Organic Chemistry: Structure, Mechanism and Synthesis

Organic Chemistry: Structure, Mechanism and Synthesis

Edited by Francesca Hopkins

CLANRYE INTERNATIONAL
www.clanryeinternational.com

Clanrye International,
750 Third Avenue, 9th Floor,
New York, NY 10017, USA

ISBN: 978-1-64726-599-1

Cataloging-in-publication Data

Organic chemistry : structure, mechanism and synthesis / Francesca Hopkins.
p. cm.
Includes bibliographical references and index.
ISBN 978-1-64726-599-1
1. Chemistry, Organic. 2. Organic compounds--Synthesis. 3. Organic reaction mechanisms.
4. Chemistry. I. Hopkins, Francesca.
QD251.3 .O65 2023
547--dc23

For information on all Clanrye International publications
visit our website at www.clanryeinternational.com

CLANRYE
INTERNATIONAL

Contents

Preface

The purpose of the book is to provide a glimpse into the dynamics and to present opinions and studies of some of the scientists engaged in the development of new ideas in the field from very different standpoints. This book will prove useful to students and researchers owing to its high content quality.

Organic chemistry is the branch of chemistry that deals with the scientific study of organic compounds that contain covalently bonded carbon atoms. It is primarily involved in the study of properties, reactions and synthesis of organic compounds. Carbon is an element that exhibits the unique property of catenation wherein it is able to form stable bonds with other carbon atoms. This ability helps in the formation of stable molecules with relatively complex structures. Therefore, the magnitude of organic chemistry can be attributed to this property of carbon. Advancements in this field have made numerous contributions to the human society, such as the synthesis of several drugs, polymers and other natural products. Some materials that are composed of organic compounds are agrichemicals, coatings, cosmetics, detergent, food, fuel, petrochemicals, pharmaceuticals, plastics and rubber. This book provides significant information for developing a good understanding of organic chemistry. It will prove to be immensely beneficial to students and researchers interested in the study of the structure, mechanism and synthesis of organic compounds.

At the end, I would like to appreciate all the efforts made by the authors in completing their chapters professionally. I express my deepest gratitude to all of them for contributing to this book by sharing their valuable works. A special thanks to my family and friends for their constant support in this journey.

Editor

1

C–H Activation Strategies for Heterofunctionalization and Heterocyclization on Quinones: Application in the Synthesis of Bioactive Compounds

Andivelu Ilangovan and
Thumadath Palayullaparambil Adarsh Krishna

Abstract

Quinone moieties in general and heterofunctionalized or heterofused quinones in particular find application in several fields such as medicinal chemistry, natural products, and functional materials. Due to its striking applications, scientists developed useful methods for the synthesis of quinone derivatives. C–H activation strategy is a fast-developing and straightforward concept, used in the construction of a diverse variety of bonds such as carbon–carbon (C–C) and carbon–hetero (C–O/N/S/P) bonds and also used is the heterofunctionalization/heterocyclization of quinones. Such approaches are useful in making use of unfunctionalized quinones for the synthesis of heterofunctionalized or heterocycle-fused quinones. The redox active nature and ligand-like properties make it difficult to carryout C–H activation on quinones. In this chapter we summarized recent developments on strategies used for C–hetero atom bond formation on quinones via C–H activation, leading to heterofunctionalization and synthesis of heterofused quinones.

Keywords: quinone moiety, C–H activation strategy, heterofunctionalization approaches, biomolecules

1. Introduction

Inspired by quinone's reactive electrophilic character, easily accessible oxidation states [1], ubiquitous natural presence [2], and important roles played in living systems (phosphorylation to electron transfer process) [3], chemists tried to mimic its acts through synthetic equivalents consisting of biologically active compounds [4], natural product analogs [5], and functional materials [6]. Consequently, several methods were developed for the synthesis of quinone derivatives. Depending upon the basic subunits (**Figure 1**), quinones are classified as benzoquinone (BQ), naphthoquinone (NQ), anthraquinone (AQ), and polyquinones (PQ).

Current, statistics on number of publications (**Figure 2**) appeared during the past two decades, ever growing research highlights, interests, importance, and applications of quinone chemistry [18].

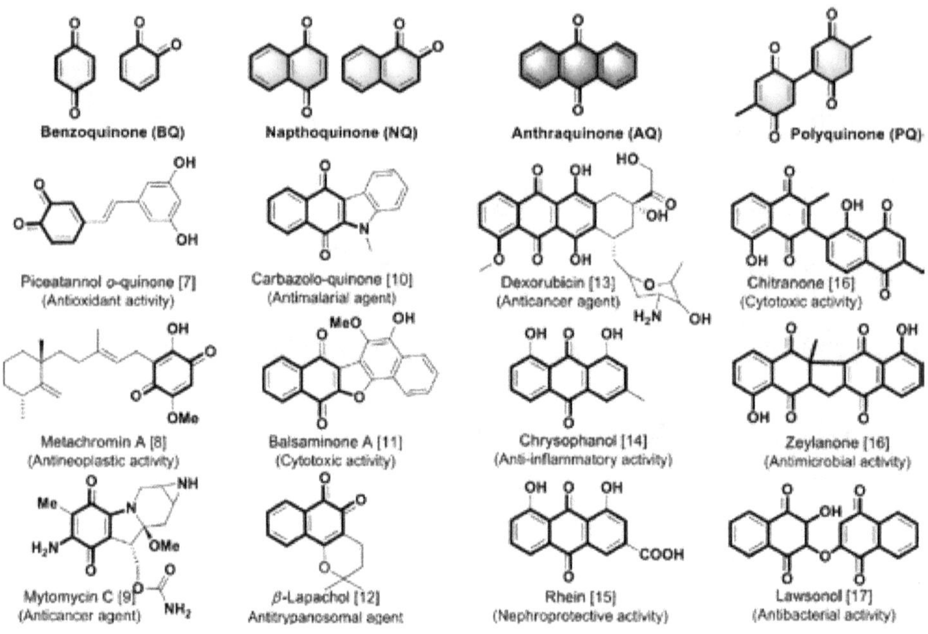

Figure 1.
Selected biologically important quinone molecules [7–17].

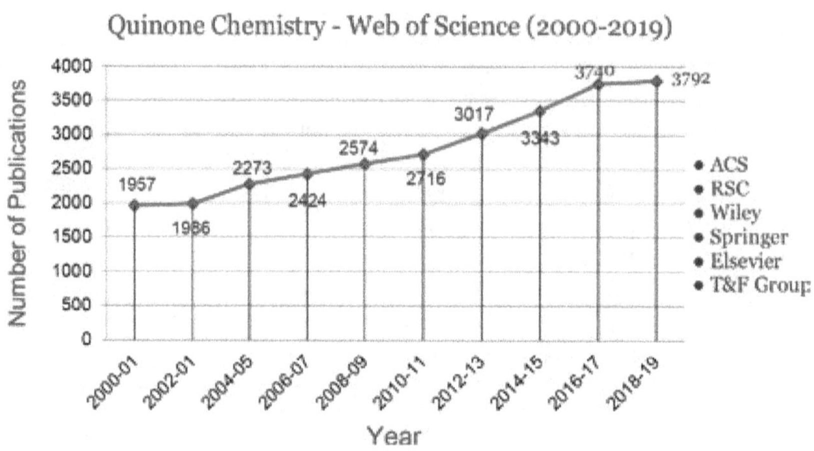

Figure 2.
Trends in the number of publications on quinones in the last 20 years based on the web of science data.

2. C–H activation and heterofunctionalization of quinones

The construction of carbon–carbon (C–C) bond or carbon–hetero atom (C–X) bond on quinone has been reported either using pre-functionalized starting materials or direct functionalization of C–H bonds [19–24]. The first step in C–H functionalization is activation, followed by the formation of an intermediate carbon-metal (C–M) bond, and final replacement with a functional group (FG). C–H activation reaction is advantageous as it is straightforward and atom economic and does not require pre-functionalization [31]. Some typical steps involved in C–H activation reaction mechanisms are oxidative addition, σ-bond metathesis, electrophilic activation, 1,2-addition, and metalloradical. C–H activation is a difficult process as it involves breaking of C–H bond having high energy (CH_4, 100 kcal/mol; benzene, 110 kcal/mol) and high pKa value (>40). In case of quinones, it is further more difficult, [32–34] as it interacts with transition metal reagents, such as Pd (Heck-type reaction), through redox reaction and ligand [35, 36] formation. The report made by Baran et al., in 2011, on coupling of quinones with boronic acids [25] for the formation of C–C bond, and by Poulsen et al., in 2018, for heterofunctionalization on quinone [26], is a notable example on C–H activation reactions on quinones. Approaches for functionalization of quinone can be broadly classified as Lewis acid (MX_3)-promoted nucleophilic addition of electron-rich arenes [27–30], transition metal-catalyzed addition of aryl radicals generated from pre-functionalized starting material [31, 32], and transition metal-catalyzed cross-coupling of halo-quinones [33–37].

Conventional methods for the heterofunctionalization (HF) of quinones involves pre-functionalization of C–H bond to form organo-halide [33–37] (Cl, Br, and I) or organo-boronic acid (–$B(OH)_2$) [38] or organo-metallic ($SnBu_3$) [39, 40] starting materials and finally to heterofunctionalization (**Figure 3**). Pre-functionalization combined with separation and purification leads to additional steps, generates waste, and lowers the efficiency drastically.

The arylation (C–C bond formation) of quinone is one of the thoroughly studied reactions using several aryl coupling partners with and without a directing group [32]. Poulsen and coworker's [26] demonstration of the synthesis of natural product stronglylophorine-26, an inhibitor of cancer cell invasion, via C–H heterofunctionalization of quinone (**Figure 4**) sets a good example on the importance of direct C–H heterofunctionalization reaction of quinones.

The undirected C–H functionalizations are much common in quinone chemistry. The presence of a directing group (DG) is helpful in achieving site-specific C–H functionalization of quinone. However, the development of efficient synthetic approaches for site-specific C–H functionalization of quinones is challenging [41–47]. This could be achieved either by manipulation of the reagent used or the presence of a directing group. For example, Junior and coworkers [41] demonstrated Rh-catalyzed C5 and C2 site-selective C–H halogenation of naphthoquinone (**Figure 5**). Similarly, by changing the type of reagent TBAI-TBHP [43] or $RuCl_2$(*p*-cymene)-PIFA [47], hydroxyl group was introduced on quinone at C-2 or C-5 position (**Figure 5**) site specifically.

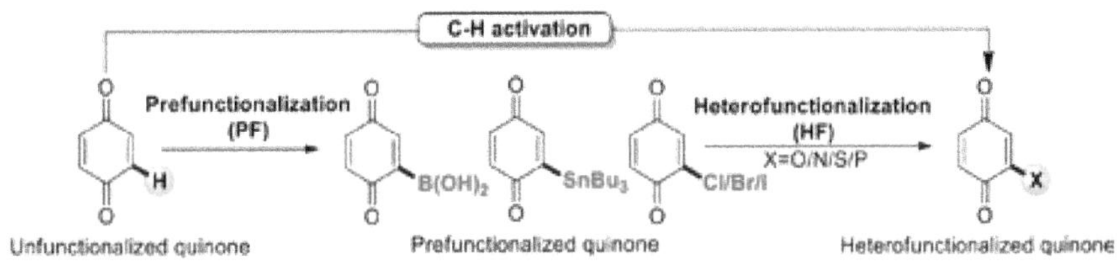

Figure 3.
Conventional heterofunctionalization vs. C–H activation approach.

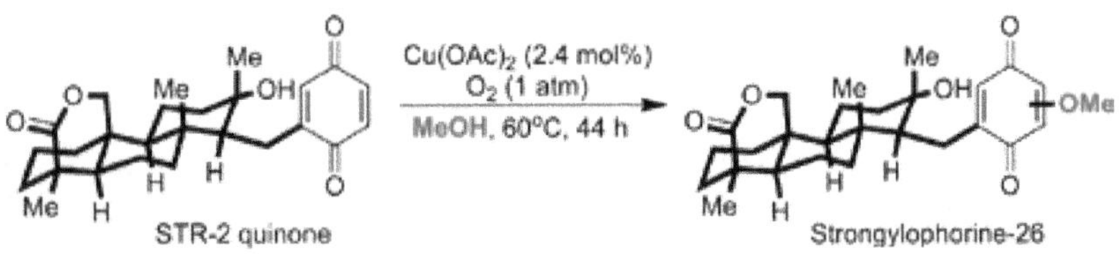

Figure 4.
Example on direct C–H heterofunctionalization of quinone.

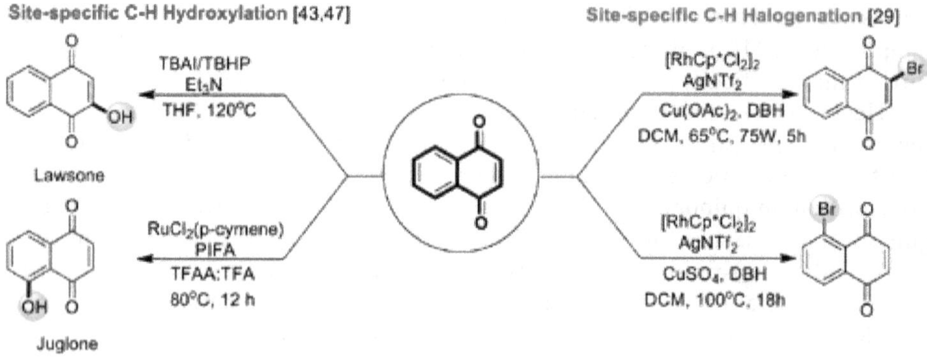

Figure 5.
Site-selective C–H functionalization on quinone or aryl ring.

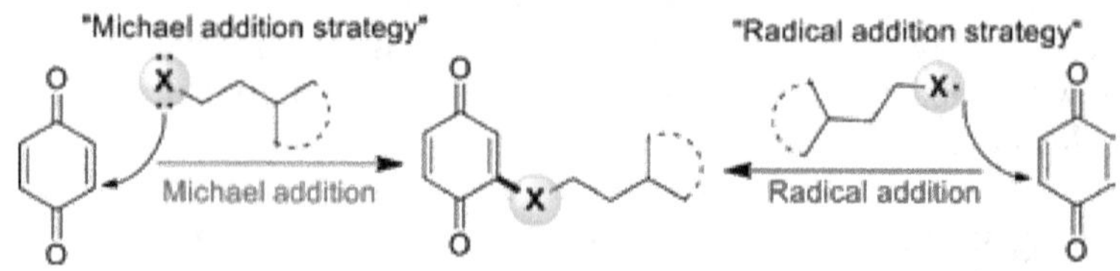

Figure 6.
Heterofunctionalization strategies.

The electrophilic character of quinone enables it to undergo facile nucleophilic attack using electron-rich nucleophilic species such as amino (R-NH_2), hydroxyl (R-OH), and thiol (R-SH) groups, as in the case of classical Michael addition [48]. Using *p*-benzoquinone most of the nucleophilic reaction leads to the forms mono-, di-, tri-, and tetra-substituted benzoquinone and most of the times hydroquinones.

In continuation of our interest on the development of C–H activation methodologies [49–55], we have developed methods for C–H functionalization of quinones [51–55]. A review article covering C–H activation of quinone with main emphasis on C–C bond-forming reactions has been reported [32]. C–H heterofunctionalization of quinone has been carried out using various catalytic systems, consisting of metal/nonmetal catalysts, organocatalyst, photocatalyst, etc. By choosing appropriate catalysts/reagents/additives, we can change the reaction pathway like radical/electrophilic/nucleophilic, etc. (**Figure 6**). For example, recently we developed an I_2-DMSO system [54] for C–H/S–H and $FeCl_3$-$K_2S_2O_8$ system [55] for C–H/C–H radical cross-coupling reactions, which normally occurs via Michael and Friedel-Crafts pathway.

Under this chapter we summarized C–H activation strategies used in heterofunctionalization and heterocyclization of quinones and its application in the synthesis of bioactive heterocycles during the past decade.

2.1 C–H activation and C–N bond formation on quinones

Aminoquinone derivatives find prominent application in medicinal chemistry and are good building blocks for many heterocyclic compounds [56]. C–N bond-forming reactions are of great importance in quinone chemistry, and in general, oxidative coupling and nucleophilic substitution reactions are involved [57–61]. It has been intensively studied using pre-functionalized quinones [62–68]. Hence, we covered some of the important C–N bond formation methodologies through C–H activation strategies which are given below.

Amines undergo smooth conjugate addition to *p*-quinones in polar solvents at ambient temperature in the absence of a catalyst and additives. Baruah et al., in 2007, synthesized a series of 2,5-bis(alkyl/arylamino)1,4-quinones from the reaction of 1,4-benzoquinone (BQ) with different amines under aerobic condition [69]. The reaction was found to be exceptionally selective and leads to only 2,5-bis(alkyl/arylamino)1,4-benzoquinones of the corresponding amine (**Figure** 7). 2,5-Isomer is formed exclusively due to electrostatic reasons. This is further evident from the fact that 1,4-naphthoquinone (NQ) on reaction with amines gives monosubstituted derivatives. In another study, Yadav et al. [70] studied H_2O-accelerated C–H amination to form highly substituted benzoquinone. In this reaction, water played a dual role of simultaneously activating the *p*-quinone and amine.

Molecular iodine-promoted direct C–H amination of NQ under ultrasonic irradiation was developed by Liu and Ji [71]. The method employs cheap, nontoxic molecular iodine as the catalyst; the desired products were obtained in moderate to excellent yield (**Figure 8**). In mechanism, molecular iodine activates the carbonyl group of the NQ to give intermediate (**A**) and is followed by amine attack at unsaturated position to give the initial addition product (**B**) which tautomerizes to form hydroquinone (**C**), which subsequently undergoes rapid oxidation finally to form quinone system [71].

Garden et al., in 2011, developed Cu(II)-catalyzed amination of NQ by oxidative coupling with derivatives of aniline (**Figure 9**). The best isolated yield was obtained in the presence of catalytic amount of copper, and the hydrated Cu(II) acetate shortens reaction time and reduces side-product formation. The study on the mechanism shows that Michael addition of anilines to NQ is facilitated by Cu(II)

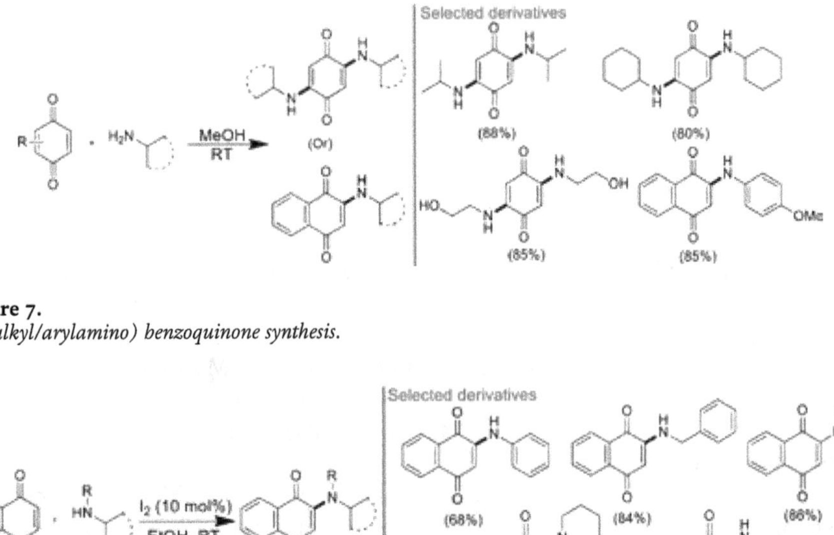

Figure 7.
Bis(alkyl/arylamino) benzoquinone synthesis.

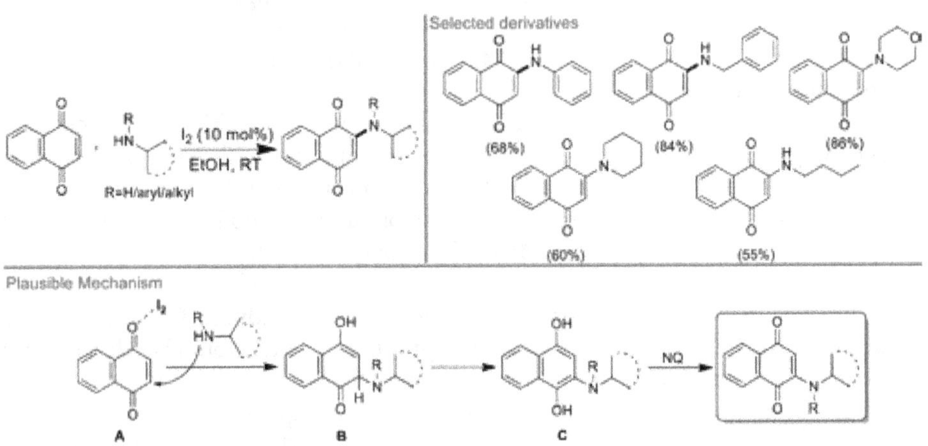

Figure 8.
Iodine-promoted direct C–H amination on NQ.

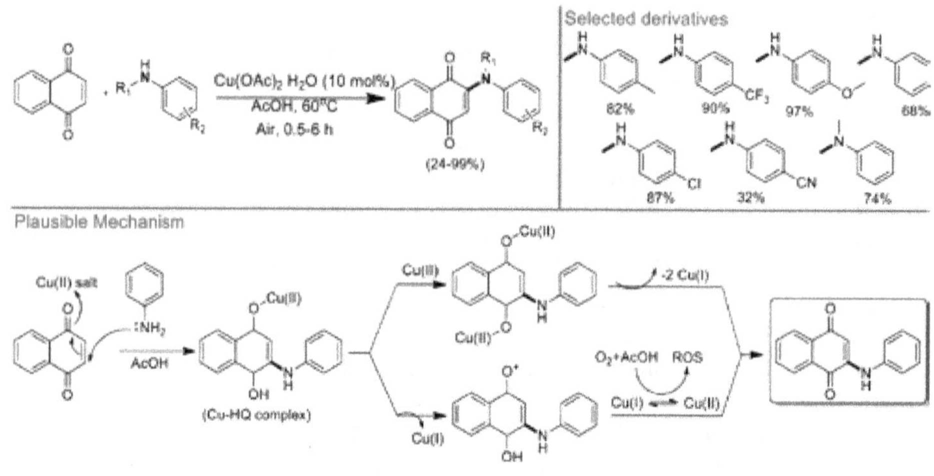

Figure 9.
Cu(II)-catalyzed oxidative coupling with anilines.

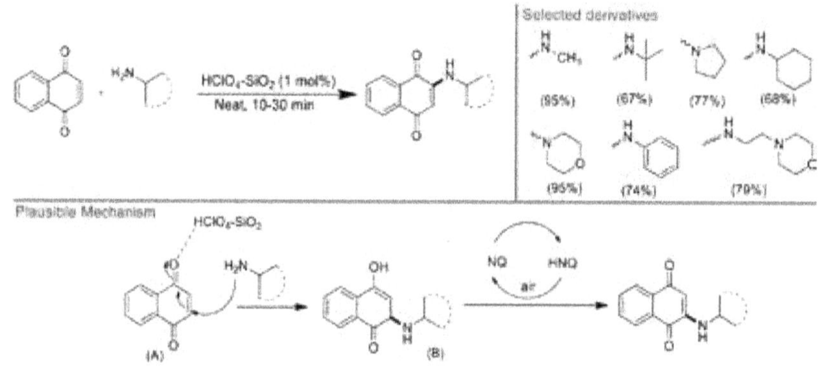

Figure 10.
$HClO_4$-SiO_2-supported C–N bond on naphthoquinone.

salt. The copper–hydroquinone (Cu–HQ) complex interacts directly with oxygen to give the quinone product or could pass through sequential one electron oxidation steps where the resulting Cu(I) species would then be reoxidized to Cu(II) by oxygen. The mechanistic proposal was supported by ESI-MS experiment, to find that the only copper species reliably observed was the copper cation as the isotopologues $Cu(I)(ACN)_2$ + (m/z 145) and $Cu(I)(ACN)_2$ + (m/z 147) in an approximately 2:1 ratio [72].

Heterogeneous SiO_2-supported $HClO_4$ catalyst promoting highly efficient and clean conjugate addition of primary and secondary amines with NQ was described by Upendra et al. [73]. Under the catalytic-ultrasonication condition, corresponding 2-amino-1,4-naphthoquinone derivatives were obtained in moderate to high yields without using any solvent (**Figure 10**). The proposed mechanism of this reaction includes two steps such as addition and oxidation. Nucleophilic addition of amines to $HClO_4$-SiO_2-activated naphthoquinone (**A**) leads to adduct (**B**). Further, the adduct (**B**) oxidized to afford NQ as final product. The authors also described a possibility of aerobic oxidation of hydroquinone (HNQ).

A base-promoted $C(sp^2)$-H sulfonamidation of 1,4-naphthoquinones via [3 + 2] cycloaddition reaction using sulfonyl azides was reported by Ramanathan and Pitchumani [74]. The straightforward, atom, and step-economic protocol provided desired product in moderate to good yield (**Figure 11**). The active alkene moiety of

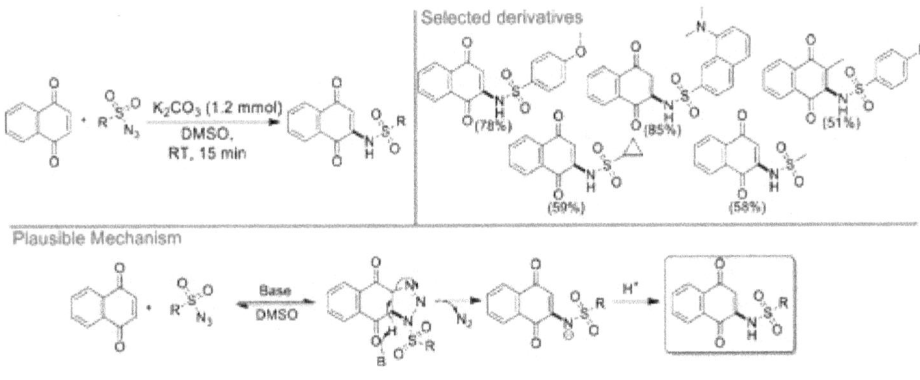

Figure 11.
K_2CO_3 promoted 1,4-quinone sp^2 C–H sulfonamidation.

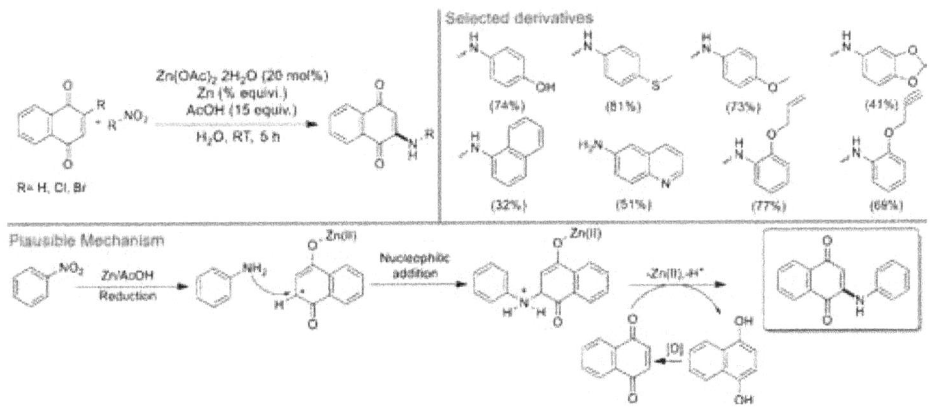

Figure 12.
Zn(II) catalyzed amination using nitro compounds.

quinone undergoes a thermal azide-alkene [3 + 2] cycloaddition followed by proton abstraction, ring opening, and elimination of a nitrogen molecule to form sulfonamidation products. Moreover, they successfully used phosphoryl azide for $-NH_2$ transfer on NQ and Menadione under optimal condition.

Recently, Chen et al. [75] developed an efficient protocol for the preparation of aminated naphthoquinone starting from NQ and nitro compounds. In the presence of Zn/AcOH system, the nitro compounds were reduced to the corresponding amines (**Figure 12**). Lewis acid $Zn(OAc)_2{\cdot}2H_2O$, 1,4-naphthoquinone is activated to generate the complex, and the intermediate reacts with aniline through 1,4-nucleophilic addition to give the adduct (**A**). Then, compound (**A**) can be oxidized to afford product in the presence of molecular oxygen along with losing a proton and the Lewis acid.

Some of the amination reactions, including multicomponent reactions, which lead to the formation of quinone-fused nitrogen heterocycles are described under the Section 3.1.

2.2 C–H activation and C–S bond formation on quinones

Thioethers are common building blocks, found in numerous biologically active compounds and in medicinally useful natural products [76]. The C–S bond construction via direct functionalization of C–H bond with sulfenylating reagents is an important reaction. Several metal and metal-free catalysts are developed for coupling of quinones with various sulfenylating reagents.

Coupling of arylsulfonyl salts with quinones in the presence of $Pd(OAc)_2$-K_2CO_3 system was developed by Ge et al. [77]. Pd directed C–sulfone to form quinone by C–S coupling (**Figure 13**). Mechanistic study shows that initially oxidative addition of Pd with sulfonyl chloride affords intermediate species **A** which is followed by carbopalladation to form intermediate **B**. After β-H elimination, intermediate **B** released the coupling product to complete the catalytic cycle.

In another study, Huang et al., in 2016, developed reaction with $[Cp^{*}IrCl_2]_2$-$AgSbF_6$ [78] system. Like palladium-catalyzed carbopalladation on sulfonyl chloride, here Ir(I) to form carboiridation (**Figure 14**). Further similar way, β-H elimination leads to the final product.

CuI-PPh_3 catalytic system was used for the synthesis of quinonyl thioethers [79]. It was reported to produce sulfonyl-quinones when palladium catalyst was used [77]. In this reaction arylsulfonyl chloride (PhSCl) was formed on reaction with PPh_3 (**Figure 15**) which on reaction with intermediate **B** (which might have been

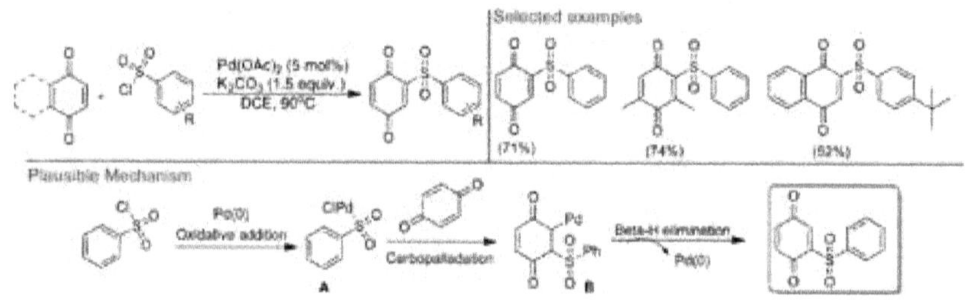

Figure 13.
Pd-catalyzed direct C–sulfone formation on quinone.

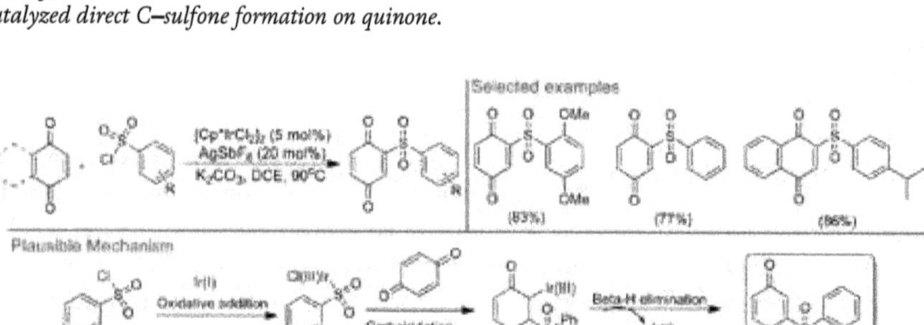

Figure 14.
Ir-catalyzed C–S coupling of quinones with sulfonyl chloride.

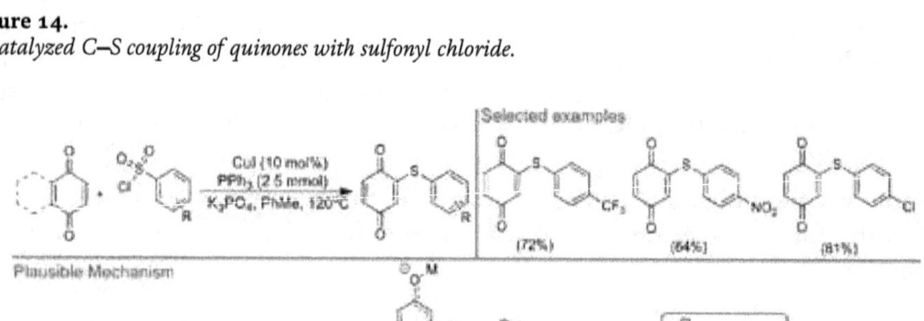

Figure 15.
Cu-PPh_3-promoted sulfenylation of quinones.

formed with the help of base through Baylis-Hillman process) produced arylthioquinone derivatives.

In 2015 Chou et al. [80] used silver catalyst system for the reaction of various aryl disulfides to synthesize a variety of quinonyl aryl thioether moderate to high yields. The authors carried out some control experiments to predict the plausible mechanism. Studies indicate that the reaction is initiated by active disulfide-silver intermediates formed through interactions of the silver with aryl disulfides in DMSO (**Figure 16**).

Furthermore, under metal-free conditions, various sulfenylating reagents such as [bmim]BF_4-arylsulfinic acids [81], NH_4I-sodium arylsulfinates [82], and H_2O-arylsulfonyl hydrazides [83] systems gave sulfonyl hydroquinones.

Notably, I_2-DMSO system [54] for the thiomethylation of quinone was recently developed by us (**Figure 17**). Based on the verification experiments, we proposed plausible radical pathway. At 120°C, DMSO decomposes to CH_3SH and CH_2O. Meanwhile, iodine releases two iodine radicals at high temperature that reacts with CH_3SH to yield methylthiyl radical (**A**). The addition of methylthiyl radical (**A**) to naphthoquinone results in the formation of radical intermediate (**B**) which should loose H•to another iodine radical leading to the formation of the product.

Moreover, very recently, CuI-O_2 [84] and $Co(OAc)_2$-O_2 [26] systems were utilized for direct thiol addition to quinone to form ether. In addition, there are limited reports available for the conversion of hydroquinone to quinone followed by in situ C–S bond formation. Notably, under metal-free condition Runtao et al. [85] utilized S-alkylisothiouronium salts on hydroquinone for the synthesis of quinonyl

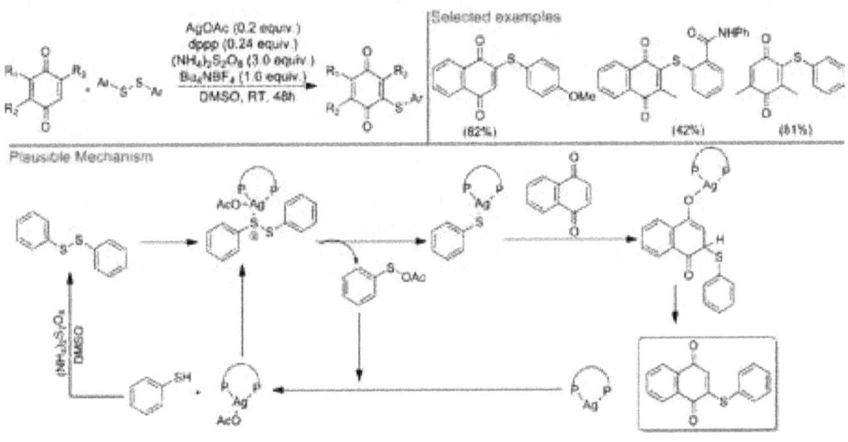

Figure 16.
Silver-catalyzed direct thiolation of quinones.

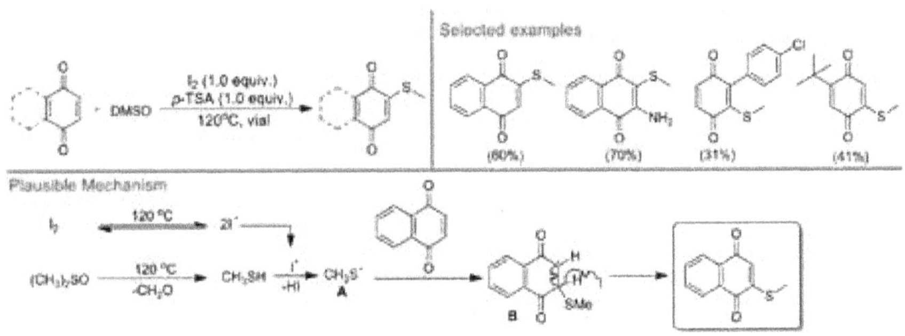

Figure 17.
I_2-DMSO-promoted thiomethylation on quinone.

thioether. In another study, laccase-catalyzed thiol Michael addition on naphthohydroquinone [86] and hydroquinone [87, 88] was observed. Less selectivity and poor yield are the main drawbacks of these enzymatic reactions.

2.3 C–H activation and C–O bond formation on quinones

Naturally occurring quinone molecules, containing C–O link, such as byrsonimaquinone, balsaminone A, maturone and lambertelinare, are biologically important. Several methods for the construction of C–O bond through the activation of C–H bonds on quinone have developed rapidly. However, this research area is less explored than C–N and C–S bond formation as oxygen has lower nucleophilicity than nitrogen and sulfur. In this section, we discuss the formation of the C–O bond through C–H functionalization.

In 2007, Tamura et al. [89] developed a simple method for the synthesis of dibenzofuranquinones, which is the core structure of the natural products balsaminone A, utilizing a novel oxidative cyclization of the quinone-arenols under the special condition (**Figure 18**). As an application of this method to natural product synthesis, a facile synthesis of violet-quinone was demonstrated.

Coupling of propargyl carbonate with quinone through Claisen rearrangement to furanonaphthoquinones (FNQ) was recently established by Zhiyu et al. [90] (**Figure 19**). Though two groups have reported the synthesis of FNQ, both of these methods had several disadvantages. The first method reported by Perez et al. [91] needs use of Cs_2CO_3, CsI, and CuI as mediator. The second method reported by da Silva Emery et al. [92] employs CuI as catalyst, which still required rigorous condition of refluxing for 24 h.

Weitz reported a useful method for the introduction of hydroxy group through a sequence of in situ Weitz-Scheffer-type epoxidation/epoxide cleavage reaction with $H_2O_2/Na_2CO_3/H_2SO_4$ [93]. In 2013, Schwalbe showed that brominated naphthoquinones could be hydroxylated with nucleophilic substitution under KOH/MeOH [94]. In 2016, Martins has accomplished the Suzuki coupling reactions between 2-hydroxy-3-iodo-1,4-naphthoquinone and boronic acids to prepare several 2-hydroxy-3-aryl-1,4-naphthoquinones by palladium catalyst [37]. In general

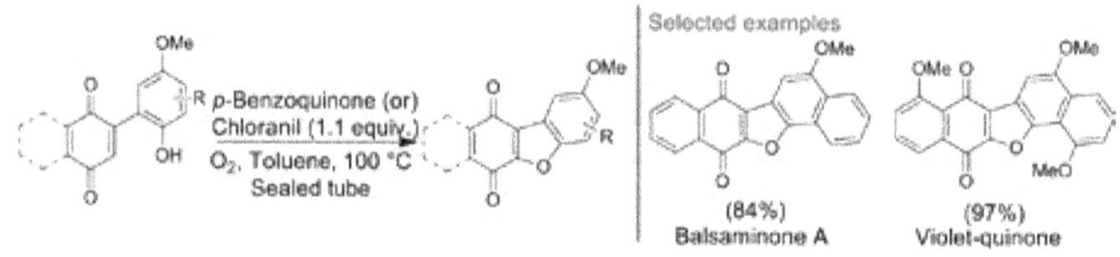

Figure 18.
Oxidative cyclization of quinone-arenols.

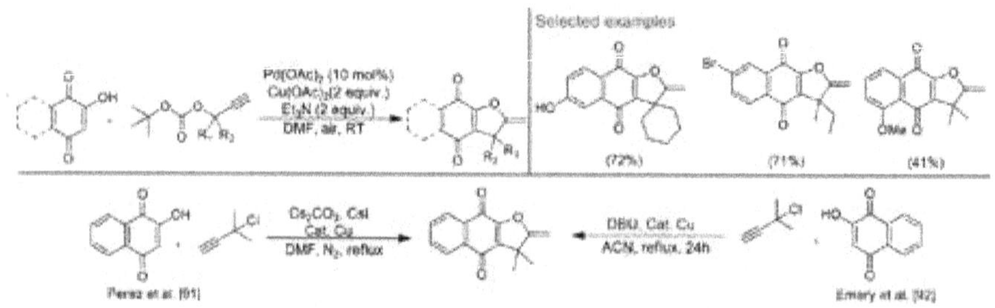

Figure 19.
Synthesis of Furano-naphthoquinone.

most of the existing methods suffer from the requirement for strong alkaline or acidic conditions, metal catalysts, pre-halogenation, and fairly limited substrate scope.

Recently, hydroxylation of naphthoquinone derivatives using tetrabutylammonium iodide (TBAI) as a catalyst and *tert*-butyl hydroperoxide (TBHP) as an oxidant was disclosed by Wang and coworkers [35]. This methodology allowed direct installation of hydroxyl groups on the quinone ring which was used for the synthesis of the corresponding substituted lawsone derivatives (**Figure 20**). Interestingly, parvaquone and lapachol were synthesized by this methodology.

Poulsen et al. [26] disclosed powerful methods for oxidative *p*-quinone functionalization using $Co(OAc)_2$ and $Mn(OAc)_3{\cdot}2H_2O$ with a collection of O, N, and S-nucleophiles, wherein oxygen was used as the terminal oxidant (**Figure 21**). Preliminary mechanistic observations and synthesis of the cytotoxic natural product strongylophorine-26 for the first time were presented.

2.4 C–H activation for multiple heterofunctionalization of quinones

Multicomponent reactions (MCRs) constitute one of the most efficient tools in modern synthetic organic chemistry, since they have all features that contribute to an ideal synthesis. Features of this type of reaction are (i) high atom efficiency, (ii) quick and simple implementation, (iii) time and energy saving, (iv) environment friendly, and (v) offer a target and diversity-oriented synthesis. Under this section we have classified some of the multicomponent reaction which let the formation of multiple heterofunctionalization of quinones but not heterocyclization.

Hong et al., in 2017, reported Ag(I)-mediated one-pot multicomponent reaction in which BQ, diarylphosphine oxides, and imines underwent regioselective CDC reaction to undergo dual C–H/P–H (phosphination) and C–H/N–H (amination) on 1,4-benzoquinone (BQ), and the desired products were obtained in moderate yield (**Figure 22**). Under the optimized condition when 1,4-naphthoquinone (NQ) instead of BQ, and aniline instead of corresponding imine was used, lowering of yield of the desired product was observed. Moreover, interestingly a competitive side reaction, namely, hydrophosphinylation reaction was observed in the absence of Ag(I). In this strategy Ag(I) plays versatile role such as a mediator and oxidant. The authors characterized the X-ray crystal structures of several new functionalized quinone derivatives [95].
Based on the control experiments, Ag(I)-mediated mechanism was proposed. Firstly, Ag(I) ions coordinate with BQ oxygen atom, rendering BQ to act as a better

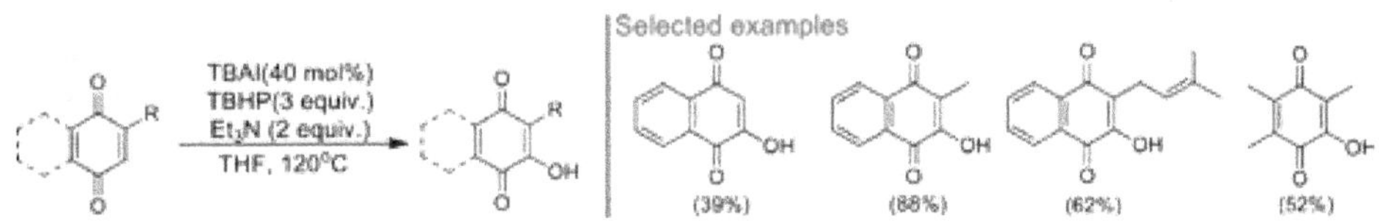

Figure 20.
Hydroxylation of naphthoquinone derivatives.

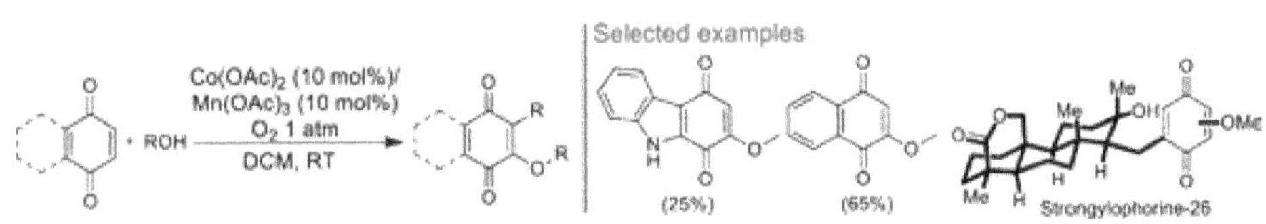

Figure 21.
Oxidative C–H functionalization of p*-quinone with alcohols.*

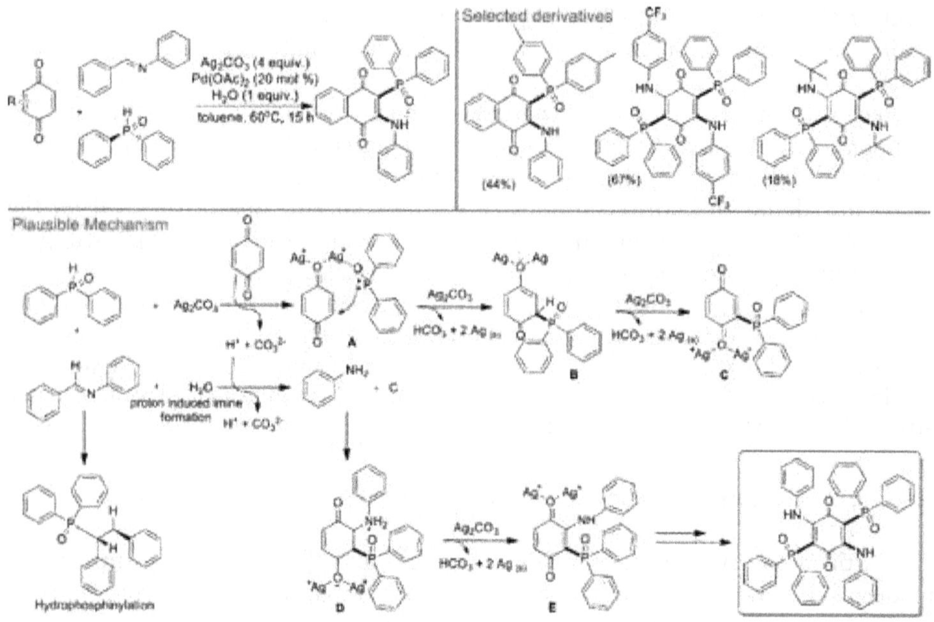

Figure 22.
Ag-mediated regioselective phosphination and amination.

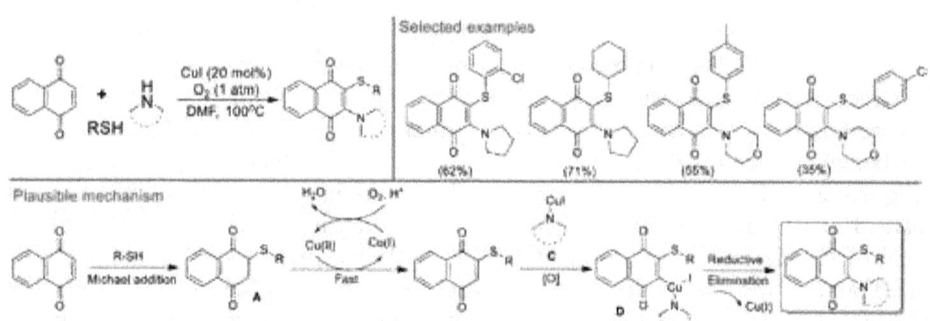

Figure 23.
Cu-catalyzed thioamination on quinone.

electrophile for diarylphosphine oxides, which is presumably released from the adduct (**A**). After the formation of intermediate (**B**), deprotonation takes place with the assistance of CO_3^{2-}, and then two electrons transfer from the intermediate to two AgI ions to give monosubstituted product (**C**) along with two equivalents of AgO species. Further, nucleophilic attack of aniline at the three positions of (**C**) forms intermediate (**D**) which subsequently forms 2,3-disubstituted intermediate (**E**) by deprotonation with the assistance of CO_3^{2-}. Yuan et al. [96] also reported C–P and C–N bond formation on quinone under the same conditions.

One-pot three-component strategy for the direct thioamination of 1,4-naphthoquinone with thiols and amines was recently disclosed by Bing et al. [84]. This approach employed a catalytic amount of CuI as a catalyst and molecular oxygen as a green oxidant. Various 2-amino-3-thio-1,4-naphthoquinones products could be synthesized in moderate to good yields. This catalytic method represents a step-economic and convenient method for the difunctionalization of 1,4-naphthoquinone. Based on the systematic control experiment, the authors proposed the plausible mechanism shown in **Figure 23**. First, the Michael addition of 1,4-naphthoquinone and thiol gave intermediate (**A**), which was immediately oxidized to intermediate (**B**) by Cu(I)/O_2. Further, the oxidative addition of amine and CuI afforded the Cu(II) species (**C**), which then reacted with intermediate (**B**) giving Cu(III) species (**D**). Finally, intermediate (**D**) underwent reductive elimination producing the desired product.

3. C–H activation for the synthesis of quinone-heterocycle-fused hybrids

Heterocyclic compounds having oxygen (O), nitrogen (N), or sulfur (S) atoms are of tremendous importance [97, 98]. C–X bond formation on quinone gives heterofunctionalized quinones which are very important in organic chemistry and medicinal chemistry, especially due to their striking biological activities [1]. Mitomycin C is an approved quinone-based anticancer drug having pyrrolidine ring [99]. Several other heterofused/linked quinone molecules show good pharmacological properties [100–102]. Structure activity relationship studies from quinonoid compounds showed that the position and increasing the number of heteroatoms are important factors to achieve biological activities [103]. In general, heterocyclization strategies on quinone is mainly classified into three, namely, C–X bond formation, C–C bond formation, and cascade C–C and C–X bonds formations (**Figure 24**).

Selections of suitable intermediates for the synthesis of heterocyclic compounds are very important. Quinones are important intermediate for the assembly of heterocycles. There are several C–H activation methods which are disclosed for the synthesis of valuable heterocyclic compounds such as phenazine, carbazole, indole, phenothiazine, benzothiophene, benzofuran, cumarin, chromene, etc.

Hybrid molecules are based on the principle of combining partial or whole structures in order to create new and possibly more active molecular entities [104–106]. Hybrid molecules can incorporate two or more pharmacophore which lead to the generation of new bioactive compound which show both the activities or altogether a new kind of bioactivity. This is useful to achieve activity on "multiple targets" of a biological system, and this is called multicomponent therapeutic strategy [107].

To achieve synthesis of hybrid organic molecules, different strategies have been adopted time to time [105]. Quinones display wide variety of biological activity, hence combining quinone skeleton with another bioactive heterocycle should basically provide a hybrid organic molecules which may show some valuable biological activity profiles. Some of the interesting quinone-heterocycle-fused hybrid molecules found in the literature are shown in **Figure 25**. There were several strategies developed for the synthesis of quinone-based hybrid molecules [108–111].

Recently, Mancini et al. [112] selected different compounds acting as inhibitors of the cancer protein targets tubulin, human topoisomerase II, and ROCK1 (**Figure 26**).

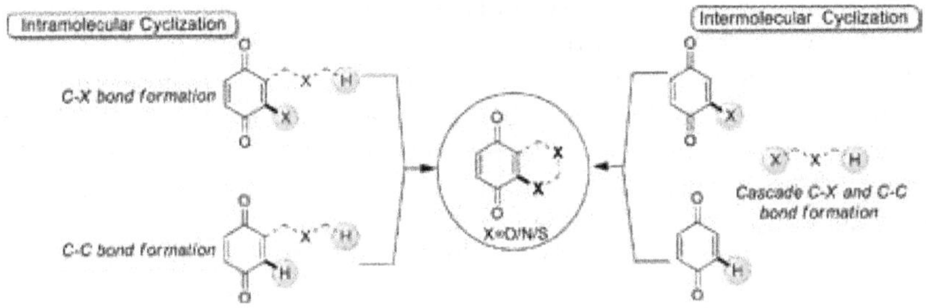

Figure 24.
Heterocyclization approaches on quinone.

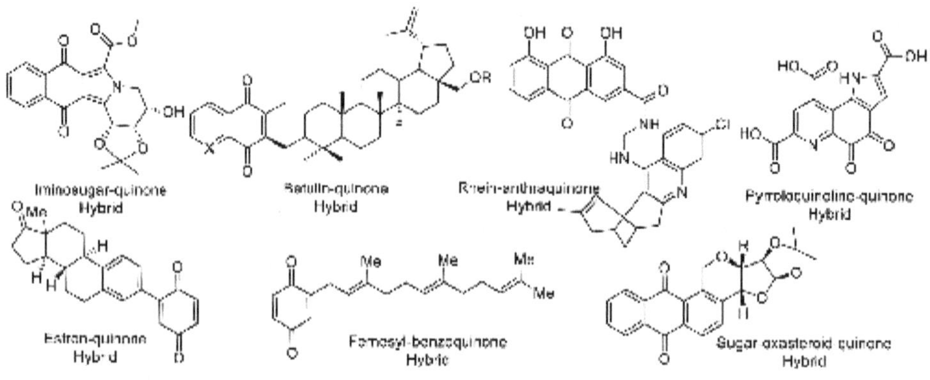

Figure 25.
Selected quinone-heterocycle hybrid molecules.

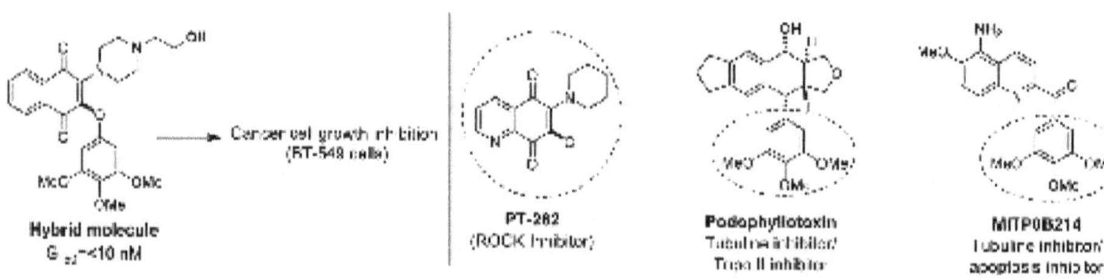

Figure 26.
Quinone-heterocycle hybrids showing anticancer activity.

The synthesized quinone-hybrid molecules displayed good and sometimes better growth inhibition GI_{50} than the ROCK inhibitor Y-27632, the Topo II inhibitor podophyllotoxin, and the tubulin inhibitor combretastatin A-4.

In this direction in the forthcoming sections, we have listed out methods known for the synthesis of quinone-heterocycle hybrid molecules and some of its importance.

3.1 C–H activation for the synthesis of quinone-fused heterocycle hybrids through two component reaction

The oxidative coupling reactions of NH isoquinolones with 1,4-benzoquinone proceeded efficiently to form spiro compounds through C–C and C–N bond in the presence of an Ir(III) catalyst (**Figure 27**) [113]. $Cu(OAc)_2{\cdot}H_2O$ was used as external oxidant for substrates such as NQ and other substituted 1,4-benzoquinone. The authors performed preliminary mechanistic experiments and a catalytically competent five-membered iridacycle was isolated and structurally characterized, thus revealing a key intermediate in the catalytic cycle. The first step of mechanism is likely to be a C(sp^2)-H activation process affording a five-membered iridacycle intermediate **A**. The coordination of BQ to **A** delivers intermediate **B**. The migratory insertion of the coordinated BQ into the Ir–C bond leads to intermediate **C** which on protonation by HOAc forms intermediate **D**. Subsequent iridation occurs at the α-position to afford iridacycle I, which undergoes a C–N reductive elimination to afford the final product and Cp*Ir(I). Cp*Ir(I) is oxidized by BQ in the presence of HOAc to $Cp^*Ir(Oac)_2$ for the next catalytic cycle.

In another study, an Rh-catalyzed substrate-tunable oxidative annulation and spiroannulation reaction of 2-arylindoles with benzoquinone was reported by

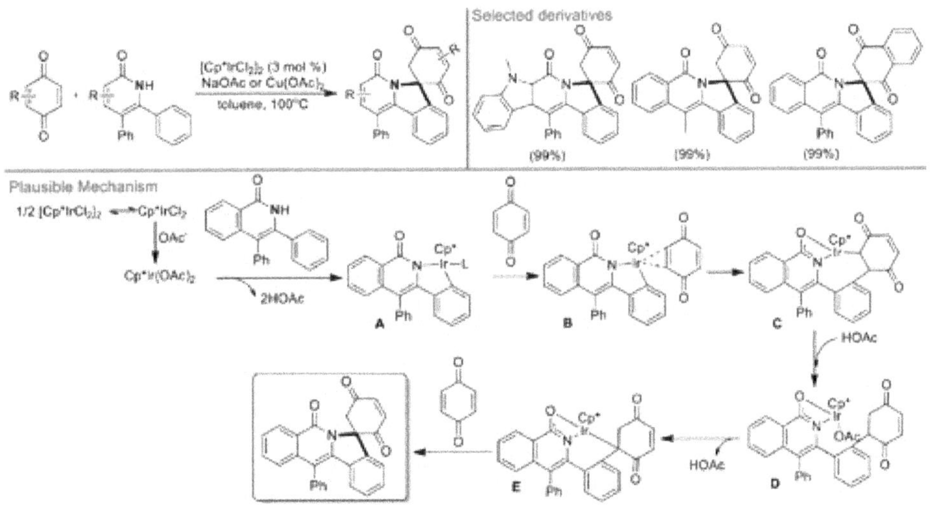

Figure 27.
Ir(III)-catalyzed oxidative coupling of NH isoquinolones.

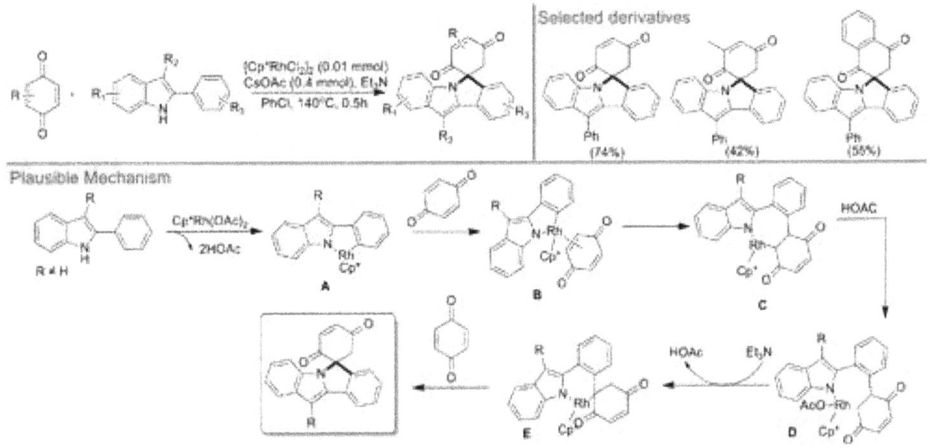

Figure 28.
Rh(III)-catalyzed oxidative coupling of NH indoles.

Shenghai et al. [114]. Mechanistic study revealed that Rh(III)-catalyzed dual N–H/C–H bond cleavage of indole occurs to afford a rhodacycle (**A**). Further, the coordination of BQ to A yields (**B**), which undergoes a migratory insertion of the coordinated BQ into the Rh–C bond to furnish (**C**). Further (**C**) undergoes a selective Rh–C protonolysis with one equivalent of HOAc to afford the key intermediate (**D**).

Subsequently, the promotion of nucleophilic attack by Et_3N, the tertiary α-C atom on the Rh center, generates I, which undergoes a C–N reductive elimination to give the desired product and a Rh(I) species. The Rh(I) species is oxidized to the active Rh(III) catalyst by BQ in the presence of HOAc (**Figure 28**).

Cu(II)-catalyzed sequential C,N-difunctionalization reaction between naphthoquinone and β-enaminones [115] which leads to the formation of indaloquinone. New C–C and C–N bonds are easily formed in the reaction course. Cu(II) salt plays a dual role as Lewis acid and oxidative catalyst, and O_2 acts as the terminal oxidant. Based on the experimental results, a plausible reaction pathway was suggested by the authors as shown in **Figure 29**.

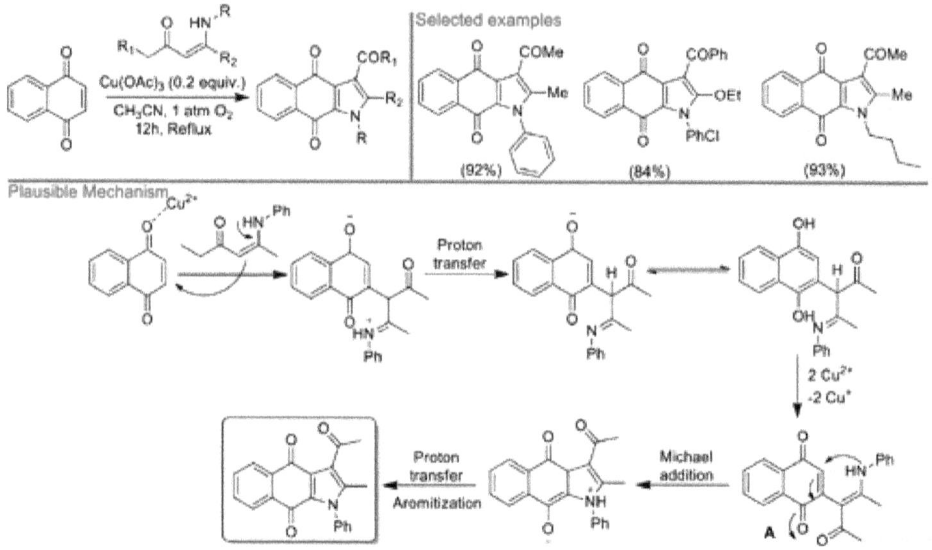

Figure 29.
Cu(II)-catalyzed NQ sequential C,N-difunctionalization.

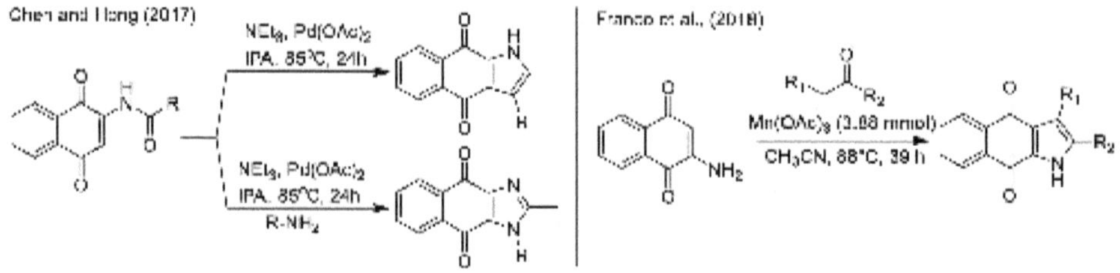

Figure 30.
Pd/Mn-catalyzed C–H functionalization of amino NQ derivatives.

First, nucleophilic attack of α-carbon atom of β-enaminone to Cu^{2+} complexed NQ followed by tautomerization and oxidation by Cu^{2+} results in the formation of intermediate (**A**). Further, intramolecular Michael addition takes place. Finally, oxidative aromatization affords cyclic product. To complete the catalytic cycle, molecular oxygen was involved for the oxidation of Cu(I) and regeneration of Cu(II).

Chen and Hong [116] reported Pd(II)-catalyzed *ortho*-CH functionalization of amido-substituted 1,4-naphthoquinone with primary and tertiary amines (**Figure 30**). The reaction occurred through an intramolecular rearrangement followed by oxidation process, which lead to the formation of imidazole and pyrrole ring-fused quinone derivatives. In another study, Franco and coworkers [117] performed oxidative free-radical reaction of quinone with either aldehydes or simple ketones in the presence of $Mn(OAc)_3$ to afford a series of indole-naphthoquinone-fused heterocycles [117].

Mito et al., in 2016, developed a method for benzo[f]indole-4,9-diones from inactivated naphthoquinone with α-aminoacetals [118]. This reaction underwent via intramolecular nucleophilic attack of aminoquinones to aldehydes. Based on the detailed mechanistic studies, the authors proposed the plausible mechanism represented in **Figure 31**.

Haiming and coworkers [119] developed a simple protocol for the synthesis of highly functionalized 3-hydroxycarbazoles by acetic acid-promoted annulation of electron-rich anilines and quinones (**Figure 32**). This chemistry, although tolerant

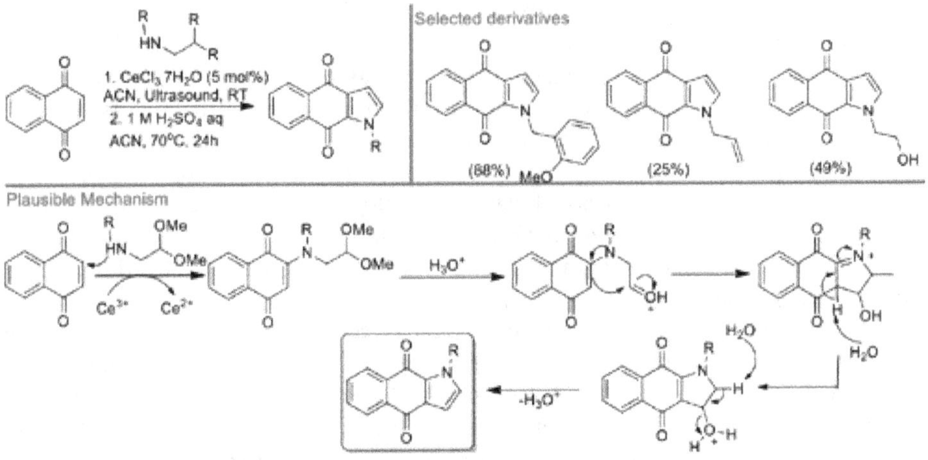

Figure 31.
Ultrasound-assisted one-pot synthesis.

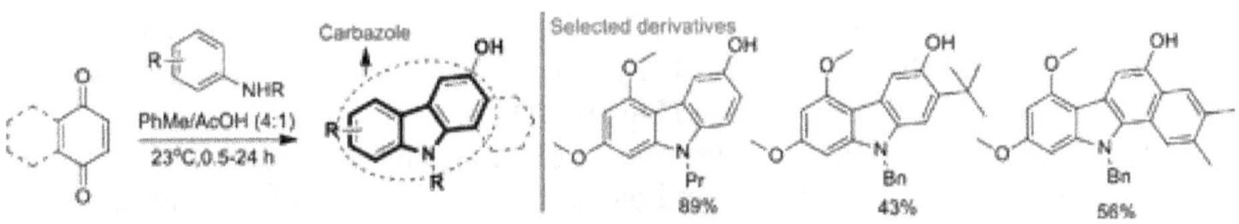

Figure 32.
Annulation of electron-rich anilines and quinones.

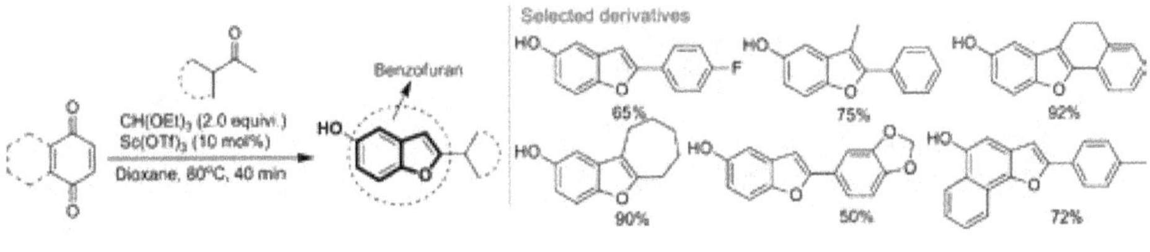

Figure 33.
Synthesis of benzofurans from ketones and 1,4-benzoquinones.

of various quinones, is sensitive to both steric and electronic elements on the anilines, as well as the steric hindrance introduced to the quinones. Although the yields are generally moderate, this reaction nevertheless provides a single-step alternative to prepare various otherwise difficult to make densely substituted 3-hydroxycarbazoles under mild conditions. Similarly to Nenitzescu indole synthesis, the mechanism of this carbazole formation is believed to involve a C–C bond formation by a Michael-type nucleophilic addition of aniline to quinone, followed by intramolecular cyclization and dehydration.

In another study, a sequential Michael addition and intramolecular cyclization reaction of ketones and 1,4-benzoquinones by using triethyl orthoformate as an additive (**Figure 33**). In the presence of $Sc(OTf)_3$ as catalyst, triethyl orthoformate may be utilized to convert enolizable ketone into ethyl vinyl ether. As a result, nucleophilicity increases. This reaction is a simple way to obtain 5-hydroxybenzofurans. The authors used this methodology to synthesize some important 2-phenylbenzofuran derivatives [120].

Wang et al. [121] developed $Pd(OAc)_2$/BQ catalytic system for ring contraction reactions which allow 2-hydroxyl-1,4-naphthoquinones to convert into various phthalides. The significance of phthalide and fulvene scaffolds as structural units should render this method attractive for both medicinal chemistry and synthetic ring contraction reactions chemistry, paving the way for efficient synthesis of other complex cyclic systems (**Figure 34**). Moreover, they utilized phthalides as versatile synthetic intermediates toward many other useful synthetic building blocks.

Peddinti et al., in 2014, reported [122] Michael addition of the 1,4-benzoxazinone derivatives, a novel class of vinylogous carbamates to the Michael acceptors. 1,4-Benzoxazinone derivative undergoes Michael addition with *p*-quinone in the presence of trifluoroacetic acid, and subsequent cyclization affords corresponding products (**Figure 35**).

A nucleophilic addition of terminal alkynes to 2-methoxy-1,4-benzoquinone afforded the corresponding quinols containing an alkyne unit [123], which were converted to phenols via mild Zn-mediated reduction. After proper protection of the free phenolic OH group, under metal-free system, 5-endo-dig iodocyclization allowed facile access to a number of 3-iodobenzofurans (**Figure 36**).

After successful establishment of kinetic controlled, Rh(III)-catalyzed annulation of C–H bonds with quinones for chemo-selective synthesis of dibenzo[*b*,*d*]pyran-6-ones [124] and phenanthridinones [125], Yang and coworkers [126] demonstrated a three-component cascade reaction for 6H-benzo[c]chromenes. Similarly, this reaction involved Rh(III)-catalyzed annulation of aryl ketone O-acyloximes, quinones, and acetone (**Figure 37**).

In another report [127], the synthesis of diverse dihydronaphtho[1,2-*b*]furans starting from 1,4-naphthoquinones and olefins in the presence of ceric ammonium nitrate (CAN) was reported. The reaction was based on the CAN-catalyzed [3 + 2] cycloaddition of 1,4-naphthoquinones. This methodology was also used to synthesize the biologically important natural product furomollugin in only two steps (**Figure 38**).

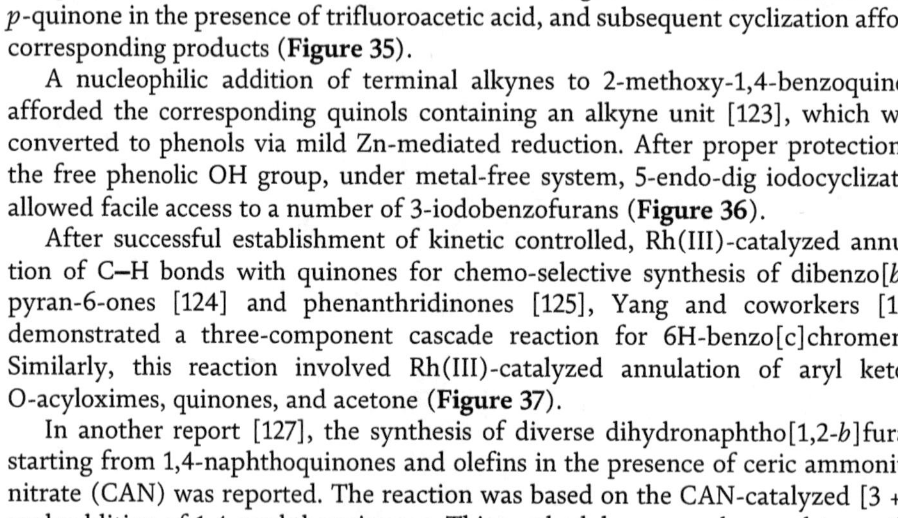

Figure 34.
Synthesis of phthalides via ring contraction reactions.

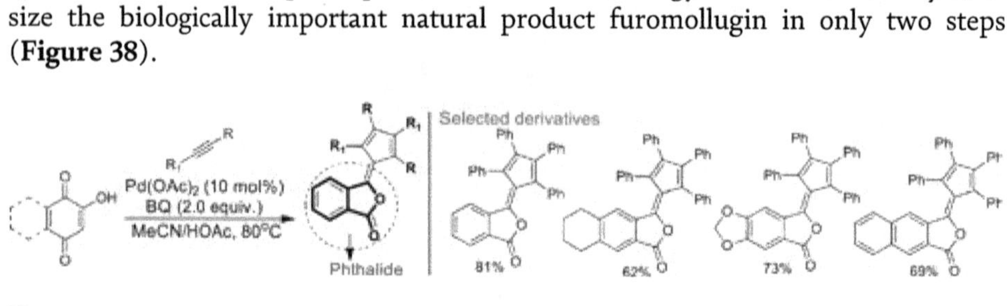

Figure 35.
TFA-catalyzed Michael addition reaction for cumarin.

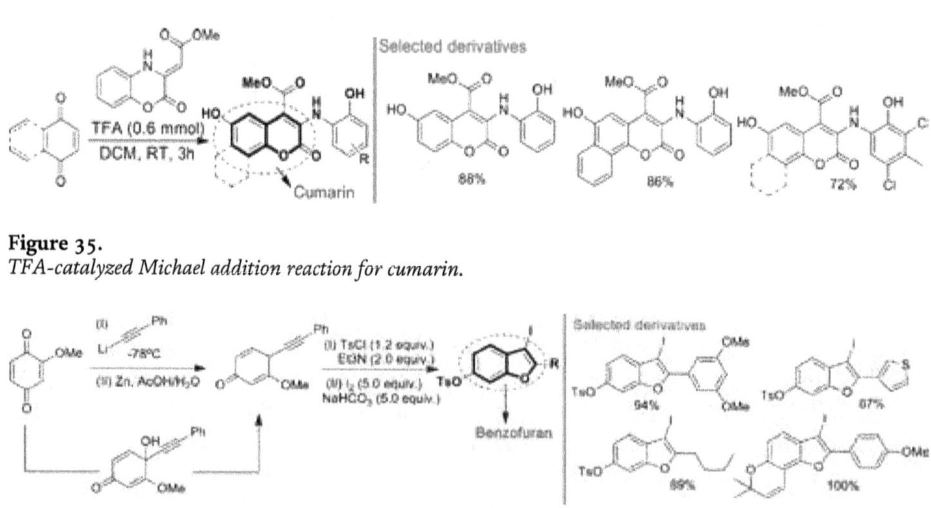

Figure 36.
5-Endo-dig iodocyclization reaction for 3-iodobenzofurans.

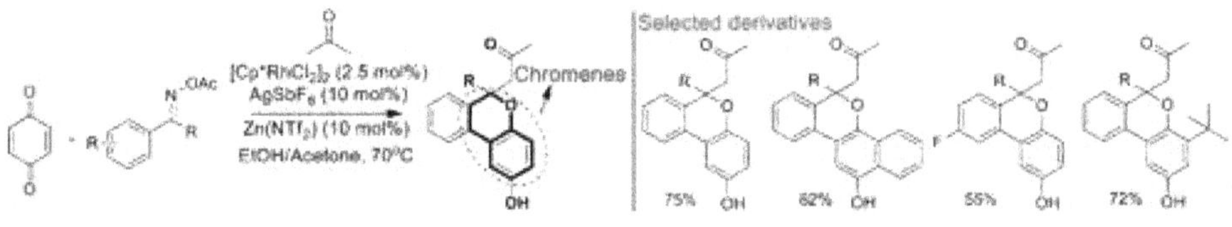

Figure 37.
Rh(III)-catalyzed annulation for benzochromenes.

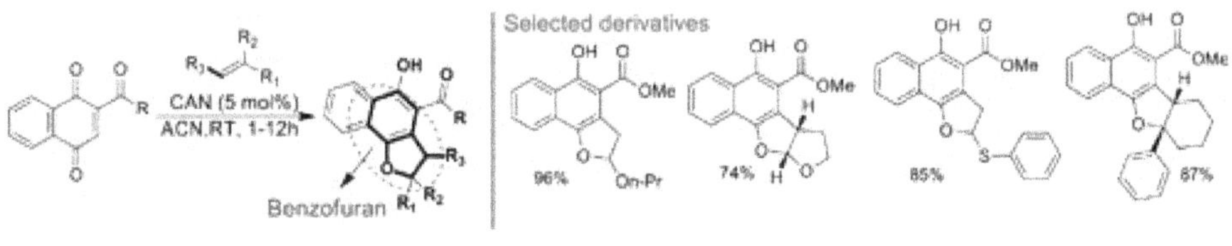

Figure 38.
CAN-catalyzed [3 + 2] cycloaddition of 1,4-naphthoquinones.

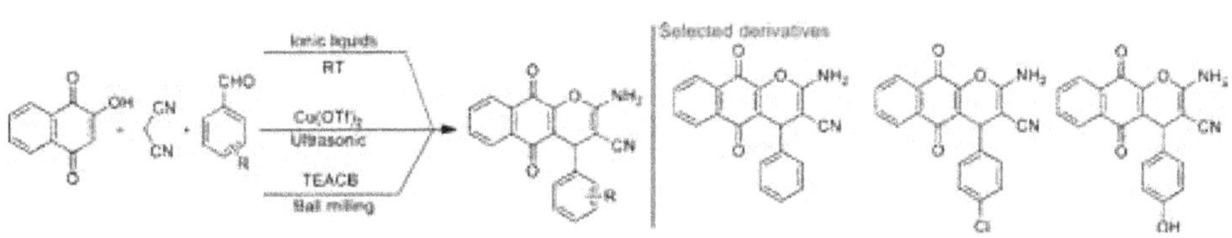

Figure 39.
Three-component synthesis of functionalized 4H-pyrans.

3.2 C–H activation for the synthesis of quinone-fused heterocycle hybrids through multicomponent reaction

The applications of MCRs have been sequenced with multiple ring-forming reactions that leads thereby to the synthesis of diverse heterocyclic scaffolds. MCRs on quinones were used for the generation of quinone-fused heterocycles.

Seven mild basic ionic liquids [128] made out of 1,8-diazabicyclo[5.4.0]-undec-7-en-8-ium acetate, pyrrolidinium acetate, pyrrolidinium formate, piperidinium acetate, piperidinium formate, *N*-methylimidazolium formate, and 3-hydroxypropanaminium acetate were used as catalyst for three-component coupling of aldehyde, malononitrile, and 2-hydroxynaphthoquinone for the formation of 2-amino-3-cyano-4-aryl-5,10-dioxo-5,10-dihydro-4H-benzo[g]chromene and hydroxyl naphthalene-1,4-dione derivatives under ambient and solvent-free conditions (**Figure 39**). The main advantages of this protocol are mild, solvent-free conditions, ecofriendly catalysts and easy to work-up procedure. Similar reaction was also reported by Javanshir et al. [129] and Manisankar et al. [130] using organocatalyst and copper catalyst, respectively.

Notably, Cao and coworkers [131] developed one-pot, pseudo-four-component reaction of 2-hydroxy-1,4-naphthoquinone, aromatic amine, and formaldehyde in aqueous media under ultrasound irradiation (**Figure 40**) naphthoquinone-fused oxazine derivatives under this operationally simple and efficient condition.

A proposed mechanism shows that amination reaction occurred first between the formaldehyde and amine, followed by H_2O elimination to furnish intermediate **A** which was then attacked by 2-hydroxy-1,4-naphthoquinone to furnish an intermediate **B**, which further reacted with formaldehyde and eliminated H_2O to produce intermediate **C**. At last the final product was formed through an intramolecular cyclization process.

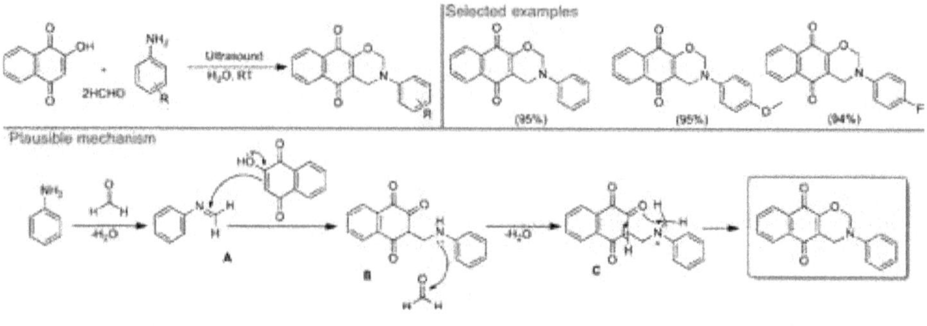

Figure 40.
Synthesis of naphthoquinone-fused oxazine derivatives.

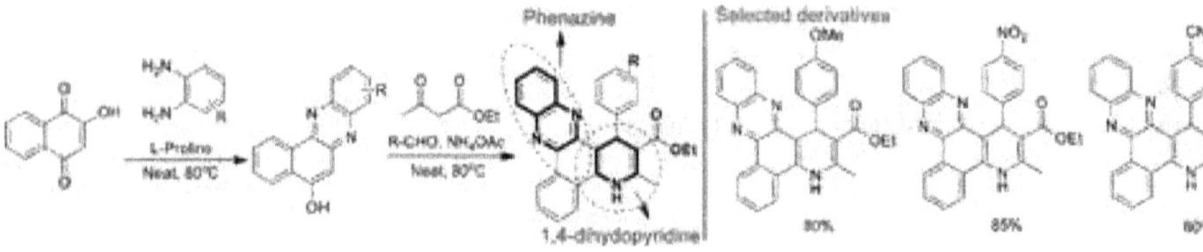

Figure 41.
Synthesis of phenazine-dihydropyridine molecules.

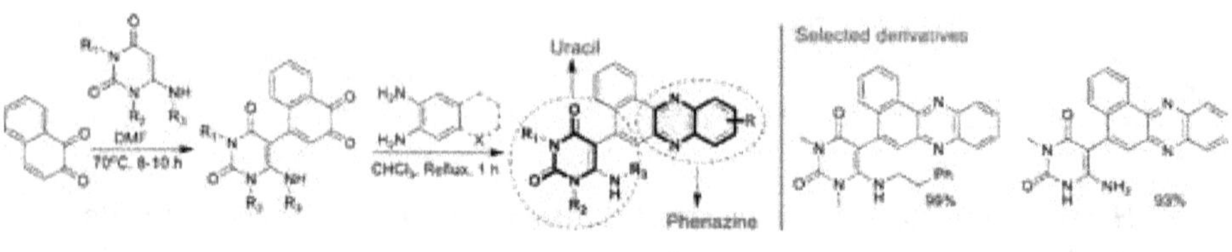

Figure 42.
Synthesis of uracil-phenazine molecules.

Afshin and coworkers [132] developed L-proline-catalyzed one-pot, two-step, five-component reaction for the synthesis of novel 1,4-dihydrobenzo[a]pyrido[2,3-c]phenazines by the condensation reaction of 2-hydroxynaphoquinone, aromatic 1,2-diamines, aldehydes, ammonium acetate, and ethyl acetoacetate under conventional heating in solvent-free conditions. In this domino transformation, six bonds and two new rings such as phenazine and 1,4-dihydropyridine are efficiently formed (**Figure 41**). High yields, short reaction time, operational simplicity, easy work-up procedure, avoidance of hazardous or toxic catalysts, and organic solvents are the main advantages of this green methodology.

In another study [133], catalyst-free synthesis of aminouracils bearing naphthoquinone in DMF system was developed by Jamaledini et al. [133]. Further it was used as intermediate for the synthesis of uracil-phenazine linked heterocycles via condensation reaction with various vicinal diamines, in chloroform under reflux condition (**Figure 42**).

Copper-catalyzed, TEMPO-mediated straightforward synthesis of 2,3-disubstituted naphtho[2,1-b]thiophene-4,5-diones via cross-dehydrogenative thienannulation was reported [134]. The reaction proceeded via in situ generated naphthalene-1,2-diones by dearomatization of β-naphthols, followed by oxidative heteroannulation with α-enolic dithioesters chemoselectively (**Figure 43**).

Further, the naphtho[2,1-b]thiophene-4,5-diones undergo L-proline-catalyzed cross-dehydrogenative coupling (CDC) with *ortho* -phenylenediamine enabling

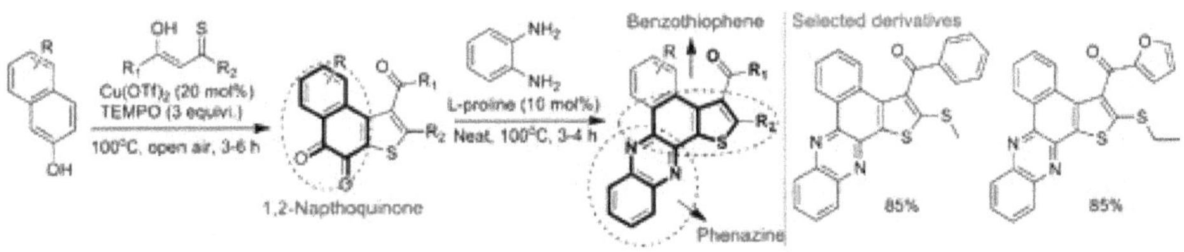

Figure 43.
Synthesis of benzothiophene-phenazine molecules.

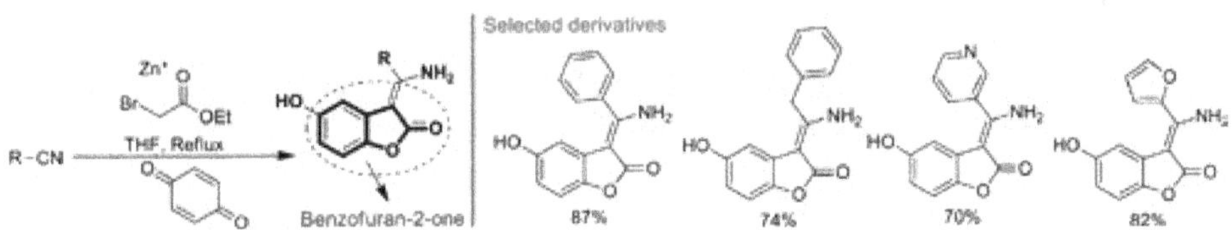

Figure 44.
Tandem Blaise-Nenitzescu reaction forbenzofuran-2(3H)-ones.

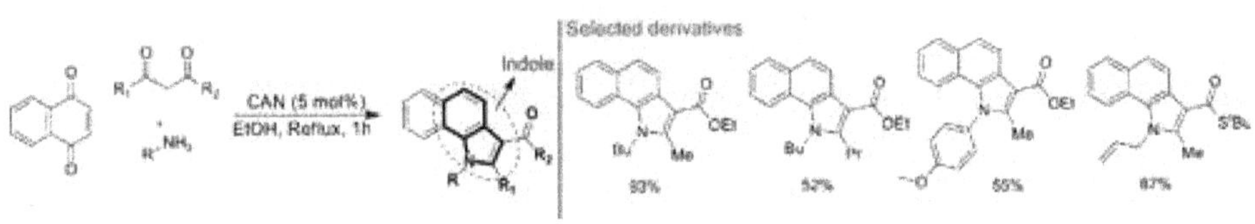

Figure 45.
CAN-catalyzed three-component domino sequences.

formation of pentacyclic benzo[a]thieno[3,2-c]phenazine derivatives in good yields under solvent-free conditions.

Interestingly, Lee and coworkers [134] reported one-pot synthesis of benzofuran-2(3H)-one derivatives from nitriles. This result underscore the high potential of the Blaise reaction intermediate as an amphiphilic organozinc complex for forming carbon—carbon bonds and provides a divergent synthetic platform toward heterocycles (**Figure 44**).

CAN-catalyzed three-component reaction between primary amines, β-dicarbonyl compounds, and functionalized or unfunctionalized naphthoquinones was reported by Menendez et al. [62]. The enamine formation Michael addition-intramolecular imine formation domino sequence starting from amines, β-dicarbonyl compounds, and quinones, in a three-component variation of the Nenitzescu indole synthesis (**Figure 45**). Further, protocol was extended to the synthesis of linear benzo[f]indolequinones by using pre-functionalized quinones as the starting materials. Moreover, the benzo[g]indole derivatives were transformed into 9,12-dihydro-8H-azepino[1,2-a]benzo[g]-indoles, a new class of fused indole derivatives, using a C-alkylation/ring-closing metathesis strategy.

4. Conclusion

Recent advances in the direct heterofunctionalization and heterocyclization of quinones were summarized in this chapter. Most of the C—hetero bond formation on quinone occurred via Michael addition in the presence/absence of a metal catalyst. Transition metal-catalyzed cross-coupling reactions were another important strategy for the direct functionalization of quinones. These reactions allowed for the construction of not only simple coupling products but also many important biologically active compounds. Moreover, the formation of C–O bond on quinone was less explored than C–N and C–S bond formation; it may be due to the fact that oxygen has lower nucleophilicity than nitrogen and sulfur, and lack of suitable synthetic reagents that can tolerate the presence of oxygen functional groups. However, due to the unique electronic property of quinones, the types of direct functionalization remain limited, and great efforts are still needed in the future.

Acknowledgements

AKTP would like to thank UGC-RFSMS (F.No.25-1/2014-2015 (BSR)/7-22/2007-(BSR) Dated: 13.03.2015), New Delhi, for the award of the fellowship for Ph.D.

Author details

Andivelu Ilangovan* and Thumadath Palayullaparambil Adarsh Krishna
School of Chemistry, Bharathidasan University, Tiruchirappalli, Tamilnadu, India

*Address all correspondence to: ilangovanbdu@yahoo.com

References

[1] Judy LB, Tareisha D. Formation and biological targets of quinones: Cytotoxic versus cytoprotective effects. Chemical Research in Toxicology. 2017;**30**(1): 13-37. DOI: 10.1021/acs.chemrestox. 6b00256

[2] Thomson RH. Naturally Occurring Quinones IV. London: Blackie Academic; 1997

[3] Marcin S, Artur O. Electronic connection between the quinone and cytochrome redox pools and its role in regulation of mitochondrial electron transport and redox signaling. Physiological Reviews. 2015;**95**:219-224. DOI: 10.1152/physrev.00006.2014

[4] Judy LB, Michael AT, Trevor MP, Glenn D, Terrence JM. Role of quinones in toxicology. Chemical Research in Toxicology. 2000;**13**(3):135-160. DOI: 10.1021/tx9902082

[5] Azadeh G, Jayne G, Jennifer RB, Cecilia CR, Jennette AS, Adam MC. A focused library synthesis and cytotoxicity of quinones derived from the natural product bolinaquinone. Royal Society Open Science. 2018;**5**(4): 1-23. DOI: 10.1098/rsos.171189

[6] Eun JS, Jae HK, Kayoung K, Chan BP. Quinone and its derivatives for energy harvesting and storage materials. Journal of Materials Chemistry A. 2016; **4**:11179-11202. DOI: 10.1039/C6TA03123D

[7] Bolton JL, Dunlap TL, Dietz BM. Formation and biological targets of botanical o-quinones. Food and Chemical Toxicology. 2018;**120**: 700-707. DOI: 10.1016/j.fct.2018.07.050

[8] Almeida WP, Correia CR. A total synthesis of the sesquiterpene quinone metachromin-A. Tetrahedron Letters. 1994;**35**(9):1367-1370. DOI: 10.1016/S0040-4039(00)76220-4

[9] Crooke ST, Bradner WT. Mitomycin C: A review. Cancer Treatment Reviews. 1976;**3**(3):121-139. DOI: 10.1016/S0305-7372(76)80019-9

[10] Norman AR, Norcott P, McErlean CS. Overview of the synthesis of carbazoloquinone natural products. Tetrahedron Letters. 2016;**57**(36): 4001-4008. DOI: 10.1016/j.tetlet.2016.07.092

[11] Sharna-kay AD, Downer-Riley NK. An improved synthesis of balsaminone A. Synlett. 2019;**30**(03):325-328. DOI: 10.1055/s-0037-1611975

[12] Hussain H, Krohn K, Ahmad VU, Miana GA, Green IR. Lapachol: An overview. ARKIVOC Journal. 2007;**2**: 145-171

[13] Ojha S, Al Taee H, Goyal S, Mahajan UB, Patil CR, Arya DS, et al. Cardioprotective potentials of plant-derived small molecules against doxorubicin associated cardiotoxicity. Oxidative Medicine and Cellular Longevity. 2016;**2016**:1-19. DOI: 10.1155/2016/5724973. Article ID 5724973

[14] Yusuf MA, Singh BN, Sudheer S, Kharwar RN, Siddiqui S, Abdel-Azeem AM, et al. Chrysophanol: A natural anthraquinone with multifaceted biotherapeutic potential. Biomolecules. 2019;**9**(2):68. DOI: 10.3390/biom9020068

[15] Zhou YX, Xia W, Yue W, Peng C, Rahman K, Zhang H. Rhein: A review of pharmacological activities. Evidence-Based Complementary and Alternative Medicine. 2015;**2015**:1-10. DOI: 10.1155/2015/578107

[16] Gu JQ, Graf TN, Lee D, Chai HB, Mi Q, Kardono LB, et al. Cytotoxic and antimicrobial constituents of the bark of *Diospyros maritima* collected in two

geographical locations in Indonesia. Journal of Natural Products. 2004;**67**(7): 1156-1161. DOI: 10.1021/np040027m

[17] Akhter S, Rony SR, Al-Mansur MA, Hasan CM, Rahman KM, Sohrab MH. Lawsonol, a new bioactive naphthoquinone dimer from the leaves of *Lawsonia alba*. Chemistry of Natural Compounds. 2018;**54**(1):26-29. DOI: 10.1007/s10600-018-2251-0

[18] Quinone Chemistry, Web of Science-Search for peer-reviewed journals, articles, book chapters and open access content. 2019

[19] Zhengkai C, Binjie W, Jitan Z, Wenlong Y, Zhanxiang L, Yuhong Z. Transition metal-catalyzed C–H bond functionalizations by the use of diverse directing groups. Organic Chemistry Frontiers. 2015;**2**:1107-1295. DOI: 10.1039/C5QO00004A

[20] Alison EW, Shannon S. Quinone-catalyzed selective oxidation of organic molecules. Angewandte Chemie International Edition. 2015;**54**(49): 14587-14977. DOI: 10.1002/anie.201505017

[21] Ruipu Z, Sanzhong L. Bio-inspired quinone catalysis. Chinese Chemical Letters. 2018;**29**(8):1193-1200. DOI: 10.1016/j.cclet.2018.02.009

[22] Huaisu G, Weilin G, Yang L, Xiaohua R. Quinone-modified metal-organic frameworks MIL-101(Fe) as heterogeneous catalysts of persulfate activation for degradation of aqueous organic pollutants. Water Science and Technology. 2019;**79**(12):2357-2365. DOI: 10.2166/wst.2019.239

[23] Lysons TW, Sanford MS. Palladium catalyzed ligand directed C–H functionalization reactions. Chemical Reviews. 2010;**110**:1147-1169. DOI: 10.1021/cr900184e

[24] Liu C, Zhang H, Shi W, Lei A. Bond formations between two nucleophiles: Transition metal catalyzed oxidative cross-coupling reactions. Chemical Reviews. 2011;**111**:1780-1842. DOI: 10.1021/cr100379j

[25] Yuta F, Victoriano D, Ian BS, Ryan G, Matthew DB, Baran PS. Practical C–H functionalization of quinones with boronic acids. Journal of American Chemical Society. 2011; **133**(10):3292-3295. DOI: 10.1021/ja111152z

[26] Yu W, Hjerrild P, Jacobsen KM, Tobiesen HN, Clemmensen L, Poulsen TB. A catalytic oxidative quinone heterofunctionalization method: Synthesis of strongylophorine-26. Angewandte Chemie International Edition. 2018;**57**(31):9805-9809. DOI: 10.1002/anie.201805580

[27] Engler TA, Reddy JP. An unusual gamma-silyl effect in titanium tetrachloride catalyzed arylation of 1,4-benzoquinones. Journal of Organic Chemistry. 1991;**56**:6491. DOI: 10.1021/jo00023a005

[28] Pirrung MC, Liu Y, Deng D, Halstead DK, Li Z, May JF, et al. Methyl scanning: Total synthesis of demethylasterriquinone B1 and derivatives for identification of sites of interaction with and isolation of its receptors. Journal of American Chemical Society. 2005;**127**:4609-4626. DOI: 10.1021/ja044325h

[29] Zhang HB, Liu L, Chen YJ, Wang D, Li CJ. Synthesis of aryl-substituted 1,4-benzoquinone via water-promoted and In(OTf)3-catalyzed in situ conjugate addition-dehydrogenation of aromatic compounds to 1,4-benzoquinone in water Advance Synthesis and Catalysis. 2006;**348**:229-235. DOI: 10.1002/adsc.200505248

[30] Katarzyna K, Oleg MD, Marta W, Pietrusiewicz KM. Brönsted acid catalyzed direct oxidative arylation of 1,4-naphthoquinone. Current

Chemistry Letters. 2014;**3**:23-36. DOI: 10.5267/j.ccl.2013.10.001

[31] Li BJ, Yang SD, Shi ZJ. Recent advances in direct arylation via palladium-catalyzed aromatic CH activation. Synlett. 2008;**07**:949-957. DOI: 10.1055/s-2008-1042907

[32] Yijun W, Shuai Z, Liang HZ. Recent advances in direct functionalization of quinones. European Journal of Organic Chemistry. 2019;**12**(31):2179-2201. DOI: 10.1002/ejoc.201900028

[33] Nuria T, Echavarren AM, Paredes MC. Palladium catalyzed coupling of 2-bromonaphthoquinones with stannanes: A concise synthesis of antibiotics WS 5995 A and C and related compounds. Journal of Organic Chemistry. 1991;**56**:6490-6494

[34] Gan X, Jiang W, Wang W, Hu L. An approach to 3,6-disubstituted 2,5-dioxybenzoquinones via two sequential suzuki couplings: Three-step synthesis of Leucomelone. Organic Letters. 2009;**11**(3):589-592. DOI: 10.1021/ol802645f

[35] Hadden MK, Hill SA, Davenport I, Matts RL, Blagg BS. Synthesis and evaluation of Hsp90 inhibitors that contain the 1,4-naphthoquinone scaffold. Bioorganic and Medicinal Chemistry. 2009;**17**:634-640. DOI: 10.1016/j.bmc.2008.11.064

[36] Hassan Z, Ullah I, Ali I, Khera RA, Knepper I, Ali A, et al. Synthesis of tetra aryl-p-benzoquinones and 2,3-diaryl-1,4-naphthoquinones via Suzuki-Miyaura cross-coupling reactions. Tetrahedron. 2013;**69**:460-469. DOI: 10.1016/j.tet.2012.11.040

[37] Louvis AR, Silva NAA, Semaan FS, Da-Silva FC, Saramago G, Souza LC, et al. Synthesis, characterization and biological activities of 3-aryl 1,4-naphthoquinones green palladium-catalysed Suzuki cross coupling. New Journal of Chemistry. 2016;**40**: 7643-7656. DOI: 10.1039/C6NJ00872K

[38] Redondo MC, Veguillas M, Ribagorda M, Carreno MC. Control of the regio-and stereoselectivity in Diels Alder reactions with quinone boronic acids. Angewandte Chemie International Edition. 2009;**48**(2): 370-374. DOI: 10.1002/anie.200803428

[39] Liebeskind LS, Foster BS. Stannylquinones synthesis and utilization as quinone carbanion synthetic equivalents. Journal of the American Chemical Society. 1990; **112**(23):8612-8613. DOI: 10.1021/ja00179a072

[40] Lanny SL, Steven WR. Substituted quinone synthesis by palladium-copper cocatalyzed cross-coupling of stannylquinones with aryl and heteroaryl iodides. Journal of Organic Chemistry. 1993;**58**:408-413. DOI: 10.1021/jo00054a025

[41] Liu RH, He YH, Yu W, Zhou B, Bing H. Silver catalyzed site-selective ring-opening and C–C bond functionalization of cyclic amines: Access to distal aminoalkyl-substituted quinones. Organic Letters. 2019;**21**: 4590-4594. DOI: 10.1021/acs.orglett.9b01496

[42] Jardim GA, Silva TL, Goulart MO, de Simone CA, Barbosa JM, Salomão K, et al. Rhodium-catalyzed CH bond activation for the synthesis of quinonoid compounds: Significant anti-trypanosoma cruzi activities and electrochemical studies of functionalized quinones. European Journal of Medicinal Chemistry. 2017; **136**:406-419. DOI: 10.1016/j.ejmech.2017.05.011

[43] Dias GG, Rogge T, Kuniyil R, Jacob C, Menna-Barreto RF, da Silva Júnior EN, et al. Ruthenium-catalyzed C–H oxygenation of quinones by weak O-coordination for potent trypanocidal

agents. Chemical Communications. 2018;**54**(91):12840-12843. DOI: 10.1039/C8CC07572G

[44] Jardim GA, Bozzi ÍA, Oliveira WX, Mesquita-Rodrigues C, Menna-Barreto RF, Kumar RA, et al. Copper complexes and carbon nanotube–copper ferrite-catalyzed benzenoid A-ring selenation of quinones: An efficient method for the synthesis of trypanocidal agents. New Journal of Chemistry. 2019;**43**(35): 13751-13763. DOI: 10.1039/C9NJ02026H

[45] Yakkala PA, Giri D, Chaudhary B, Auti P, Sharma S. Regioselective C–H alkylation and alkenylation at the C5 position of 2-amino-1, 4-naphthoquinones with maleimides under Rh (III) catalysis. Organic Chemistry Frontiers. 2019;**6**(14): 2441-2446. DOI: 10.1039/C9QO00538B

[46] Dias GG, Nascimento TA, de Almeida AK, Bombaça AC, Menna-Barreto RF, Jacob C, et al. Ruthenium (II)-catalyzed C–H Alkenylation of quinones: Diversity-oriented strategy for trypanocidal compounds. European Journal of Organic Chemistry. 2019; **9**(13):2344-2353. DOI: 10.1002/ejoc.201900004

[47] Yu D, Chen XL, Ai BR, Zhang XM, Wang JY. Tetrabutylammonium iodide catalyzed hydroxylation of naphthoquinone derivatives with tert-butyl hydroperoxide as an oxidant. Tetrahedron Letters. 2018;**59**(40): 3620-3623. DOI: 10.1016/j.tetlet.2018.08.052

[48] Castellano S, Bertamino A, Gomez-Monterrey I, Santoriello M, Grieco P, Campiglia P, et al. A practical, green, and selective approach toward the synthesis of pharmacologically important quinone-containing heterocyclic systems using alumina-catalyzed Michael addition reaction. Tetrahedron Letters. 2008;**49**(4): 583-585. DOI: 10.1016/j.tetlet.2007.11.148

[49] Satish G, Polu A, Ramar T, Ilangovan A. Iodine-mediated C–H functionalization of sp, sp2, and sp3 carbon: A unified multisubstrate domino approach for isatin synthesis. The Journal of Organic Chemistry. 2015; **80**(10):5167-5175. DOI: 10.1021/acs.joc.5b00581

[50] Ilangovan A, Satish G. Copper-mediated selective C–H activation and cross-dehydrogenative C–N coupling of 2′-aminoacetophenones. Organic Letters. 2013;**15**(22):5726-5729. DOI: 10.1021/ol402750r

[51] Ilangovan A, Polu A, Satish G. $K_2S_2O_8$-mediated metal-free direct C–H functionalization of quinones using arylboronic acids. Organic Chemistry Frontiers. 2015;**2**(12):1616-1620. DOI: 10.1039/C5QO00246J

[52] Ashok P, Ilangovan A. Transition metal mediated selective C vs N arylation of 2-aminonaphthoquinone and its application toward the synthesis of benzocarbazoledione. Tetrahedron Letters. 2018;**59**(5):438-441. DOI: 10.1016/j.tetlet.2017.10.075

[53] Ilangovan A, Saravanakumar S, Malayappasamy S. γ-Carbonyl quinones: Radical strategy for the synthesis of evelynin and its analogues by C–H activation of quinones using cyclopropanols. Organic Letters. 2013; **15**(19):4968-4971. DOI: 10.1021/ol402229m

[54] Rajasekar S, Krishna TA, Tharmalingam N, Andivelu I, Mylonakis E. Metal-free C–H thiomethylation of quinones using iodine and DMSO and study of antibacterial activity. ChemistrySelect. 2019;**4**(8):2281-2287. DOI: 10.1002/slct.201803816

[55] Krishna TPA, Sakthivel P, Ilangovan A. Iron-mediated site-selective oxidative C–H/C–H cross-coupling of aryl radicals with quinones:

Synthesis of β-secretase-1 inhibitor B and related arylated quinones. Organic Chemistry Frontiers. 2019;**6**:3244-3251. DOI: 10.1039/c9qo00623k

[56] Ramos-Peralta L, López-López LI, Silva-Belmares SY, Zugasti-Cruz A, Rodríguez-Herrera R, Aguilar-González CN. Naphthoquinone: Bioactivity and green synthesis. The Battle Against Microbial Pathogens: Basic Science, Technological Advances and Educational Programs. 2015:542-550

[57] Lavergne O, Fernandes AC, Bréhu L, Sidhu A, Brézak MC, Prévost G, et al. Synthesis and biological evaluation of novel heterocyclic quinones as inhibitors of the dual specificity protein phosphatase CDC25C. Bioorganic and Medicinal Chemistry Letters. 2006;**16**(1):171-175. DOI: 10.1016/j.bmcl.2005.09.030

[58] Arnone A, Merlini L, Nasini G, de Pava OV. Direct amination of naphthazarin, juglone, and some derivatives. Synthetic Communications. 2007;**37**(15):2569-2577. DOI: 10.1080/00397910701462864

[59] Bukhtoyarova AD, Rybalova TV, Ektova LV. Amination of 5-hydroxy-1,4-naphthoquinone in the presence of copper acetate. Russian Journal of Organic Chemistry. 2010;**46**(6):855-859. DOI: 10.1134/S1070428 01006012

[60] Tuyun AF, Bayrak N, Yıldırım H, Onul N, Mataraci Kara E, Ozbek CB. Synthesis and in vitro biological evaluation of aminonaphthoquinones and benzo[b]phenazine-6,11-dione derivatives as potential antibacterial and antifungal compounds. Journal of Chemistry. 2015;**2015**:1-8. DOI: 10.1155/2015/645902

[61] Janeczko M, Demchuk OM, Strzelecka D, Kubiński K, Masłyk M. New family of antimicrobial agents derived from 1,4-naphthoquinone. European Journal of Medicinal Chemistry. 2016;**29**(124):1019-1025. DOI: 10.1016/j.ejmech.2016.10.034

[62] Suryavanshi PA, Sridharan V, Menéndez JC. Expedient, one-pot preparation of fused indoles via CAN-catalyzed three-component domino sequences and their transformation into polyheterocyclic compounds containing pyrrolo [1,2-a] azepine fragments. Organic and Biomolecular Chemistry. 2010;**8**(15):3426-3436. DOI: 10.1039/C004703A

[63] Tapia RA, Cantuarias L, Cuéllar M, Villena J. Microwave-assisted reaction of 2,3-dichloronaphthoquinone with aminopyridines. Journal of the Brazilian Chemical Society. 2009;**20**(5):999-1002. DOI: 10.1590/S0103-50532009 000500027

[64] Gouda MA, Eldien HF, Girges MM, Berghot MA. Synthesis and antioxidant activity of novel series of naphthoquinone derivatives attached to benzothiophene moiety. Medicinal Chemistry. 2013;**3**(2):2228-2232. DOI: 10.4172/2161-0444.1000143

[65] Brandy Y, Brandy N, Akinboye E, Lewis M, Mouamba C, Mack S, et al. Synthesis and characterization of novel unsymmetrical and symmetrical 3-halo- or 3-methoxy-substituted 2-dibenzoylamino-1, 4-naphthoquinone derivatives. Molecules. 2013;**18**(2):1973-1984. DOI: 10.3390/molecules18021973

[66] Mital A, Sonawane M, Bindal S, Mahlavat S, Negi V. Substituted 1,4-naphthoquinones as a new class of antimycobacterial agents. Der Pharma Chemica. 2010;**2**(3):63-73

[67] Tran NC, Le MT, Nguyen DN, Tran TD. Synthesis and biological evaluation of halogen substituted 1,4-naphthoquinones as potent

antifungal agents. Molecular Diversity Preservation International. 2009:1-7

[68] Liu R, Li H, Ma WY. A new method for the biomimetic synthesis of 2-hydroxy-3-amino-1,4 naphthoquinone. Advanced Materials Research. 2013;**781**: 287-290. DOI: 10.4028/www.scientific.net/AMR.781-784.287

[69] Bayen S, Barooah N, Sarma RJ, Sen TK, Karmakar A, Baruah JB. Synthesis, structure and electrochemical properties of 2, 5-bis (alkyl/arylamino) 1,4-benzoquinones and 2-arylamino-1, 4-naphthoquinones. Dyes and Pigments. 2007;**75**(3):770-775. DOI: 10.1016/j. dyepig.2006.07.033

[70] Yadav JS, Reddy BV, Swamy T, Shankar KS. Green protocol for conjugate addition of amines to p-quinones accelerated by water. Monatshefte für Chemie-Chemical Monthly. 2008;**139**(11):1317. DOI: 10.1007/s00706-008-0917-1

[71] Liu B, Ji SJ. Facile synthesis of 2-amino-1,4-naphthoquinones catalyzed by molecular iodine under ultrasonic irradiation. Synthetic Communications. 2008;**38**(8):1201-1211. DOI: 10.1080/00397910701866254

[72] Lisboa CD, Santos VG, Vaz BG, de Lucas NC, Eberlin MN, Garden SJ. C–H functionalization of 1,4-naphthoquinone by oxidative coupling with anilines in the presence of a catalytic quantity of copper (II) acetate. The Journal of Organic Chemistry. 2011; **76**(13):5264-5273. DOI: 10.1021/jo200354u

[73] Sharma U, Katoch D, Sood S, Kumar N, Singh B, Thakur A, et al. Synthesis, antibacterial and antifungal activity of 2-amino-1,4-naphthoquinones using silica-supported perchloric acid ($HClO_4$-SiO_2) as a mild, recyclable and highly efficient heterogeneous catalyst. Indian Journal of Chemistry. 2013;**54**:1431-1440

[74] Devenderan R, Kasi P. Metal-free, base promoted sp2 C–H functionalization in the sulfonamidation of 1, 4-naphthoquinones. Organic and Biomolecular Chemistry. 2018;**16**(29): 5294-5300. DOI: 10.1039/C8OB00818C

[75] Chen XL, Dong Y, He S, Zhang R, Zhang H, Tang L, et al. A one-pot approach to 2-(N-substituted amino)-1,4-naphthoquinones with use of nitro compounds and 1,4-naphthoquinones in water. Synlett. 2019;**30**(05):615-619. DOI: 10.1055/s-0037-1610689

[76] Feng M, Tang B, Liang SH, Jiang X. Sulfur containing scaffolds in drugs: Synthesis and application in medicinal chemistry. Current Topics in Medicinal Chemistry. 2016;**16**(11): 1200-1216

[77] Ge B, Wang D, Dong W, Ma P, Li Y, Ding Y. Synthesis of arylsulfonyl-quinones and arylsulfonyl-1,4-diols as FabH inhibitors: Pd-catalyzed direct C-sulfone formation by C–S coupling of quinones with arylsulfonyl chloride. Tetrahedron Letters. 2014;**55**(40): 5443-5446. DOI: 10.1016/j.tetlet.2014.08.023

[78] Wang L, Xie YB, Yang QL, Liu MG, Zheng KB, Hu YL, et al. Ir-catalyzed C–S coupling of quinones with sulfonyl chloride. Journal of the Iranian Chemical Society. 2016;**13**(10):1797-1803. DOI: 10.1007/s13738-016-0897-8

[79] YuX,WuQ,WanH,XuZ,XuX, Wang D. Copper and triphenylphosphine-promoted sulfenylation of quinones with arylsulfonyl chlorides. RSC Advances. 2016;**6**(67):62298-62301. DOI: 10.1039/C6RA11301J

[80] Zhang C, McClure J, Chou CJ. Silver-catalyzed direct thiolation of quinones by activation of aryl disulfides to synthesize quinonyl aryl thioethers. The Journal of Organic Chemistry. 2015; **80**(10):4919-4927. DOI: 10.1021/acs.joc.5b00247

[81] Yadav JS, Reddy BV, Swamy T, Ramireddy N. Ionic liquids-promoted addition of arylsulfinic acids to p-quinones: A green synthesis of diaryl sulfones. Synthesis. 2004;**2004**(11): 1849-1853. DOI: 10.1055/s-2004-829145

[82] Yuan JW, Liu SN, Qu LB. Ammonium iodide-promoted unprecedented arylsulfonylation of quinone with sodium arylsulfinates. Tetrahedron. 2017;**73**(48):6763-6772. DOI: 10.1016/j.tet.2017.10.022

[83] Tandon VK, Maurya HK. 'On water': Unprecedented nucleophilic substitution and addition reactions with 1,4-quinones in aqueous suspension. Tetrahedron Letters. 2009;**50**(43): 5896-5902. DOI: 10.1016/j.tetlet.2009.07.149

[84] Zeng FL, Chen XL, He SQ, Sun K, Liu Y, Fu R, et al. Copper-catalyzed one-pot three-component thioamination of 1,4-naphthoquinone. Organic Chemistry Frontiers. 2019;**6**(9):1476-1480. DOI: 10.1039/C9QO00091G

[85] Lu Y, Zhao Y, Wang S, Wang X, Ge Z, Li R. An efficient synthesis of 2-thio-5-amino substituted benzoquinones via KI catalyzed cascade oxidation/michael addition/oxidation starting from hydroquinone. RSC Advances.2016;**6**(14):11378-11381. DOI: 10.1039/C5RA26524J

[86] Wellington KW, Gordon GE, Ndlovu LA, Steenkamp P. Laccase-catalyzed C–S and C–C coupling for a one-pot synthesis of 1,4-naphthoquinone sulfides and 1, 4-naphthoquinone sulfide dimers. ChemCatChem. 2013;**5**(6):1570-1577. DOI: 10.1002/cctc.201200606

[87] Wellington KW, Bokako R, Raseroka N, Steenkamp P. A one-pot synthesis of 1,4-naphthoquinone-2, 3-bis-sulfides catalysed by a commercial laccase. Green Chemistry. 2012;**14**(9): 2567-2576. DOI: 10.1039/C2GC35926J

[88] Schlippert M, Mikolasch A, Hahn V, Schauer F. Enzymatic thiol Michael addition using laccases: Multiple CS bond formation between *p*-hydroquinones and aromatic thiols. Journal of Molecular Catalysis B: Enzymatic. 2016;**126**:106-114. DOI: 10.1016/j.molcatb.2015.12.012

[89] Takeya T, Kondo H, Otsuka T, Tomita K, Okamoto I, Tamura O. A novel construction of dibenzofuran-1,4-diones by oxidative cyclization of quinone-arenols. Organic Letters. 2007; **9**(15):2807-2810. DOI: 10.1021/ol070951i

[90] Feng X, Qiu X, Huang H, Wang J, Xu X, Xu P, et al. Palladium (II)-catalyzed reaction of lawsones and propargyl carbonates: Construction of 2,3-furanonaphthoquinones and evaluation as potential indoleamine 2,3-dioxygenase inhibitors. The Journal of Organic Chemistry. 2018;**83**(15): 8003-8010. DOI: 10.1021/acs.joc.8b00872

[91] Perez AL, Lamoureux G, Sánchez-Kopper A. Efficient syntheses of streptocarpone and (±)-α-dunnione. Tetrahedron Letters. 2007;**48**(21): 3735-3738. DOI: 10.1016/j.tetlet.2007.03.090

[92] Da Silva Júnior PE, de Araujo NM, da Silva Emery F. Claisen rearrangement of hydroxynaphthoquinones: Selectivity toward naphthofuran or α-xiloidone using copper salts and iodine. Journal of Heterocyclic Chemistry. 2015;**52**(2): 518-521. DOI: 10.1002/jhet.2087

[93] Weitz E, Scheffer A. Über die Einwirkung von alkalischem Wasserstoffsuperoxyd auf ungesättigte Verbindungen. Berichte der Deutschen Chemischen Gesellschaft (A and B Series). 1921;**54**(9):2327-2344. DOI: 10.1002/cber.19210540922

[94] Nasiri HR, Madej MG, Panisch R, Lafontaine M, Bats JW, Lancaster CR,

et al. Design, synthesis, and biological testing of novel naphthoquinones as substrate-based inhibitors of the quinol/fumarate reductase from Wolinella succinogenes. Journal of Medicinal Chemistry. 2013;**56**(23):9530-9541. DOI: 10.1021/jm400978u

[95] Chang YC, Yuan PT, Hong FE. C–H bond functionalization of 1, 4-benzoquinone by silver-mediated regioselective phosphination and amination reactions. European Journal of Organic Chemistry. 2017;**2017**(17): 2441-2450. DOI: 10.1002/ejoc.201700109

[96] Yuan PT, Pai CH, Huang SZ, Hong FE. Making CN and CP bonds on the quinone derivatives through the assistance of silver-mediated CH functionalization processes. Tetrahedron. 2017;**73**(48):6786-6794. DOI: 10.1016/j.tet.2017.10.037

[97] Kalaria PN, Karad SC, Raval DK. A review on diverse heterocyclic compounds as the privileged scaffolds in antimalarial drug discovery. European Journal of Medicinal Chemistry. 2018; **158**:917-936. DOI: 10.1016/j.ejmech.2018.08.040

[98] Buntrock RE. Review of heterocyclic chemistry. Journal of Chemical Education. 2012;**89**(11):1349-1350. DOI: 10.1021/ed300616t

[99] Verweij J, Pinedo HM. Mitomycin C: Mechanism of action, usefulness and limitations. Anti-Cancer Drugs. 1990; **1**(1):5-13

[100] Tisšler M. Heterocyclic quinones. In: Advances in Heterocyclic Chemistry. Vol. 45. Academic Press, Elsevier; 1989. pp. 37-150. DOI: 10.1016/S0065-2725(08)60329-3

[101] Garuti L, Roberti M, Pizzirani D. Nitrogen-containing heterocyclic quinones: A class of potential selective antitumor agents. Mini Reviews in Medicinal Chemistry. 2007;7(5): 481-489. DOI: 10.2174/ 13895570778061 9626

[102] Take Y, Oogose K, Kubo T, Inouye Y, Nakamura S, Kitahara Y, et al. Comparative study on biological activities of heterocyclic quinones and streptonigrin. The Journal of Antibiotics. 1987;**40**(5):679-684. DOI: 10.7164/antibiotics.40.679

[103] Deniz NG, Ibis C, Gokmen Z, Stasevych M, Novikov V, Komarovska-Porokhnyavets O, et al. Design, synthesis, biological evaluation, and antioxidant and cytotoxic activity of heteroatom-substituted 1,4-naphtho- and benzoquinones. Chemical and Pharmaceutical Bulletin. 2015; **63**(12):1029-1039. DOI: 10.1248/cpb.c15-00607

[104] Maier ME. Design and synthesis of analogues of natural products. Organic and Biomolecular Chemistry. 2015; **13**(19):5302-5343. DOI: 10.1039/C5OB00169B

[105] Choudhary S, Singh PK, Verma H, Singh H, Silakari O. Success stories of natural product-based hybrid molecules for multi-factorial diseases. European Journal of Medicinal Chemistry. 2018; **151**:62-97. DOI: 10.1016/j.ejmech.2018.03.057

[106] Ramsay RR, Popovic-Nikolic MR, Nikolic K, Uliassi E, Bolognesi ML. A perspective on multi-target drug discovery and design for complex diseases. Clinical and Translational Medicine. 2018;**7**(1):3. DOI: 10.1186/s40169-017-0181-2

[107] Wink M. Evolutionary advantage and molecular modes of action of multi-component mixtures used in phytomedicine. Current Drug Metabolism. 2008;**9**(10):996-1009. DOI: 10.2174/138920008786927794

[108] Kaliappan KP, Ravikumar V. Design and synthesis of novel sugar-oxasteroid-

quinone hybrids. Organic and Biomolecular Chemistry. 2005; **3**(5): 848-851. DOI: 10.1039/B418659A

[109] Nepovimova E, Uliassi E, Korabecny J, Pena-Altamira LE, Samez S, Pesaresi A, et al. Multitarget drug design strategy: Quinone–tacrine hybrids designed to block amyloid-β aggregation and to exert anticholinesterase and antioxidant effects. Journal of Medicinal Chemistry. 2014;**57**(20):8576-8589. DOI: 10.1021/jm5010804

[110] Mallavadhani UV, Prasad CV, Shrivastava S, Naidu VG. Synthesis and anticancer activity of some novel 5,6-fused hybrids of juglone based 1,4-naphthoquinones. European Journal Of Medicinal Chemistry. 2014;**83**: 84-91. DOI: 10.1016/j.ejmech.2014. 06.012

[111] Frenkel-Pinter M, Tal S, Scherzer-Attali R, Abu-Hussien M, Alyagor I, Eisenbaum T, et al. Naphthoquinone-tryptophan hybrid inhibits aggregation of the tau-derived peptide PHF6 and reduces neurotoxicity. Journal of Alzheimer's Disease. 2016; **51**(1): 165-178. DOI: 10.3233/JAD-150927

[112] Defant A, Mancini I. Design, synthesis and cancer cell growth inhibition evaluation of new aminoquinone hybrid molecules. Molecules. 2019;**24**(12):2224. DOI: 10.3390/molecules24122224

[113] Zhou T, Li L, Li B, Song H, Wang B. Ir (III)-catalyzed oxidative coupling of NH isoquinolones with benzoquinone. Organic Letters. 2015; **17**(17):4204-4207. DOI: 10.1021/acs. orglett.5b01974

[114] Guo S, Liu Y, Zhao L, Zhang X, Fan X. Rhodium-catalyzed selective oxidative (spiro) annulation of 2-arylindoles by using benzoquinone as a C2 or C1 synthon. Organic Letters. 2019; **21**(16):6437-6441. DOI: 10.1021/ acs. orglett.9b02336

[115] Sun JW, Wang XS, Liu Y. Copper (II)-catalyzed sequential C,N-difunctionalization of 1,4-naphthoquinone for the synthesis of benzo[f]indole-4,9-diones under base-free condition. The Journal of Organic Chemistry. 2013;**78**(20):10560-10566. DOI: 10.1021/jo401842d

[116] Chen SW, Hong FE. Palladium-catalyzed C–H functionalization of amido-substituted 1, 4-napthoquinone in the presence of amines toward the formation of pyrroles and imidazoles. ChemistrySelect. 2017;**2**(31): 10232-10238. DOI: 10.1002/ slct.201702173

[117] Acuña J, Piermattey J, Caro D, Bannwitz S, Barrios L, López J, et al. Synthesis, anti-proliferative activity evaluation and 3D-QSAR study of naphthoquinone derivatives as potential anti-colorectal cancer agents. Molecules. 2018;**23**(1):186. DOI: 10.3390/ molecules23010186

[118] Luu QH, Guerra JD, Castaneda CM, Martinez MA, Saunders J, Garcia BA, et al. Ultrasound assisted one-pot synthesis of benzo-fused indole-4,9-dinones from 1,4-naphthoquinone and α-aminoacetals. Tetrahedron Letters. 2016; **57**(21):2253-2256. DOI: 10.1016/j. tetlet.2016.04.031

[119] Pushkarskaya E, Wong B, Han C, Capomolla S, Gu C, Stoltz BM, et al. Single-step synthesis of 3-hydroxycarbazoles by annulation of electron-rich anilines and quinones. Tetrahedron Letters. 2016;**57**(50): 5653-5657. DOI: 10.1016/j. tetlet.2016.11.009

[120] Wu F, Bai R, Gu Y. Synthesis of benzofurans from ketones and 1,4-benzoquinones. Advanced Synthesis and Catalysis. 2016;**358**(14):2307-2316. DOI: 10.1002/adsc.201600048

[121] Wang L, Zhang J, Lang M, Wang J. Palladium-catalyzed ring contraction reaction of naphthoquinones upon

reaction with alkynes. Organic Chemistry Frontiers. 2016;**3**(5): 603-608. DOI: 10.1039/C6QO00045B

[122] Naganaboina RT, Nayak A, Peddinti RK. Trifluoroacetic acid-promoted Michael addition–cyclization reactions of vinylogous carbamates. Organic and Biomolecular Chemistry. 2014;**12**(21):3366-3370. DOI: 10.1039/c4ob00437j

[123] Jung Y, Kim I. Chemoselective reduction of quinols as an alternative to Sonogashira coupling: Synthesis of polysubstituted benzofurans. Organic and Biomolecular Chemistry. 2016; **14**(44):10454-10472. DOI: 10.1039/c6ob01941b

[124] Yang W, Wang S, Zhang Q, Liu Q, Xu X. Rh (iii)-catalyzed oxidative C–H bond arylation with hydroquinones: Sustainable synthesis of dibenzo[b,d]pyran-6-ones and benzo[d]naphtho[1,2-b]pyran-6-ones. Chemical Communications. 2015;**51**(4):661-664. DOI: 10.1039/C4CC08260E

[125] Yang W, Wang J, Wei Z, Zhang Q, Xu X. Kinetic control of Rh (III)-catalyzed annulation of C–H bonds with quinones: Chemoselective synthesis of hydrophenanthridinones and phenanthridinones. The Journal of Organic Chemistry. 2016;**81**(4): 1675-1680. DOI: 10.1021/acs.joc.5b02903

[126] Yang W, Wang J, Wang H, Li L, Guan Y, Xu X, et al. Rhodium (iii)-catalyzed three-component cascade synthesis of 6H-benzo[c]chromenes through C–H activation. Organic and Biomolecular Chemistry. 2018;**16**(38): 6865-6869. DOI: 10.1039/C8OB01938J

[127] Xia L, Lee YR. A novel and efficient synthesis of diverse dihydronaphtho[1,2-b] furans using the ceric ammonium nitrate-catalyzed formal [3 + 2] cycloaddition of 1,4-naphthoquinones to olefins and its application to furomollugin. Organic and Biomolecular Chemistry. 2013; **11**(36): 6097-6107. DOI: 10.1039/c3ob40977e

[128] Shaterian HR, Mohammadnia M. Effective preparation of 2-amino-3-cyano-4-aryl-5,10-dioxo-5,10-dihydro- 4H-benzo[g]chromene and hydroxyl naphthalene-1,4-dione derivatives under ambient and solvent-free conditions. Journal of Molecular Liquids. 2013;**177**:353-360. DOI: 10.1016/j.molliq.2012.10.012

[129] Dekamin MG, Alikhani M, Javanshir S. Organocatalytic clean synthesis of densely functionalized 4H-pyrans by bifunctional tetraethylammonium 2-(carbamoyl) benzoate using ball milling technique under mild conditions. Green Chemistry Letters and Reviews. 2016; **9**(2):96-105. DOI: 10.1080/17518253.2016.1139191

[130] Perumal M, Sengodu P, Venkatesan S, Srinivasan R, Paramsivam M. Environmentally benign copper triflate-mediated multicomponent one-pot synthesis of novel benzo[g]chromenes possess potent anticancer activity. ChemistrySelect. 2017;**2**(18):5068-5072. DOI: 10.1002/slct.201700170

[131] Cao YQ, Li XR, Wu W, Zhang D, Zhang ZH, Mo LP. A green approach for synthesis of naphthoquinone-fused oxazine derivatives in water under ultrasonic irradiation. Research on Chemical Intermediates. 2017;**43**(7): 3745-3755. DOI: 10.1007/s11164-016-2854-7

[132] Yazdani-Elah-Abadi A, Pour SA, Kangani M, Mohebat R. L-Proline catalyzed domino cyclization for the green synthesis of novel 1,4-dihydrobenzo[a]pyrido[2,3-c] phenazines. Monatshefte für Chemie-Chemical Monthly. 2017;**148**(12): 2135-2142. DOI: 10.1007/s00706-017-2008-7

[133] Jamaledini A, Mohammadizadeh MR, Mousavi SH. Catalyst-free, efficient, and green procedure for the synthesis of 5-heterocyclic substituted 6-aminouracils. Monatshefte für Chemie-Chemical Monthly. 2018; **149**(8): 1421-1428. DOI: 10.1007/s00706-018-2164-4

[134] Shukla G, Srivastava A, Yadav D, Singh MS. Copper-catalyzed one-pot cross-dehydrogenativethienannulation: Chemoselective access to naphtho[2,1-b]thiophene-4,5-diones and subsequent transformation to benzo[a]thieno[3,2-c] phenazines. The Journal of Organic Chemistry. 2018;**83**(4):2173-2181. DOI: 10.1021/acs.joc.7b03092

2

Catalytic Activity of Iron N-Heterocyclic Carbene Complexes

Badri Nath Jha, Nishant Singh and Abhinav Raghuvanshi

Abstract

Recent research towards development of more efficient as well as cost effective catalyst as a substitute to traditional precious metal catalysts has witnessed significant growth and interest. Importance has been given to catalyst based on 3d-transition metals, especially iron because of the broad availability and environmental compatibility which allows its use in various environmentally friendly catalytic processes. N-Heterocyclic carbene (NHC) ligands have garnered significant attention because of their unique steric and electronic properties which provide substantial scope and potential in organometallic chemistry, catalysis and materials sciences. In the context of catalytic applications, iron-NHC complexes have gained increasing interest in the past two decades and could successfully be applied as catalysts in various homogeneous reactions including C–C couplings (including biaryl cross-coupling, alkyl-alkyl cross-coupling, alkyl-aryl cross-coupling), reductions and oxidations. In addition to this, iron-NHC complexes have shown the ability to facilitate a variety of reactions including C-heteroatom bond formation reactions, hydrogenation and transfer-hydrogenation reactions, polymerization reactions, etc. In this chapter, we will discuss briefly recent advancements in the catalytic activity of iron-NHC complexes including mono-NHC, bis-NHC (bidentate), tripodal NHC and tetrapodal NHC ligands. We have chosen iron-NHC complexes because of the plethora of publications available, increasing significance, being more readily available, non-toxic and economical.

Keywords: N-heterocyclic carbene (NHC), singlet carbenes, triplet carbenes, percent buried volume (% V_{bur}), *σ-donation*, *π-donation*, CO complexes, NO complexes, halide complexes, donor-substituted NHCs, pincer motifs, scorpionato motifs, macrocyclic ligands, piano stool motifs, iron-sulfur clusters, C-C bond formations, allylic alkylations, C-X (heteroatom) bond formations, reduction reactions, cyclization reactions, polymerization

1. Introduction

Story of N-heterocyclic carbene builds up from an unstable non-isolable reactive species to a stable and highly flourished ligand for the synthesis of a variety of organometallic compounds and many important catalytic reactions. Based on the orbital occupancy of the electrons, carbenes can be classified as singlet and triplet carbenes. In singlet carbene, a lone pair of electron occupies sp^2-hybrid orbital (**Figure 1A**) whereas, in triplet carbene, two single electrons occupy two different p-orbitals (**Figure 1B**). Carbenes are inherently unstable, hence highly reactive species due to incomplete electron octet. Initial reports of isolable carbene came in the late 1980s, where the carbene is stabilized by adjacent silicon and phosphorus substituents.

Credit for the discovery of stable and isolable carbene goes to Arduengo, where carbene carbon is a part of a nitrogen heterocycle and gave the first N-heterocyclic carbene (NHC) compound called 1,3-di(adamantyl)imidazol-2-ylidene briefly called IAd (**Figure 2A**) [1]. Since then NHC compounds are enjoying their success to several dimensions of synthesis and organic transformations.

1.1 Structure and general properties of NHCs

Thus, a heterocyclic compound with a carbene carbon and at least a nitrogen atom adjacent to it within the ring can be termed as NHC [2]. NHCs are singlet carbenes and their remarkable stability is contributed by both steric and electronic effects. Dimerization of carbene carbon is kinetically frustrated by keeping bulky groups on the two sides of the carbene carbon, as is the case with IAd (**Figure 2A**) where two adamantyl groups are attached to the nitrogen atoms (adjacent to the carbene center). Nolan and his co-workers have quantified the steric properties in terms of the 'buried volume' parameter (% V_{bur}) (**Figure 2B**) [3]. Metal ion of the NHC-metal complex is assumed to be at the center of a sphere and then % V_{bur} is calculated as the portion of the sphere occupied by the NHC ligand (**Figure 2B**). Larger the value of % V_{bur}, greater is the steric repulsion at the metal center. The buried volume is usually determined from crystallographic data of the NHC-metal complex [4] or directly from theoretical calculations with the free NHC.

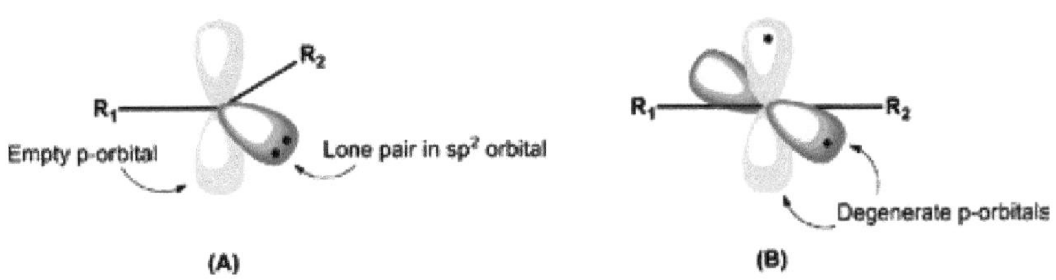

Figure 1.
(A) Singlet carbenes; (B) triplet carbenes.

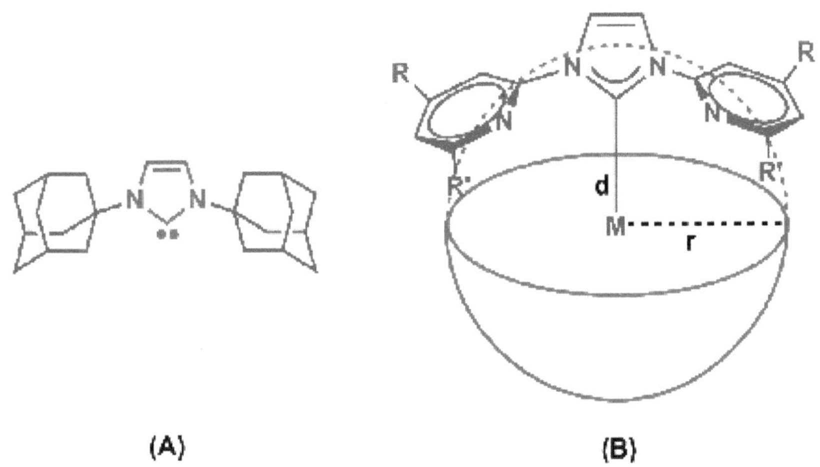

Figure 2.
(A) Structure of IAd; (B) percent buried volume (% V_{bur}).

The value of % V_{bur} is affected by both the nature of the NHC ligand as well as the geometry of the NHC-metal complex; therefore, data is useful only for the comparison within the same family of complexes. A small change in the structure of ligands may bring more than 10% increase or decrease in percent buried volume [5]. Caution should also be paid as the calculation of % V_{bur} is carried out in solid-phase through crystallographic data analysis or in gas phase by DFT calculation. In both the methods the behavior of the complexes in solution and solvation is not considered where ligand may adopt several conformations. The stability of an NHC is far more affected by the electronic factor. Carbene carbon of NHC has three sp^2-orbitals orientated in triangular planar fashion and one p-orbital (p_z) perpendicular to the plane of the NHC ring. Two sp^2-orbitals are bonded with two nitrogen atoms in the ring and one sp^2-orbital houses the lone pair of electrons. The two nitrogen atoms stabilize the carbene carbon in two ways: (i) by withdrawing the sigma-electrons through inductive effect and (ii) through a π-electron donation to the empty p_z-orbital of the carbene carbon (mesomeric effect). This π-electron donation is so strong that NHCs are also described by its zwitterionic resonance structure and is evident by the intermediate bond length of car-bene C-N bond (1.37 Å) in IAd, which falls in between C-N single bond length (1.49 Å) and C-N double bond length (1.33 Å) of the corresponding analog compounds ($IAdH_2$ and $IAdH^+$ respectively). In the molecular orbital model, sp^2 and p_z-orbital can be described as HOMO (A_1 non-bonding molecular orbital) and LUMO (B_2^* bonding molecular orbital), respectively (**Figure 3**) [6, 7]. The cyclic nature of NHCs is also an important structural aspect as it creates a preferable situation for the singlet state by forcing the carbene carbon to adopt a more sp^2-like arrangement.

Like the phosphines, the electron-donating capability of NHCs is evaluated using Tolman electronic parameter (TEP) [8]. Any build-up of electron density on the metal center of the complex $[Ni(CO)_3(NHC)]$ due to electron donation by the NHC is reflected by the decrease in the infrared-stretching frequency of CO bonded with the metal ion. Now-a-days, instead of $[Ni(CO)_3(NHC)]$, less toxic $[(NHC)\ IrCl(CO)_2]$ and $[(NHC)RhCl(CO)_2]$ are used and a correlation formula is used [Eqs. (1) and (2)], respectively [9, 10].

$$TEP = 0.847.\nu_{CO}(Ir) + 336\ cm^{-1} \quad (1)$$

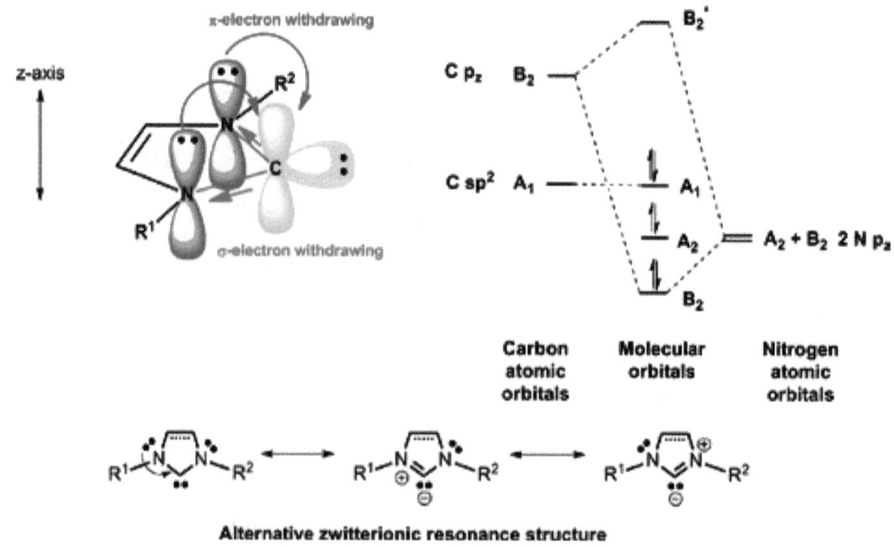

Figure 3.
Molecular orbital diagram of an NHC.

where, $\nu_{CO}(Ir)$ = average IR-stretching frequency of CO in $[(NHC)IrCl(CO)_2]$ complex.

$$\nu_{CO}(Ir) = 0.8695 . \nu_{CO}(Rh) + 250.7 \text{ cm}^{-1} \tag{2}$$

where, $\nu_{CO}(Ir)$ = average IR-stretching frequency of CO in $[(NHC)IrCl(CO)_2]$ complex, and $\nu_{CO}(Rh)$ = average IR-stretching frequency of CO in $[(NHC)RhCl(CO)_2]$ complex.

1.2 Synthesis of NHCs precursor and generation of carbene

Azolium or dihydroimidazolium salts are sufficiently stable solids and the generation of NHCs can be carried out *in situ* by their deprotonation using non-nucleophilic bases such as sodium hydride, butyllithium or *t*-butoxide. Alkoxides form an adduct with azolium salt, however, in presence of transition metal precursor, NHC is transferred to the metal and usually moves toward complex formation rather than the disruption of the azolium ring. Generation of NHCs is also carried out using mild metal oxides like silver (I) or copper (I) oxides where after generation, NHC forms NHC-silver(I) or copper(I) complexes and *in situ* transfer of NHC occurs to the desired metal center. A general protocol for the synthesis of NHCs and NHC precursor **11** is outlined below in **Figures 4** and **5**, respectively [11, 12].

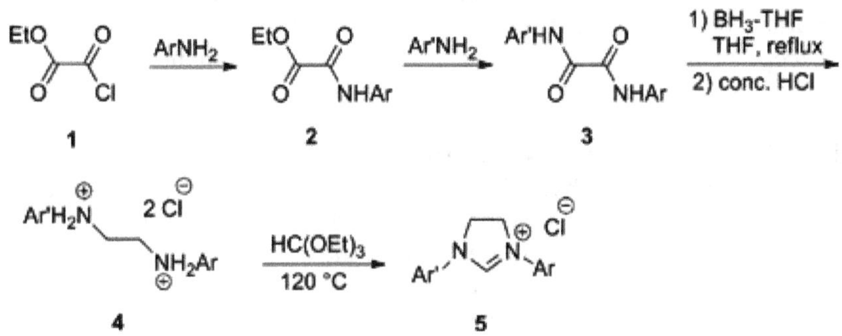

Figure 4.
General protocol for the synthesis of unsymmetrical substituted NHCs.

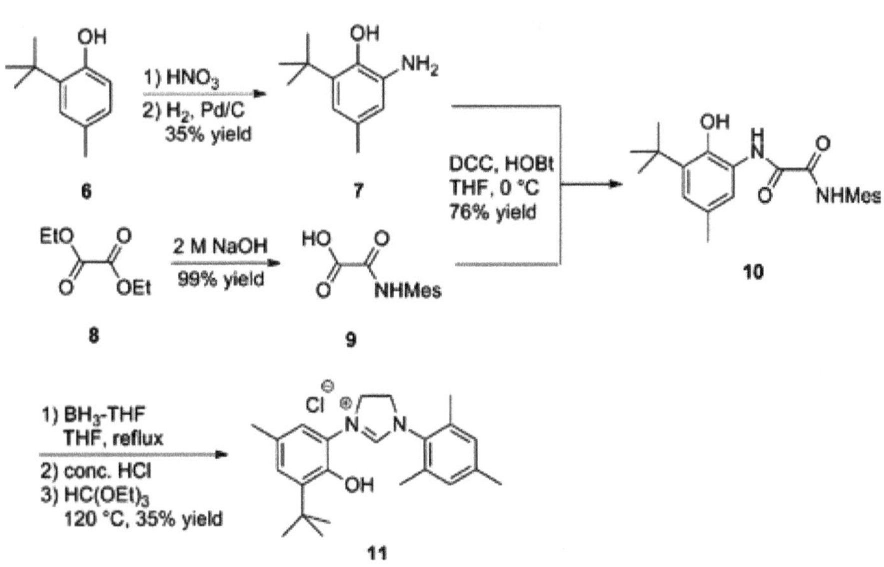

Figure 5.
*Synthesis of NHC precursor **11**.*

1.3 Generation of NHCs

Formation of saturated and unsaturated NHCs upon treatment with an alkoxide base is shown in **Figure 6A** and **B**, respectively [13].

1.4 Coordination of NHCs to transition metals

Thus, the coordination of NHC ligand to the transition metal ion occurs largely through the *strong σ-donation* of the formal sp^2-hybridized lone pair to a σ-accepting orbital of the transition metal and a *weak but not inconsiderable π-donation* [14] either in the form of π-back donation from metal to the p_z orbital of the ligand or vice versa [15, 16]. However, in practice a single bond is drawn since the free rotation energy across the M-C bond is very low (**Figures 2B** and **6**).

1.5 Phosphine versus NHCs

NHCs are being compared with strong sigma donating ligands like phosphines and cyclopentadienes. As a ligand, NHCs edge ahead of phosphines on several points:

i. *Electron donor*: NHCs are relatively stronger electron-donor than phosphines and produce thermodynamically stronger metal-ligand bonds, except when there are steric constraints interfere with metal-ligand binding [17].

ii. *Steric properties*: Whereas the spatial arrangement of steric bulk takes up a *cone-shape* due to sp^3-hybridization of phosphines; most of the NHCs results in *umbrella-shaped* steric bulk and the orientation of the substituent on the two nitrogen atoms are more toward the metal center. Thus, the steric crowd around the metal center can be tuned by changing the substituent on the two nitrogen atoms and the heterocyclic ring, if required.

iii. *Ease of varying their steric and electronic properties*: There are several well-established synthetic routes to tune the steric and electronic properties of NHCs, whereas it is usually difficult to tune the properties to the desired level for the phosphines.

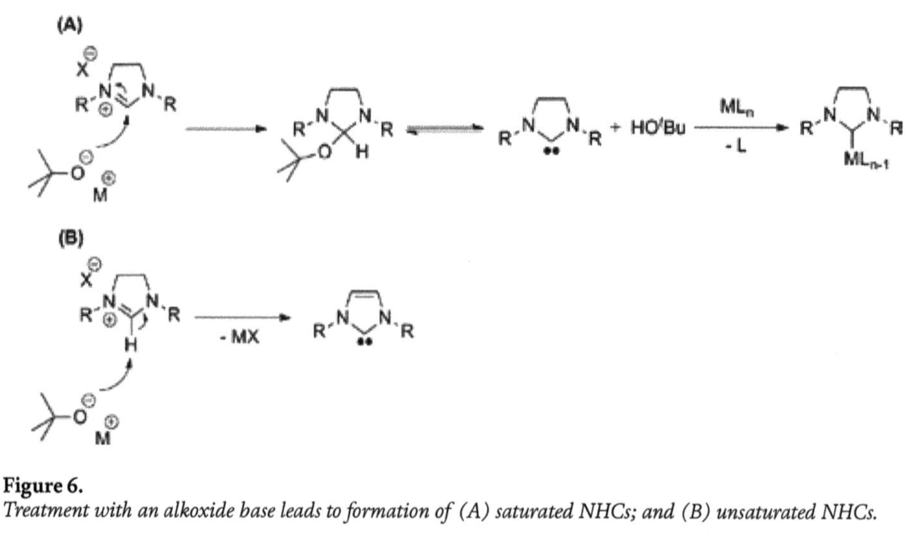

Figure 6.
Treatment with an alkoxide base leads to formation of (A) saturated NHCs; and (B) unsaturated NHCs.

iv. In the case of phosphines, changing the substituent on the phosphorus inevitably changes both steric and the electronic properties whereas each parameter can be modified independently through modifying the substituents on nitrogen, functionalities on the heterocycle and the type of heterocycle itself.

2. Various motifs of Fe-NHC complexes

The structural diversity in various motifs of Fe-NHC complexes is shown in **Figure 7** and each of them is explained below along with their known applications in different areas.

2.1 Mono- and bis-(mono- or chelating) carbene ligands

2.1.1 CO complexes

The chemistry of Fe-NHC complexes began with the synthesis of their unsaturated and saturated ligand precursors with carbonyl as their motifs, and extensive studies on molecular structure determination and reactivity (**Figure 8A–D**). These CO complexes were further subjected to substitution reaction, e.g. ligand exchange with monophos-phines and oxidation, to develop newer Fe-NHC complexes (on oxidation their geometry tends to change from trigonal bipyramidal to distorted square pyramidal). These transformations, in progression, led to the formation of new classes of complexes with novel attributes viz. monocarbene, bis-monocarbene, and chelating biscarbene ligands having variable oxidation states of iron from Fe(0) to Fe(II), which contributed to new horizons in bioinorganic chemistry and biomimetic systems e.g. Novel Fe(II) monocarbene complexes (**Figure 8C**) as models for basic structure of the monoiron hydrogenase [18].

2.1.2 NO complexes

Synthesis of novel and intriguing Fe-NHC complexes in the field of biomimetic chemistry e.g. dinitrosyliron complexes (DNICs) (**Figure 8G**) displaying a variety of vital biological functions [18], forced the scientific community to shift their attention toward novel monocarbenes and bis-monocarbene ligands having nitrosyl as their struc-tural attributes (**Figure 8E–G**). Not only as biomimetic structural models, these nitrosyl complexes can act as catalyst in chemical transformations e.g. allylic alkylation [18, 19].

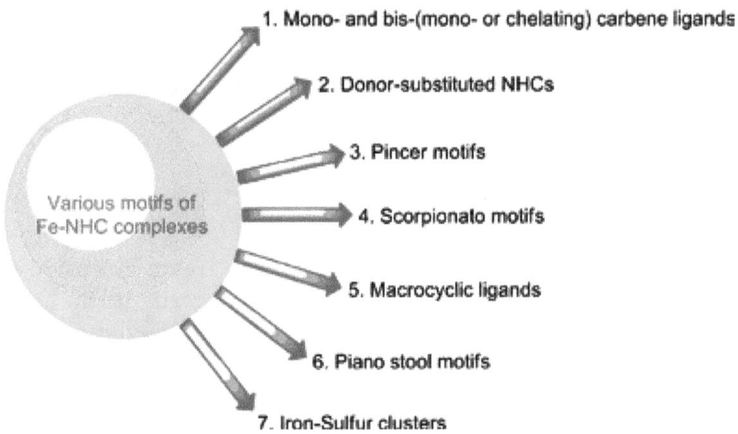

Figure 7.
Different motifs of Fe-NHC complexes.

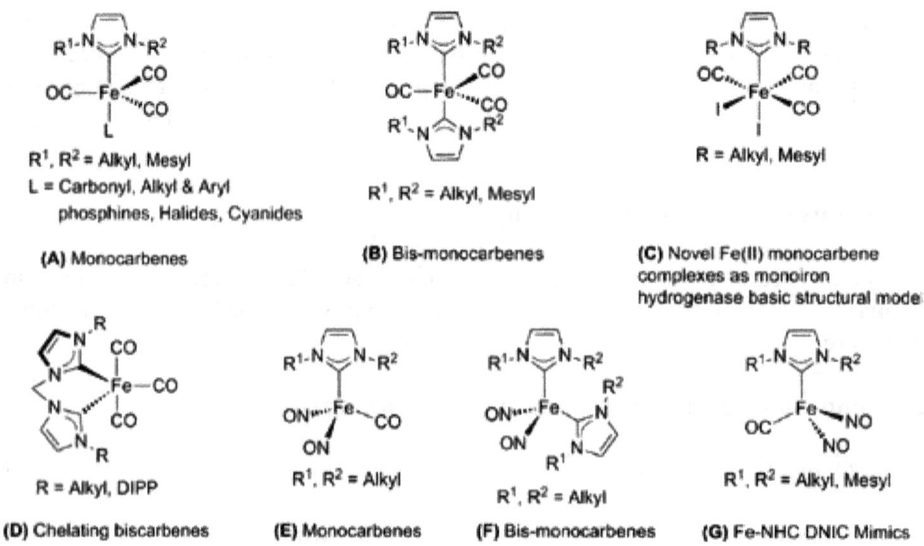

Figure 8.
(A–D) CO complexes; (E–G) NO complexes.

2.1.3 Halide complexes

Just like carbonyl and nitrosyl motifs in Fe-NHCs chemistry, halides do play a major role in influencing the role of Fe-NHC complexes in both catalysis as well as biomimetics. Halide complexes catalytic role varies from polymerization catalysis by bis-monocarbene dihalide Fe-NHC complexes [18, 20] C-C cross-coupling reactions catalyzed by dinuclear Fe-NHC imido complexes [18, 21] to catalytic hydrosilylation by ethylenediamine-derived Fe-NHC complex [18]. Depending upon the structural versatility in halide complexes, many subclasses have been synthesized and studied, namely monocarbene ligands, bis-monocarbene ligands, chelating biscarbene ligands, dinuclear Fe-NHC imido complexes, halide-bridged Fe-NHC complexes, immobilized Fe-NHC complexes, three-coordinate Fe-NHC complexes (**Figure 9A–G**).

2.2 Donor-substituted NHCs

Effects on the reactivity of organometallic iron complexes could be observed when the ligand environment changes from CO, NO, halides to donor-substituted NHC ligands (**Figure 10A**). These donor-substituted NHC ligands possess nitrogen or oxygen as heteroatoms, thus present themselves as potential coordinating "arms" attached to the NHCs and exhibit coordination from bi- to pentadentate as ligand systems. These complexes have shown their catalytic role in ring-opening polymerization of ε-caprolactone [18, 22].

2.3 Pincer motifs

Chelating biscarbene pincer ligands (**Figure 10B**) are an extension of donor-substituted NHCs in Fe-NHC chemistry, where instead of the presence of heteroatoms as "arms", two NHC units are linked by a pyridyl moiety and hence "chelation". Structurally, pincer motifs exhibit two coordination geometries predominantly, octahedral and square pyramidal, due to their strict binding mode to three adjacent coplanar centers. Catalysis by Fe-NHC complexes bearing pincer motifs has been demonstrated by their catalytic role in concerted C-H oxidation addition reaction [18], hydroboration reaction [18, 23], and hydrogenation reaction [18].

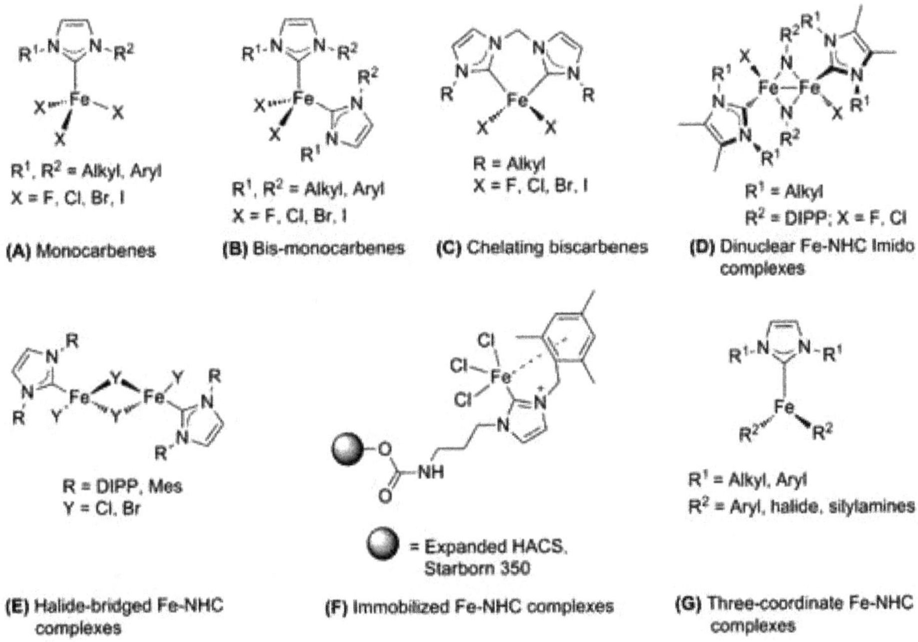

Figure 9.
(A–G) Halide complexes.

2.4 Scorpionato motifs

Scorpionato-type motifs (**Figure 10C**) means boron linked anionic chelating triscarbene ligands and on complexation with iron results in a new class of Fe-NHC complexes. Therefore, if any iron complex/compound is bearing two scorpionato-type ligands, it will be, (a) coordinated by six carbenes, (b) highly stable, and (c) showing S_6 symmetry along Fe-B-H axis [18]. Different types of scorpionato-type motifs have also been synthesized e.g. tripodal borane NHC iron complexes [18], amine-bridged scorpionato Fe-NHC motifs [18].

2.5 Macrocyclic ligands

Macrocyclic ligands, despite well-investigated other cyclic ligands such as cyclam, porphyrin, on complexation with iron developed a new class of complexes in Fe-NHC coordination chemistry (**Figure 10D**). Their catalytic aspect has been successfully employed in aziridination of alkenes with aryl azides [18, 24].

2.6 Piano stool motifs

The term "piano stool Fe-NHC complexes" states that all such complexes bear both, (a) N-heterocyclic carbene motif and (b) cyclopentadienyl (Cp) ligand. The structural variations in these complexes are well explained by (a) mono- and dimeric piano stool Fe-NHC complexes [18], (b) donor-substituted piano stool Fe-NHC complexes [18], (c) biscarbene-chelated piano stool complexes [18], (d) alkyl piano stool Fe-NHC complexes [18], (e) three coordinate piano stool Fe-NHC complexes [18], and many more [18] (**Figure 10E–G**). These have shown their catalytic activities in C-H bond activation [18], borylation reactions [18, 23], hydrosilylation [18, 25–27], transfer hydrogenation [18], C-N bond formation [18, 24].

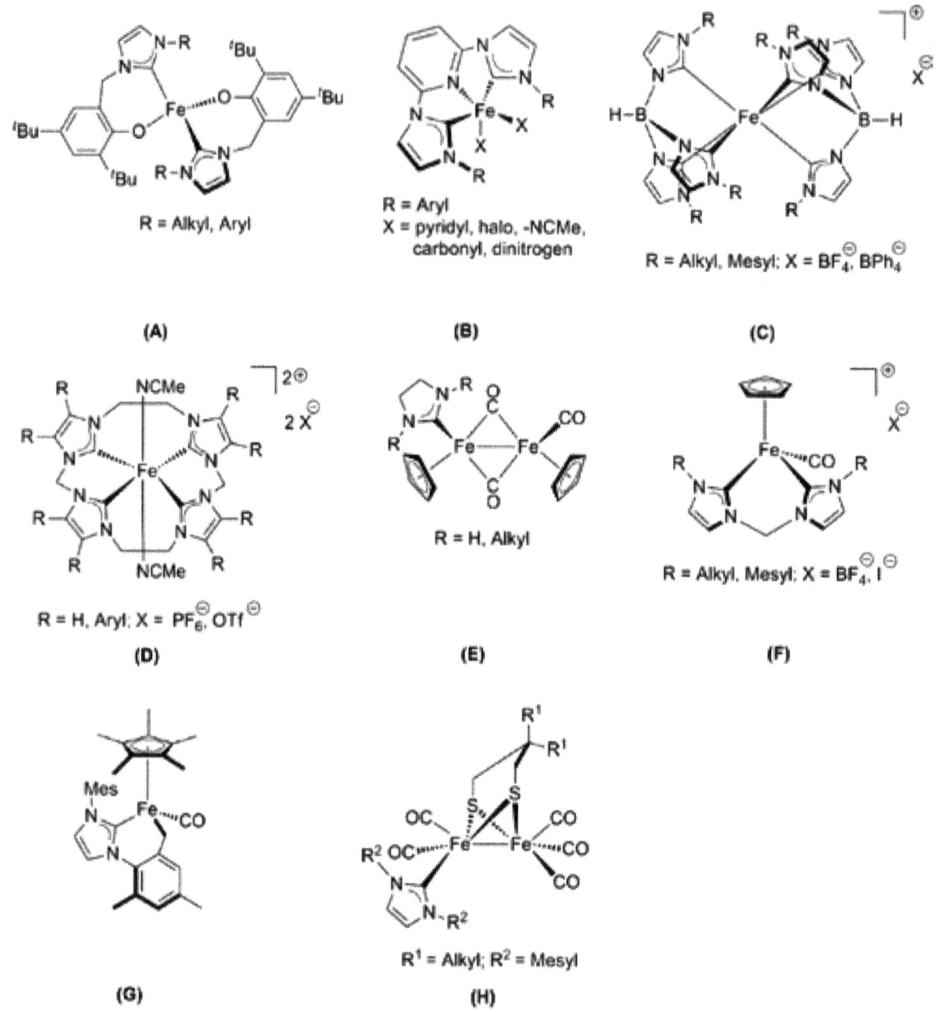

Figure 10.
(A) Donor-substitutes NHCs; (B) pincer motifs; (C) Scorpionato motifs; (D) macrocyclic ligands; (E–G) piano stool motifs; (H) iron-sulfur clusters

2.7 Iron-sulfur clusters

Diiron dithiolate complexes (**Figure 10H**) have been reported to mimic the active site of [FeFe] hydrogenase [18]. Also, the substitution of carbonyl motifs (one or more) in the diiron dithiolate complexes by σ-donor ligands (in this case NHCs) is shown to influence the redox potential of the iron center [18]. Further, donor-substituted NHCs motifs were included in the molecular framework of [FeFe] hydrogenase model compounds to extend its molecular assembly [18]. Another notable characteristic presence of Fe-NHC complexes bearing iron-sulfur clusters was demonstrated in synthesis of nitrogenase model compounds, which were based on all-ferrous $[Fe_4S_4]^0$ [18].

3. Catalysis by Fe-NHC complexes: important transformations

Even if there are a tremendous number of catalysts based on rare/heavy transition metals such as palladium, platinum, ruthenium, rhodium, iridium, and gold

[28–30] are available for various different kind of organic transformations and they are very successful; the scientific community is trying hard to replace these metals by some environment and biological friendly metals because they are highly expensive and very toxic in nature therefore not compatible with biological systems. Iron becomes the obvious choice since it is the most abundant transition metal on the earth's crust, relatively inexpensive, environmentally benign [31] and relatively less toxic to the biological systems [32, 33]. There are several very successful examples of iron-based catalysts like Fischer-Tropsch and the Haber-Bosch processes [34] and are capable of catalysis in numerous different reactions [35, 36]. Reports related to the iron-NHC complexes started coming just after the publication of first metal-NHC complex in 1968, the growth in the research was almost ceased for next three decades and picks up the pace after the success of Grubb's catalyst for various organic transformations and polymerization reactions [20, 37]. Iron-NHC complexes are reported to have found applications in different classes of reactions such as substitution, addition, oxidation, reduction, cycloaddition, isomerization, rearrangement and polymerization reactions (**Figure 11**).

3.1 C-C bond formations

Negishi, Suzuki, and Heck were awarded the Nobel Prize in 2010 for their pioneer work in the area of cross-coupling reactions, as it provides a very effective tool for C-C bond formation. Several different protocols have been reported mainly based on palladium and, to some extent, Ni and copper metal ions. Iron-NHC complex based catalysts have been used for various Kumada-type cross-couplings such as C(sp^3)-C(sp^2), C(sp^2)-C(sp^3), C(sp^2)-C(sp^2), C(sp^3)-C(sp^3) bond formations, and C(sp^2)-C(sp^2) homo-couplings. NHC can either be generated *in situ* in a reaction or a resynthesized iron-NHC complex can be used. Bedford and co-workers, in a first, introduced the NHCs ligands and iron-NHC complexes along with $FeCl_3$ to improve the yield of Kumada-type coupling reactions (**Figure 12A**) [38]. Among

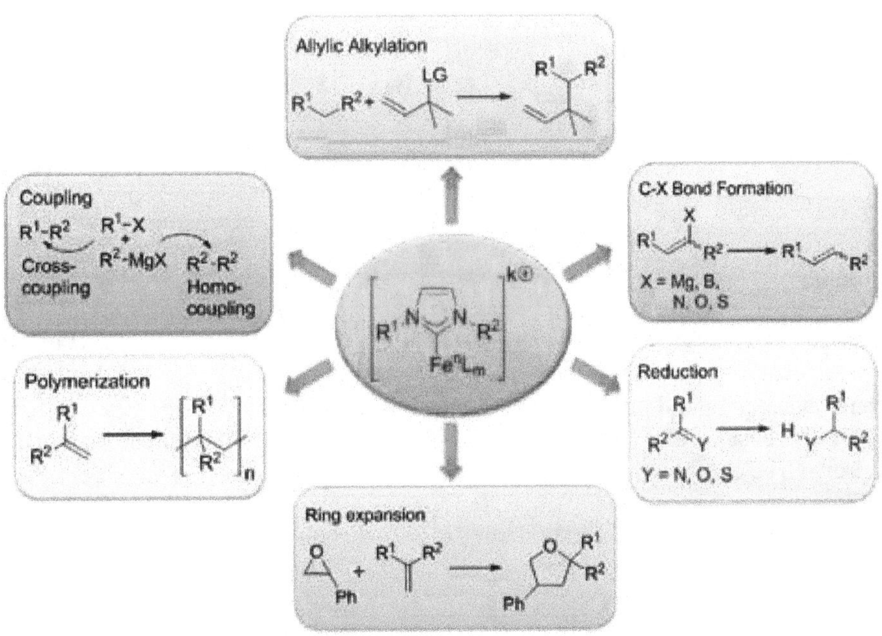

Figure 11.
Important transformations catalyzed by Fe-NHC complexes.

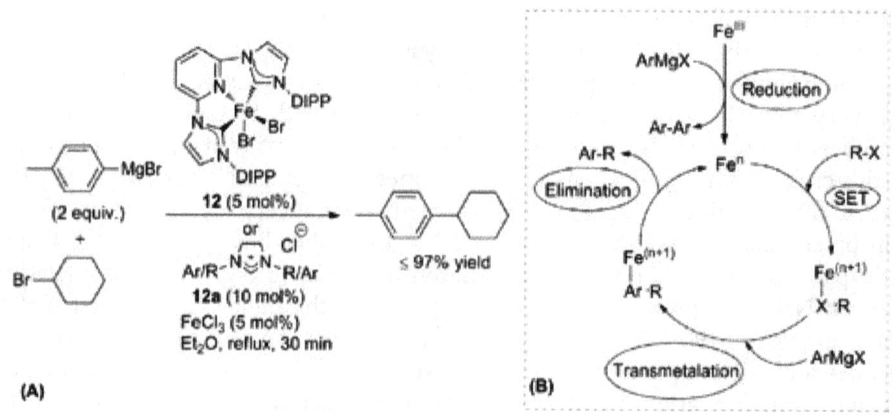

Figure 12.
(A) Aryl Grignard reagents-bromoalkanes cross-coupling [38]; (B) proposed mechanism.

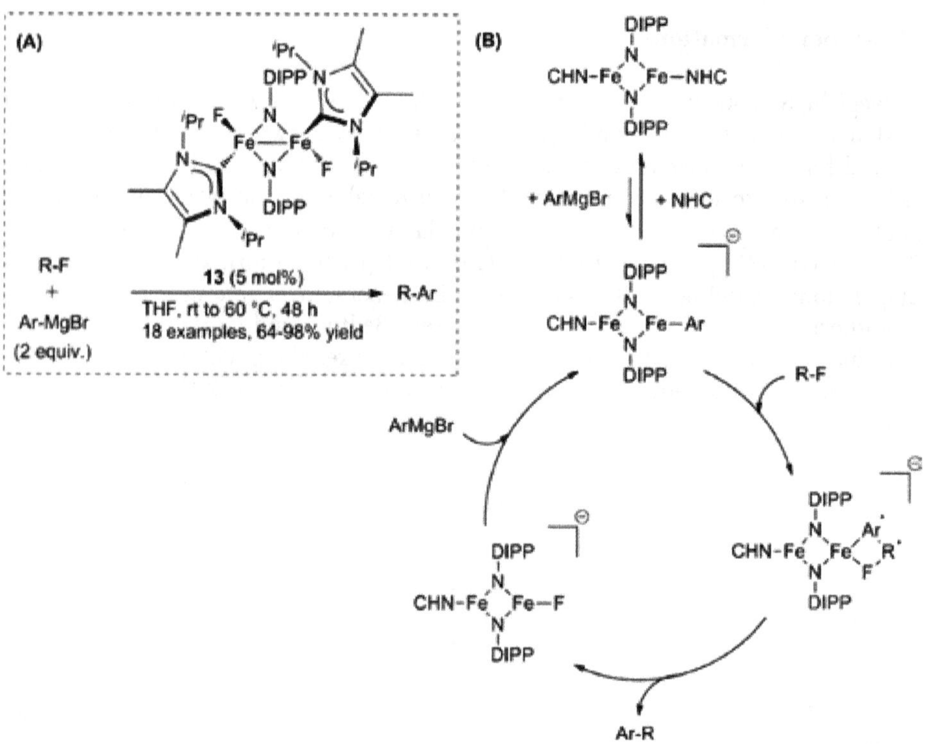

Figure 13.
(A) Primary alkyl fluorides-aryl Grignard reagents Kumada-type coupling [21]; (B) proposed mechanism.

several carbene ligand precursors, *tert*-butylimidazolinium chloride **12a** was found to give the best results (97% yield) and the performance was almost matched by the iron-NHC complex **12** (94% yield).

The proposed mechanism suggests that reaction does not follow the classical oxidative addition mechanism, but rather involves a radical intermediate produced through single electron transfer (SET) (**Figure 12B**) [39, 40]. Reaction mechanism involves the following processes: (i) generation of active catalyst through reduction of Fe(III) to Fe(II, I, or 0), (ii) generation and association (not the oxidative addition) of alkyl radical ($R^{\cdot}$) with the iron center through SET, (iii) transmetalation, where aryl group is transferred from ArMgX to the iron center, and (iv) attack of alkyl radical (R˙) to the aryl group (Ar) leading to the generation of coupled product and the catalyst [38].

It was proved through a control experiment that particularly primary and secondary alkyl halides favor iron-catalyzed reactions, in comparison to most of the Pd or Ni systems, because of their sluggish tendency toward the β-hydride elimination and hence less susceptibility to the olefin formation. Therefore, it plausibly indicated the limitations of the catalytic role of the Fe-NHC complexes, in case of *in situ* formation of an iron NHC complex or the deprotonation of the imidazolium salt. Besides Alkyl bromide, dinuclear Fe-NHC imido complexes such as **13** have been reported to be effective in activating other alkyl halides and most challenging alkyl fluoride (**Figure 13A**). Here again, the use of the substrates such as (fluoromethyl)cyclopropane suggested a radical-mediated mechanistic pathway (**Figure 13B**). The first step is the dissociation of one NHC

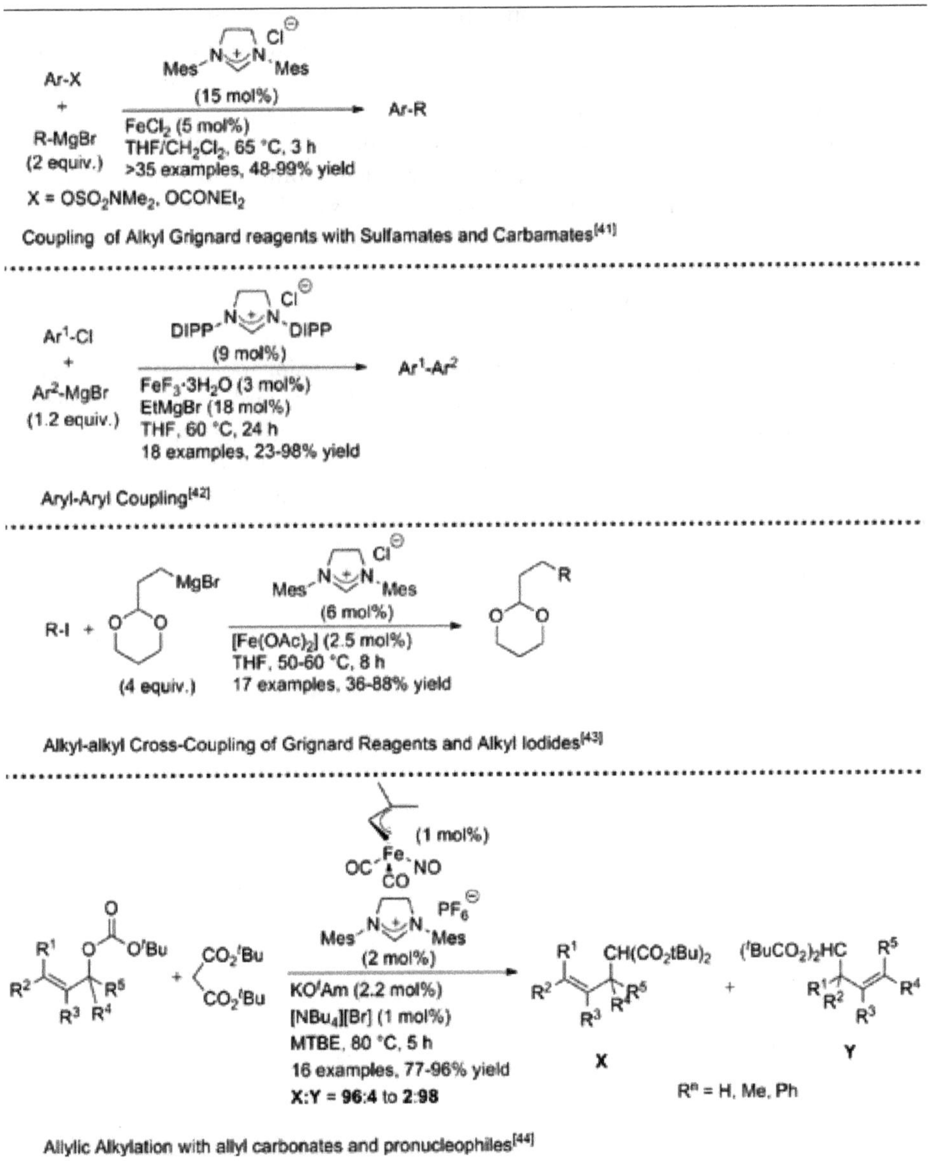

Table 1.
Other examples of C-C bond formation and allylic alkylation reactions [41–44].

ligand followed by the second step as transmetalation (note: dinuclear iron imido subunit stays intact during the process). The further mechanism involves the usual mechanistic protocol, which includes firstly the formation of radical species and secondly, attack of the radical on the aryl moiety [21]. Several more iron-NHC complex catalyzed carbon-carbon coupling reactions have been given in **Table 1**.

3.2 Allylic alkylations

In a seminal work by Plietker group [19], allylic alkylation by the catalyst **14** was shown through the reaction of allyl carbonate and a Michael donor resulting into two isomeric products, i.e. (i) Product **X**, through the *ipso* substitution, and (ii) Product **Y**, via a σ–π–σ isomerization (**Figure 14A**). Mechanistic investigation suggests that the product ratio is greatly influenced by the steric crowd around the metal center, created due to the substituents on the nitrogen atoms of NHC moiety. Increased steric crowd hinders the isomerization process and thus favoring *ipso* substitution product **X**. For example, if *tert*-butyl group is present on the N atom of NHC, *ipso* substitution is favored, on the other hand, mesitylene group, which creates less steric hindrance around the metal center, favors isomerized product **Y**. In addition, stronger nucleophilicity of Michael donor favors the *ipso*-substitution. A plausible mechanism is outlined in **Figure 14B**. Few more allylic alkylation reactions are presented in **Table 1**.

3.3 C-X (heteroatom) bond formations

Catalytic C-H bond activation has been one of the major tools to perform effective chemical transformations. Applicability of Fe-NHC complex as the catalyst for C-H bond activation has gained momentum since it can produce the formation of a range of different C-X bonds such as C-N, C-B, C-Mg, and C-S bond. Fe-NHC complex catalyzed C-N bond formation is important because of the three very basic

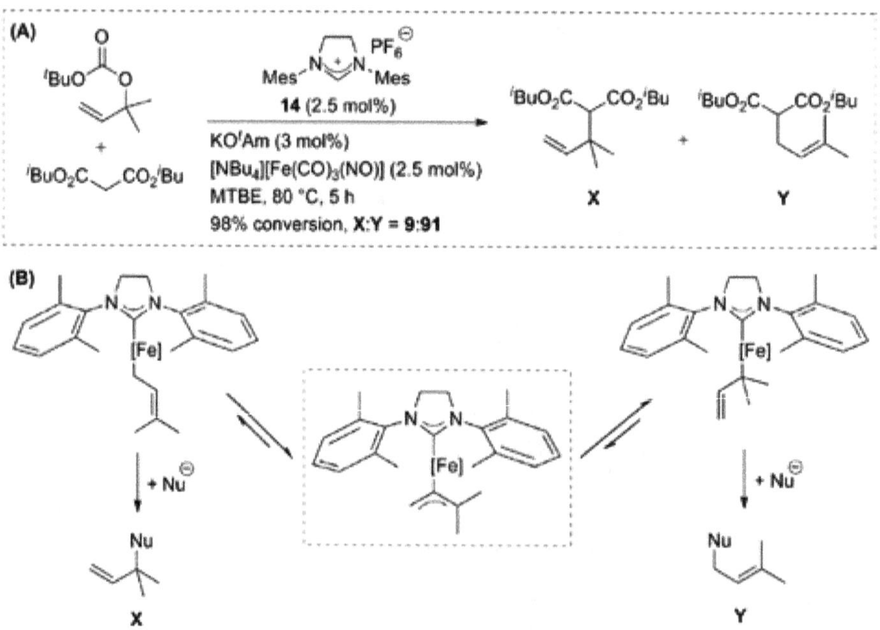

Figure 14.
(A) (TBA)Fe/NHC catalyzed allylic alkylation [19]; (B) proposed mechanism.

reasons, (a) aziridine based compounds are of medicinal importance and therefore essential for pharmaceutical industry, (b) demand of aziridine derivatives in polymer chemistry as cross-linker agents for two-component resins, and (c) relative to well-known synthesis of *O*-epoxidation analogs, it is hard to synthesize the designer *N*-building blocks. Catalytic aziridination of alkenes by using Fe-NHC complex **15**

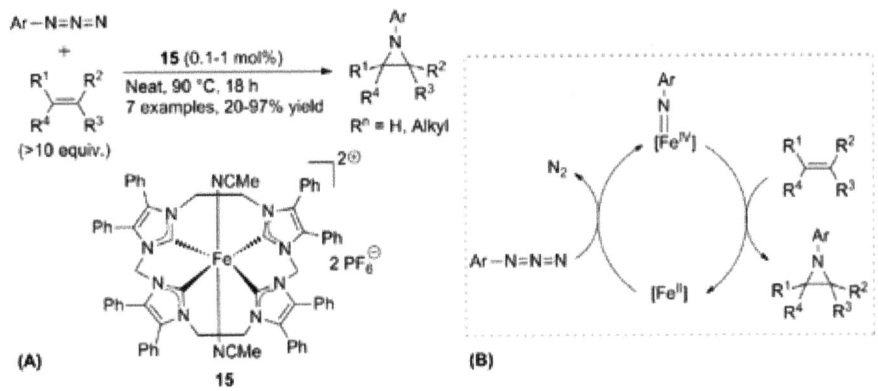

Figure 15.
(A) Fe-NHC catalyzed aziridination of alkenes [24]; (B) proposed mechanism.

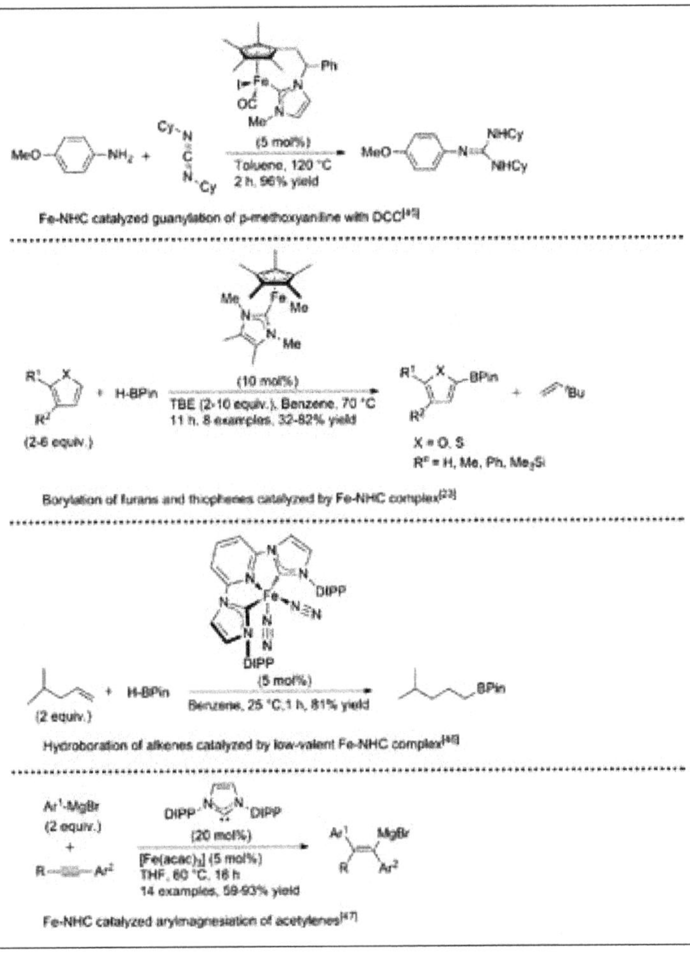

Fe-NHC catalyzed guanylation of p-methoxyaniline with DCC
Borylation of furans and thiophenes catalyzed by Fe-NHC complex
Hydroboration of alkenes catalyzed by low-valent Fe-NHC complex
Fe-NHC catalyzed arylmagnesiation of acetylenes

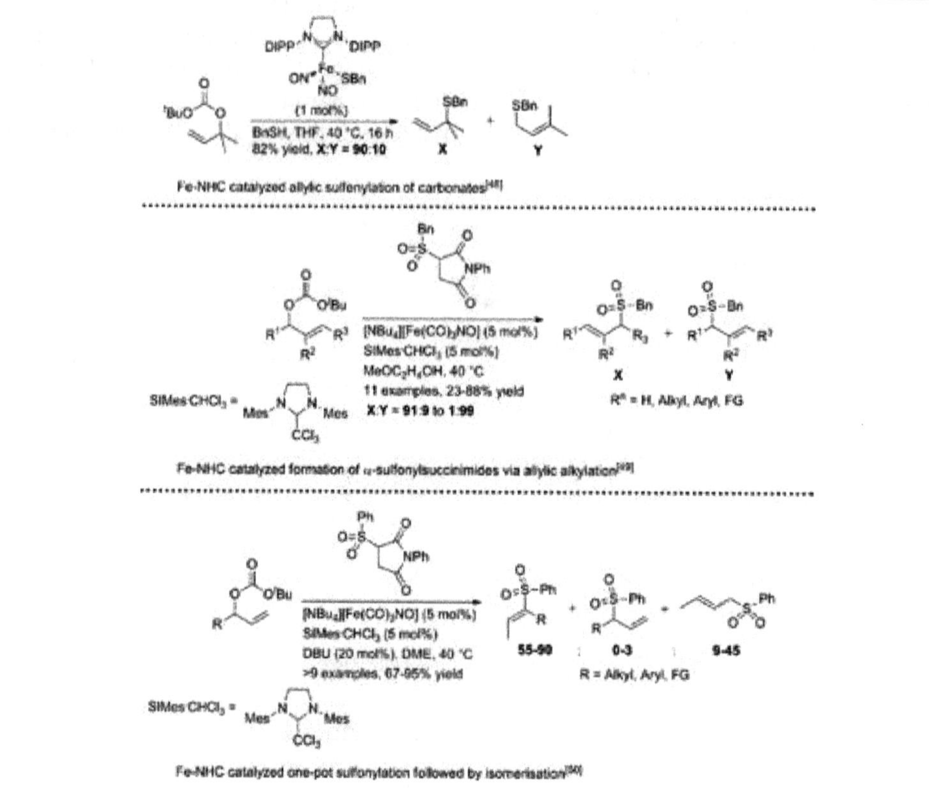

Fe-NHC catalyzed allylic sulfenylation of carbonates[48]
Fe-NHC catalyzed formation of α-sulfonylsuccinimides via allylic alkylation[49]
Fe-NHC catalyzed one-pot sulfonylation followed by isomerisation[50]

Table 2.
Other examples of C-X bond formations [23, 45–50].

(0.1–1 mol%) as the catalyst was published by Jenkins et al. [24] to form respective aryl-substituted aziridines by treating aryl azides with various substituted alkene (**Figure 15A**). As proposed, the reaction involved the formation of a key and highly reactive intermediate Fe(IV) imido complex (**Figure 15B**). Few more C-X bond formation reactions are presented in **Table 2**.

3.4 Reduction reactions

There are several reports on the reduction of alkenes via silylation using iron-NHC complexes. Royo group was first to show such conversion using piano stool type complex **16** (**Figure 16A**) [25]. The reaction is sensitive to the type of substituent present at para-position in the aromatic ring of the reactant, e.g. quantitative yields for reactions of *p*-aryl-substituted aldehydes and alkyl-substituted aldehydes or ketones remained unreactive. Another piano stool type complex **17** reduces ketones and aldehydes into the corresponding alcohols very efficiently (**Figure 16B**) [26]. Same catalyst **17** can reduce the carbonyl group of various amides in moderate to excellent yields (**Figure 16C** and **D**) [27]. In both cases, irradiation of visible light is crucial for the reported effective conversions, where $PhSiH_3$ works as the hydride source. Catalyst shows differential reactivity with the primary, secondary and tertiary amides. Secondary and tertiary amides give usual conversion of carbonyl group into alcohol, while primary amide converts into nitrile compound. Cyclic amides have to be protected before reduction; otherwise a mixture of products forms.

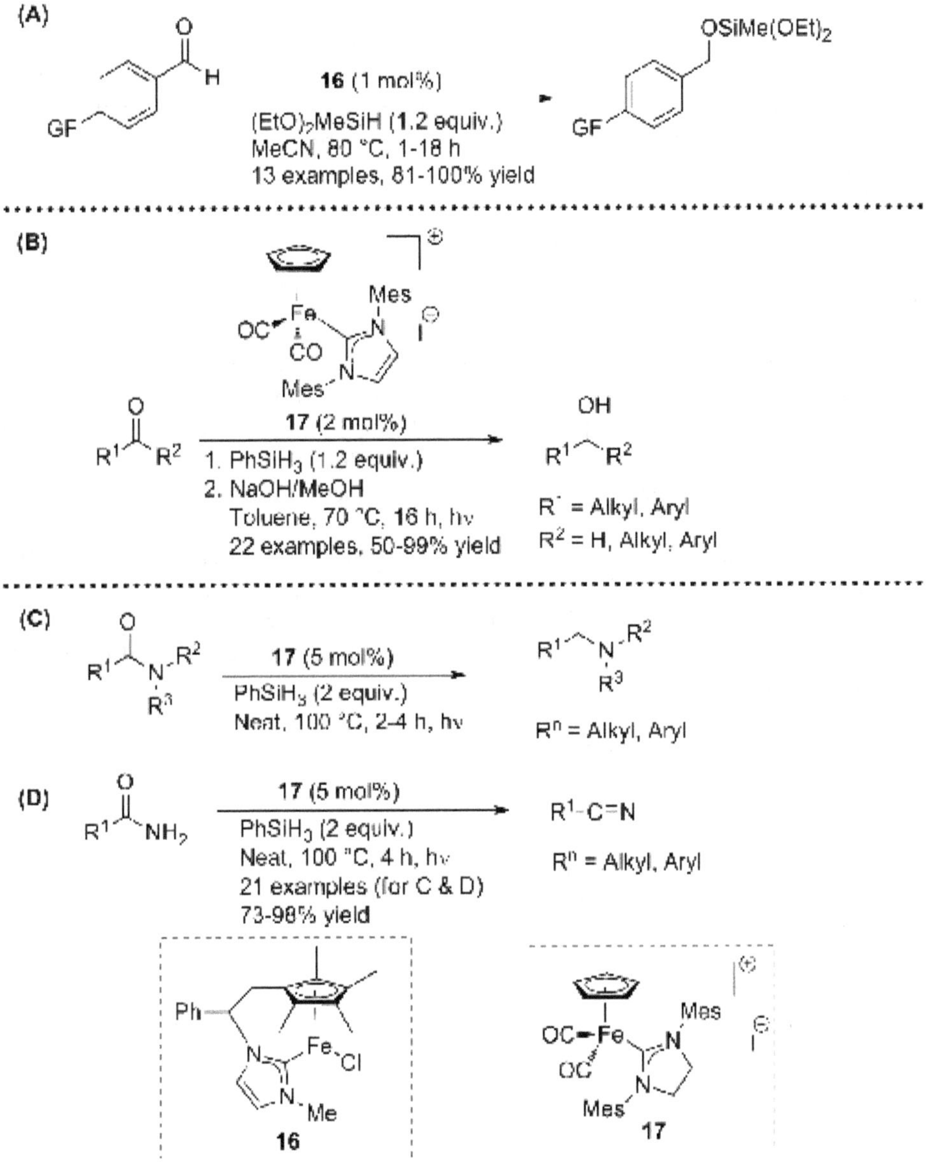

Figure 16.
Hydrosilylative reductions of (A) benzaldehyde derivatives [25]; (B and C) substituted and primary amides, respectively [27].

Various recently reported iron-NHC complex catalyzed reduction reactions are summarized in **Table 3**.

3.5 Cyclization reactions

Fe-NHC catalyzed ring expansion of the epoxides with functionalized alkenes presents a very intriguing case because cyclic structures are of great importance in various fields such as the pharmaceutical industry, fine chemicals, agriculture, etc. Fe-NHC catalyzed such reactions not only have shown functional group tolerance but also high chemo- and regioselectivity.

Hilt et al. [51] used a mixture of $FeCl_2$, phosphine ligands and *in situ* generated free NHCs, **18** and performed reaction under reductive conditions using Zn and NEt_3 (**Figure 17A**). The reaction mechanism demonstrates the first step as a SET (single-electron transfer) in epoxide ring-opening, the second step as the formation of an elongated alkoxy radical via reaction between formed radical intermediate and added alkene, and the final step as a BET (back-electron transfer), which gave the desired expanded cyclic product via a zwitterionic intermediate cyclization (**Figure 17B**).

3.6 Polymerization

So far, the application of Fe-NHC complexes have not been much explored in the area of polymerization [52]. Grubbs has first reported the use of Fe-NHC complex **19** as the catalyst in atom transfer radical polymerization (ATRP) reaction of styrene and methyl methacrylate (**Figure 18**) [20]. The reaction shows pseudo first-order kinetics, a decent control of radical concentration, and polydispersity index (PDI) near 1.1.

Shen and co-workers have reported the ring-opening polymerization (ROP) reaction of ε-caprolactone by using Fe-NHC complex **20** as the catalyst [22]. Even though reaction suffers some side reaction of transesterification, polymerization progresses with quantitative conversion and moderate number average molecular weight distribution (**Figure 19**).

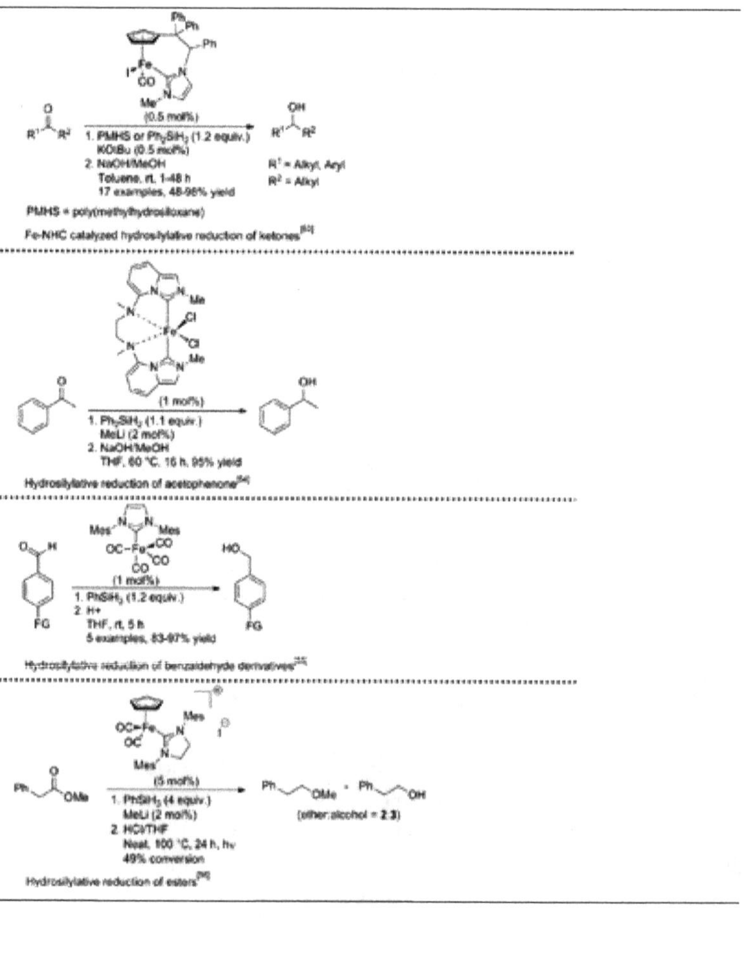

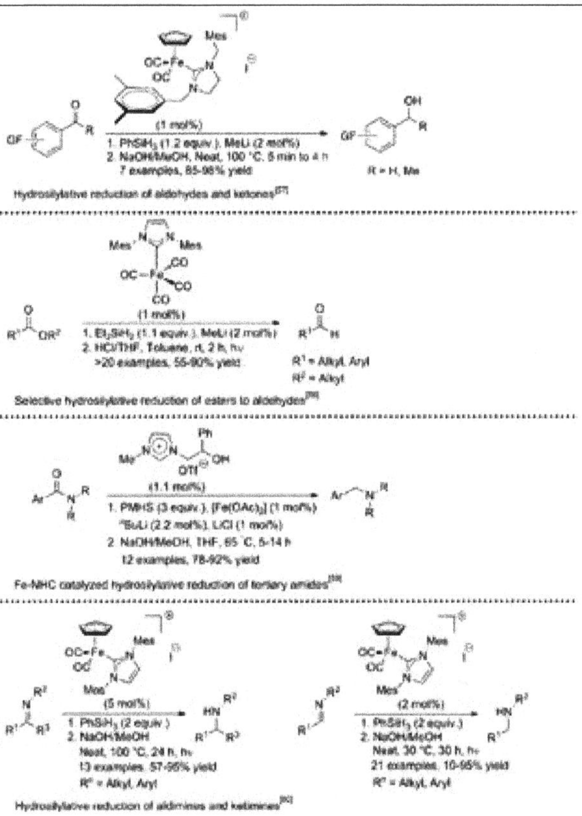

Table 3.
Other examples of reduction reactions [53–60].

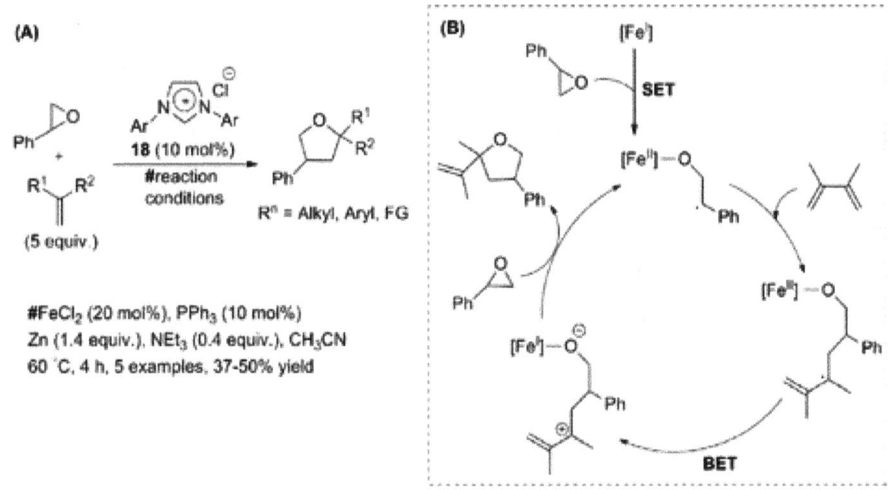

Figure 17.
(A) Fe-NHC catalyzed epoxide ring expansions [51]; (B) proposed mechanism.

4. Conclusion

Iron will remain a metal of choice for the replacement of all the heavy metal ions currently being used for the application of catalytic processes for the obvious reason

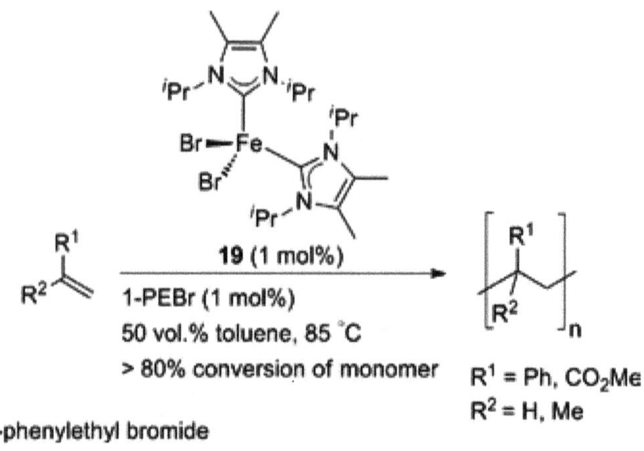

Figure 18.
Atom transfer radical polymerization (ATRP) of olefins [20].

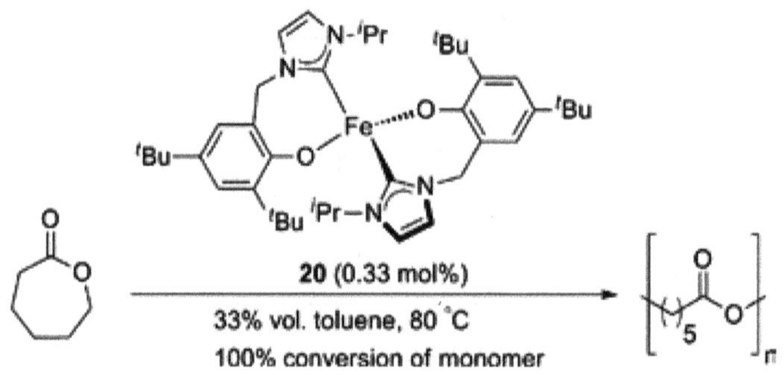

Figure 19.
Ring-opening polymerization of ε-caprolactone [22].

of it being economical, very high natural abundance, environmentally benign and more importantly biologically compatible. Earlier, several iron-based complexes have enjoyed their success in many processes like Fischer–Tropsch and the Haber – Bosch processes, but the progress of iron-NHC complexes has gained momentum only after the success of Grubb's catalyst at the onset of this century and now the number of published articles is growing with every passing year. The importance of Fe-NHC complexes can be evaluated from the aforementioned fact that they have found applicability in diverse fields from academia (e.g. biomimetic studies, various intriguing chemical transformations) to industries (e.g. pharmaceutical industry). The existing and ever possible versatility of (i) various structural motifs with different oxidation states, (ii) their flexible coordination geometries before and after the reaction, and (iii) substitution patterns in the iron N-heterocyclic carbene complexes along with their potential economic and toxicity benefits present an exciting scenario for the upcoming generation.

Acknowledgements

Badri Nath Jha is grateful to the SERB-DST (Project No. YSS/000699/2015), India, for the financial support to carry out research in the area of catalysis and cathodic materials of LIBs. B.N. Jha is also thankful to Pradeep Mathur for his continuous motivation to write book chapters/books and pursue research. Abhinav Raghuvanshi is thankful to the SERB for NPDF fellowship file no. PDF/2016/001786 for the financial support to carry out the research.

Author details

Badri Nath Jha[1*], Nishant Singh[1] and Abhinav Raghuvanshi[2]

1 University Department of Chemistry, T. M. Bhagalpur University, Bhagalpur, India

2 Department of Chemistry, Indian Institute of Technology Indore, Indore, India

*Address all correspondence to: bnjha06@gmail.com

References

[1] Arduengo AJ III, Harlow RL, Kline MA. Stable crystalline carbene. Journal of the American Chemical Society. 1991;**113**:36-363. DOI: 10.1021/ja00001a054. This is the first report of a stable, isolable NHC

[2] de Fre'mont P, Marion N, Nolan SP. Carbenes: Synthesis, properties, and organometallic chemistry. Coordination Chemistry Reviews. 2009;**253**:862-892. DOI: 10.1016/j.ccr.2008.05.018

[3] Hillier AC, Sommer WJ, Yong BS, Petersen JL, Cavallo L, Nolan SP. A combined experimental and theoretical study examining the binding of N-heterocyclic carbenes (NHC) to the Cp*RuCl (Cp* = η^5-C_5Me_5) moiety: Insight into stereoelectronic differences between unsaturated and saturated NHC ligands. Organometallics. 2003;**22**:4322-4326. DOI: 10.1021/om034016k

[4] Poater A, Cosenza B, Correa A, Giudice S, Ragone F, Scarano V, et al. SambVca: A web application for the calculation of the buried volume of N-heterocyclic carbene ligands. European Journal of Inorganic Chemistry. 2009:1759-1766. DOI: 10.1002/ejic.200801160

[5] Truscot BJ, Nelson DJ, Lujan C, Slawin AMZ, Nolan SP. Iridium(I) hydroxides: Powerful synthons for bond activation. Chemistry - A European Journal. 2013;**19**:7904-7916. DOI: 10.1002/chem.201300669

[6] Runyon JW, Steinhof O, Rasika Dias HV, Calabrese JC, Marshall WJ, Arduengo AJ III. Carbene-based Lewis pairs for hydrogen activation. Australian Journal of Chemistry. 2011;**64**:1165-1172. DOI: 10.1071/CH11246

[7] Bourissou D, Guerret O, Gabbai FP, Bertrand G. Stable carbenes. Chemical Reviews. 2000;**100**:39-92. DOI: 10.1021/cr940472u

[8] Tolman CA. Steric effects of phosphorus ligands in organometallic chemistry and homogeneous catalysis. Chemical Reviews. 1977;**77**:313-348. DOI: 10.1021/cr60307a002

[9] Kelly RA, Clavier H, Giudice S, Scott NM, Stevens ED, Bordner J, et al. Determination of N-heterocyclic carbene (NHC) steric and electronic parameters using the [(NHC)Ir$(CO)_2$Cl] system. Organometallics. 2007;**27**:202-210. DOI: 10.1021/om701001g

[10] Wolf S, Plenio H. Synthesis of (NHC)Rh(cod)Cl and (NHC)RhCl$(CO)_2$ complexes-translation of the Rh- into the Ir-scale for the electronic properties of NHC ligands. Journal of Organometallic Chemistry. 2009;**694**:1487-1492. DOI: 10.1016/j.jorganchem.2008.12.047

[11] Waltman AW, Grubb RH. A new class of chelating N-heterocyclic carbene ligands and their complexes with palladium. Organometallics. 2004;**23**:3105-3107. DOI: 10.1021/om049727c

[12] Clavier H, Coutable L, Guillemin J-C, Maudit M. New bidentate alkoxy-NHC ligands for enantioselective copper-catalysed conjugate addition. Tetrahedron: Asymmetry. 2005;**16**:921-924. DOI: 10.1016/j.tetasy.2005.01.015

[13] Trnka TM, Morgan JP, Sanford MS, Wilhelm TE, Scholl M, Choi T-L, et al. Synthesis and activity of ruthenium alkylidene complexes coordinated with phosphine and N-heterocyclic carbene ligands. Journal of the American Chemical Society. 2003;**125**:2546-2558. DOI: 10.1021/ja021146w

[14] Nemcsok D, Wichmann K, Frenking G. The significance of p interactions in group-11 complexes with N-heterocyclic carbenes. Organometallics. 2004;**23**:3640-3646. DOI: 10.1021/om049802j

[15] Díez-González S, Nolan SP. Stereoelectronic parameters associated with N-heterocyclic carbene (NHC) ligands: A quest for understanding. Coordination Chemistry Reviews. 2007;**251**:874-883. DOI: 10.1016/j.ccr.2006.10.004

[16] Jacobsen H, Correa A, Poater A, Costabile C, Cavallo L. Understanding the M-(NHC) (NHC = N-heterocyclic carbene) bond. Coordination Chemistry Reviews. 2009;**253**:687-703. DOI: 10.1016/j.ccr.2008.06.006

[17] Crudden CM, Allen DP. Stability and reactivity of N-heterocyclic carbene complexes. Coordination Chemistry Reviews. 2004;**248**: 2247-2273.DOI:10.1016/j.ccr.2004.05.013

[18] Riener K, Haslinger S, Raba A, Högerl MP, Cokoja M, Herrmann WA, et al. Chemistry of iron N-heterocyclic carbene complexes: Syntheses, structures, reactivities, and catalytic applications. Chemical Reviews. 2014;**114**:5215-5272. DOI: 10.1021/cr4006439

[19] Plietker B, Dieskau A, Möws K, Jatsch A. Ligand-Dependent mechanistic dichotomy in iron-catalyzed allylic substitutions: σ-allyl versus π-allyl mechanism. Angewandte Chemie, International Edition. 2008;**47**:198-201. DOI: 10.1002/anie.200703874

[20] Louie J, Grubbs RH. Highly active iron imidazolylidene catalysts for atom transfer radical polymerization. Chemical Communications. 2000;**16**:1479-1480. DOI: 10.1039/B003957H

[21] Mo Z, Zhang Q, Deng L. Dinuclear iron complex-catalyzed cross-coupling of primary alkyl fluorides with aryl Grignard reagents. Organometallics. 2012;**31**:6518-6521. DOI: 10.1021/om300722g

[22] Chen M-Z, Sun H-M, Li W-F, Wang Z-G, Shen Q, Zhang Y. Synthesis, structure of functionalized N-heterocyclic carbene complexes of Fe (II) and their catalytic activity for ring-opening polymerization of ε-caprolactone. Journal of Organometallic Chemistry. 2006;**691**:2489-2494. DOI: 10.1016/j.jorganchem.2006.01.031

[23] Hatanaka T, Ohki Y, Tatsumi K. C-H bond activation/borylation of furans and thiophenes catalyzed by a half-sandwich iron N-heterocyclic carbene complex. Chemistry - An Asian Journal. 2010;**5**:1657-1666. DOI: 10.1002/asia.201000140

[24] Cramer SA, Jenkins DM. Synthesis of aziridines from alkenes and aryl azides with a reusable macrocyclic tetracarbene iron catalyst. Journal of the American Chemical Society. 2011;**133**:19342-19345. DOI: 10.1021/ja2090965

[25] Kandepi VVKM, Cardoso JMS, Peris E, Royo B. Iron(II) complexes bearing chelating cyclopentadienyl-N-heterocyclic carbene ligands as catalysts for hydrosilylation and hydrogen transfer reactions. Organometallics. 2010;**29**:2777-2782. DOI: 10.1021/om100246j

[26] Jiang F, Bézier D, Sortais J-B, Darcel C. N-heterocyclic carbene piano-stool iron complexes as efficient catalysts for hydrosilylation of carbonyl derivatives. Advanced Synthesis and Catalysis. 2011;**353**:239-244. DOI: 10.1002/adsc.201000781

[27] Bézier D, Venkanna GT, Sortais J-B, Darcel C. Well-defined cyclopentadienyl NHC iron complex as the catalyst for efficient hydrosilylation of amides to amines and nitriles. ChemCatChem. 2011;**3**:1747-1750. DOI: 10.1002/cctc.201100202

[28] Beller M, Bolm C. Transition Metals for Organic Synthesis: Building Blocks and Fine Chemicals. Weinheim,

Germany: Wiley-VCH; 1998. ISBN: 978-3-527-61940-5

[29] Hartwig JF. Organotransition Metal Chemistry: From Bonding to Catalysis. Mill Valley, CA: University Science Books; 2010. ISBN: 978-1-891-38953-5

[30] Crabtree RH. The Organometallic Chemistry of the Transition Metals. Hoboken, NJ: Wiley; 2011. ISBN: 978-0-470-25762-3

[31] Huheey JE, Keiter EA, Keiter RL, Medhi OK. Inorganic Chemistry: Principles of Structure and Reactivity. Upper Saddle River, NJ: Pearson Education; 2006. ISBN: 978-8-177-58130-0

[32] Lippard SJ, Berg JM. Principles of Bioinorganic Chemistry. Mill Valley, CA: University Science Books; 1994. ISBN: 0-935702-73-3

[33] Ochiai EI. Bioinorganic Chemistry: A Survey. Amsterdam: Elsevier Science/Academic Press; 2010. ISBN: 978-0-120-88756-9

[34] Beller M, Renken A, van Santen RA. Catalysis: From Principles to Applications. Weinheim, Germany: Wiley-VCH; 2012. ISBN: 978-3-527-32349-4

[35] Gopalaiah K. Chiral iron catalysts for asymmetric synthesis. Chemical Reviews. 2013;**113**:3248-3296. DOI: 10.1021/cr300236r

[36] Plietker B. Iron Catalysis in Organic Chemistry: Reactions and Applications. Weinheim, Germany: Wiley-VCH; 2008. ISBN: 978-3-527-31927-5

[37] Lavallo V, El-Batta A, Bertrand G, Grubbs RH. Insights into the carbene-initiated aggregation of [Fe(cot)$_2$]. Angewandte Chemie, International Edition. 2011;**50**:268-271. DOI: 10.1002/anie.201005212

[38] Bedford RB, Betham M, Bruce DW, Danopoulos AA, Frost RM, Hird M. Iron-phosphine, -phosphite, -arsine, and -carbene catalysts for the coupling of primary and secondary alkyl halides with aryl grignard reagents. The Journal of Organic Chemistry. 2006;**71**:1104-1110. DOI: 10.1021/jo052250+

[39] Nakamura M, Matsuo K, Ito S, Nakamura E. Iron-catalyzed cross-coupling of primary and secondary alkyl halides with aryl grignard reagents. Journal of the American Chemical Society. 2004;**126**:3686-3687. DOI: 10.1021/ja049744t

[40] Martin R, Fürstner A. Cross-coupling of alkyl halides with aryl Grignard reagents catalyzed by a low-valent iron complex. Angewandte Chemie, International Edition. 2004;**43**:3955-3957. DOI: 10.1002/anie.200460504

[41] Silberstein AL, Ramgren SD, Garg NK. Iron-catalyzed alkylations of aryl sulfamates and carbamates. Organic Letters. 2012;**14**:3796-3799. DOI: 10.1021/ol301681z

[42] Hatakeyama T, Nakamura M. Iron-catalyzed selective biaryl coupling: Remarkable suppression of homocoupling by the fluoride anion. Journal of the American Chemical Society. 2007;**129**:9844-9845. DOI: 10.1021/ja073084l

[43] Guisán-Ceinos M, Tato F, Buñuel E, Calle P, Cárdenas D. Fe-catalysed Kumada-type alkyl-alkyl cross-coupling. Evidence for the intermediacy of Fe(I) complexes. Journal of Chemical Sciences. 2013;**4**:1098-1104. DOI: 10.1039/C2SC21754F

[44] Holzwarth M, Dieskau A, Tabassam M, Plietker B. The ammosamides: Structures of cell cycle modulators from a marine-derived *Streptomyces* species. Angewandte Chemie, International Edition.

2009;**48**:725-727. DOI: 10.1002/anie.200804890

[45] Pottabathula S, Royo B. First iron-catalyzed guanylation of amines: A simple and highly efficient protocol to guanidines. Tetrahedron Letters. 2012;**53**:5156-5158. DOI: 10.1016/j.tetlet.2012.07.065

[46] Obligacion JV, Chirik P. Highly selective bis(imino)pyridine iron-catalyzed alkene hydroboration. Journal of Organic Letters. 2013;**15**:2680-2683. DOI: 10.1021/ol400990u

[47] Yamagami T, Shintani R, Shirakawa E, Hayashi T. Iron-catalyzed arylmagnesiation of aryl(alkyl) acetylenes in the presence of an N-heterocyclic carbene ligand. Organic Letters. 2007;**9**:1045-1048. DOI: 10.1021/ol063132r

[48] Holzwarth MS, Frey W, Plietker B. Binuclear Fe-complexes as catalysts for the ligand-free regioselective allylic sulfenylation. Chemical Communications.2011;**47**:11113-11115. DOI: 10.1039/C1CC14599A

[49] Jegelka M, Plietker B. α-Sulfonyl succinimides: Versatile sulfinate donors in Fe-Catalyzed, salt-free, neutral allylic substitution. Chemistry - A European Journal. 2011;**17**:10417-10430. DOI: 10.1002/chem.201101047

[50] Jegelka M, Plietker B. Dual catalysis: Vinyl sulfones through tandem iron-catalyzed allylic sulfonation amine-catalyzed isomerization. ChemCatChem. 2012;**4**:329-332. DOI: 10.1002/cctc.201100465

[51] Hilt G, Bolze P, Kieltsch I. An iron-catalysed chemo- and regioselective tetrahydrofuran synthesis. Chemical Communications. 2005:1996-1998. DOI: 10.1039/B501100K

[52] Pintauer T, Matyjaszewski K. Atom transfer radical addition and polymerization reactions catalyzed by PPM amounts of copper complexes. Chemical Society Reviews. 2008;**37**:1087-1097. DOI: 10.1039/B714578K

[53] Bézier D, Jiang F, Roisnel T, Sortais J-B, Darcel C. Cyclopentadienyl-NHC iron complexes for solvent-free catalytic hydrosilylation of aldehydes and ketones. European Journal of Inorganic Chemistry. 2012;**2012**:1333-1337. DOI: 10.1002/ejic.201100762

[54] Grohmann C, Hashimoto T, Fröhlich R, Ohki Y, Tatsumi K, Glorius F. An Iron(II) complex of a diamine-bridged bis-N-heterocyclic carbene. Organometallics. 2012;**31**:8047-8050. DOI: 10.1021/om300888q

[55] Warratz S, Postigo L, Royo B. Direct synthesis of Iron (0) N-heterocyclic carbene complexes by using $Fe_3(CO)_{12}$ and their application in reduction of carbonyl groups. Organometallics. 2013;**32**:893-897. DOI: 10.1021/om3012085

[56] Bézier D, Venkanna GT, Misal Castro LC, Zheng J, Roisnel T, Sortais J-B, et al. Iron-catalyzed hydrosilylation of esters. Advanced Synthesis and Catalysis. 2012;**354**:1879-1884. DOI: 10.1002/adsc.201200087

[57] Demir S, Gökçe Y, Kaloğlu N, Sortais J-B, Darcel C, Özdemir İ. Synthesis of new Iron-NHC complexes as catalysts for hydrosilylation reactions. Applied Organometallic Chemistry. 2013;**27**:459- 464. DOI: 10.1002/aoc.3006

[58] Li H, Misal Castro LC, Zheng J, Roisnel T, Dorcet V, Sortais J-B, et al. Selective reduction of esters to aldehydes under the catalysis of well-defined NHC-Iron complexes. Angewandte Chemie, International Edition. 2013;**52**:8045-8049. DOI: 10.1002/anie.201303003

[59] Volkov A, Buitrago E, Adolfsson H. Direct hydrosilylation of tertiary amides

to amines by an *in situ* formed Iron/N-heterocyclic carbene catalyst. European Journal of Organic Chemistry. 2013:2066-2070. DOI: 10.1002/ejoc.201300010

[60] Misal Castro LC, Sortais J-B, Darcel C. NHC-carbene cyclopentadienyl iron based catalyst for a general and efficient hydrosilylation of imines. Chemical Communications. 2012;**48**:151-153. DOI: 10.1039/C1CC14403K

3

Synthetic Studies of Vitamin B12

David Joshua Ferguson

Abstract

Overall these are selections from the total synthesis of vitamin B12. Through the use of selected reactions in the reaction schema, hypothetical mechanisms have been provided. It is the hope of the author, that it will provide insight for students in organic chemistry. Additionally the focus was on the Eschenmoser's Variant of the total synthesis of vitamin B12. This required the reviewing of the lectures of Dr. A. Eschenmoser as well as reviews of the different mechanistic process involved. Due to constraints all of the mechanisms have not been developed, but selected ones have been provided and shown for understanding.

Keywords: mechanisms, vitamin B12

1. Section 1

1.1 Introduction

Vitamin B12 otherwise known as cyanocobalamin is a compound with synthetic elegance. Since it is composed of an aromatic macrocyclic corrin there are key features of this molecule that are observed either in its synthesis of in the biochemical reactions it plays a role in whether they be isomerization reactions or transfer reactions. In this paper the focus for the discussion will be on the history, chemical significance, and total synthesis of vitamin B12. Even more so the paper will concentrate on one of the two variants of the vitamin B12 synthesis, namely, the ETH Zurich variant spearheaded by Albert Eschenmoser. Examining the structure as a whole, it is observed that a large portion of the vitamin B12 is a corrin structure with a cobalt ion in the center of the macrocyclic part and that same cobalt ion has cyanide ligands. The general macrocyclic portion of the structure is rimmed with either methyl or amide group attachments. One of the amide groups is N-alkylated by a large isopropanol group, then a phosphate, followed by a ribose which is attached to the dimethylbenzimidazole. However, in terms of history, there were some key steps in the process of determining and synthesizing the overall structure of vitamin B12.

"We made up in our minds that we're going to specialize in research in the field of vitamins. We're going to isolate every vitamin. We're going to determine their structures if it hasn't already been done and synthesize them and make them available," Randolph Major, as told by Mac Tishler, said in 1983. Our understanding of disease in the modern world was aided by the work of Louis Pasteur and the germ theory. The issue came into being with diseases such as pellagra, anemia, and beriberi in which the origin is not pathogenic typically but based in nutrient deficiency and in this case vitamin B. This new category came into being in the 1900s.

In 1889 **Dutch physician Christiaan Eijkman** investigated beriberi. He and **Gerrit Grijins** studied the effects of dietary variations on the occurrence of beriberi. After, in 1906 English biochemist **Frederick Gowland Hopkins** suggested a connection between nutrition and diseases such as beriberi and scurvy. Following that in 1911 Casimir Funk, a Polish biochemist working in London, further advanced this idea. University of Wisconsin biochemist Elmer Mccollum was able to distinguish two different species of vitamins "fat-soluble factor A" and "water-soluble factor B." Moreover, in 1926 Dutch chemists Barend Jansen and Willem Donath isolated crystal of anti-beriberi factor from extracts of rice polishings. Chemists were arduously working with natural product chemistry, with Merck in 1930 already working on this task.

Williams of Bell Laboratories approached Merck to help isolate and make thiamine.

Randolph Major was chosen to head the new research and development laboratory Merck built as a part of its efforts to grow basic research. Eventually, Williams and Cline synthesized thiamine. As was seen in 1922, riboflavin, vitamin B2, had been discovered in 1922 by Richard Kuhn in Germany and Theodor Wagner-Jauregg in Austria. Moreover in 1933 riboflavin was isolated by Kuhn and Gyorgy in Germany. As time progressed in 1934, vitamin B6, pyridoxine, was discovered by Gyorgy and colleagues. After this in 1938, the active compound of pyridoxine was isolated by Samuel Lepovsky of U.C. Berkley, after which in 1939, Folkers and Harris along with Kuhn in Germany determined the structure of pyridoxine. Eventually in 1940, the synthesis of vitamin B5, pantothenic acid, was reported by Merck.

1.1.1 Discovery of cobalamin

Cobalamin was discovered through an interesting process. First in 1926 at Harvard University, a team of physicians found out that ingesting a half a pound of liver would prevent pernicious anemia. As time progressed liver extracts were fed to willingly participating patients. Folkers ultimately learned that Mary Shorb a microbiologist found a bacterium that reacted to liver extracts. Also it was determined that the most promising extracts were those with the "pinkish color," which implied that the vitamin being sought was a red compound. In 1947, Folkers and his team isolated vitamin B12 (cobalamin) which resulted in tiny, bright, red crystals of the vitamin [1].

1.1.2 Nominal definitions

1. Homologation

2. Corrin

3. Ammonolysis

4. Thionation

5. Methanolysis

6. Woodward-Hoffman rules

7. Protection/deprotection

1.1.2.1 Homologation

Essentially homologation is a reaction that converts the reactant into the next member of the homologous series [2]. In many cases a homologous series is a group of compounds that differ by a constant unit. Homologation occurs simply when the repeated structural unit is increased, and in the reaction above, it is a methylene (—CH2—).

1.1.2.2 Corrin

A corrin is a macrocycle. Specifically a corrin is a species consisting of four reduced pyrrole rings joined by three —CH= and one double bond [3].

A common prefix associated with corrin is "seco-" which refers to a macrocycle in which cleavage of a ring has occurred with the addition of one or more hydrogen atoms at each terminal group as indicated.

One distinction is made between the porphyrin, seen below, and the corrin, seen above, based on size in that the porphyrin is larger.

1.1.2.3 Ammonolysis

Ammonolysis is a reaction similar to hydrolysis in which ammonia reacts with another compound as a nucleophile and oftentimes the solvent usually to result in the formation of an amine functional group of the molecule [4]. An example seen above is the ammonolysis of esters which results in amides.

1.1.2.4 Thionation

Thionation is a chemical reaction in which the oxygen in a moiety (e.g., carbonyl, hydroxyl) is converted to a sulfur. In this step in the Eschenmoser variant for the total synthesis of vitamin B12, a cyclic carbonyl-containing molecule is thionated which results in the precursor to ring A for cobyric acid and vitamin B12.

1.1.2.5 Methanolysis

Methanolysis is similar to hydrolysis, but instead of water functioning as the nucleophile and solvent, methanol is functioning in that way. Overall in the reaction as seen above, the methanolysis process results in the elimination of the hydroxyl from the ester. This can also be considered a type of transesterification.

1.1.2.6 Woodward-Hoffman rules

The Woodward Hoffman rules were sorted out by Robert B. Woodward and Roald Hoffman, although further work was done by Fukui [5]. These rules involve the use of a simple procedure for determining whether a pericyclic reaction is thermally allowed. Primarily the focus is on the aromaticity of the transition state, which is understood based on orbital topology and electron count. The reaction above shows an example where these rules can be applied in this unique cycloaddition in the form of a Diels-Alder reaction. In short, these rules state that whenever possible, reactions go through aromatic transition states.

As Eschenmoser [6] wrote in his lecture, "but I should perhaps propose that we enjoy the figure just from an aesthetic point of view, by watching the corrinoid chromophore system evolve, like a bud blooming into a flower."

For the total synthesis of vitamin B12, there are two variants, both of which were accomplished in 1972. In 1960, the ETH Zurich variant was started by Albert Eschenmoser and his team. Following that in 1961, the Harvard variant was started, and after 1965 the work was collaboratively pursued. In terms of the amount of collaboration, it required the work of 91 post-doctoral fellows and 12 Ph.D. students from several different nations [7].

2. Section 2

2.1 Synthesis of the rings

Within the descriptions, both general and or mechanistic, the numbering of the compounds was based on Albert Eschenmoser's overall schema [8].

For the schema with identical steps, the mechanism and explanations are explained once. Also selected mechanisms are listed below from the overall schema.

Ring A:

1. Claisen-Schmidt condensation [9]

2. and 3. Diels-Alder

4. Oxidation

5. Arndt-Eistert

6. Ammonolysis

7. Ring opening

8. Thionation [10]

Ring B:

1. Claisen-Schmidt condensation

2. and 3. Diels-Alder [9]

4. Oxidation

5. Arndt-Eistert [11]

6. Ammonolysis

7. Thionation

Ring C:

1. Claisen-Schmidt condensation

2. Diels-Alder

3. Oxidation

4. Arndt-Eistert

5. Ammonolysis

6. Esterification and methanolysis

7. Thioesterification [12]

8. Reductive decarbonylation

Ring D:

1. Claisen-Schmidt condensation

2. Diels-Alder reaction

3. Oxidation reaction

4. Ammonolysis reaction

5. Ring opening reaction

6. Arndt-Eistert reaction

7. Hydrolysis, decarboxylation, and esterification

8. Protection/sulfonation

9. Reduction/deprotection

10. Protonation

11. Beckmann fragmentation

12. Bromination of the ketimine.

3. Section 3

Ring A:

1. Claisen-Schmidt condensation

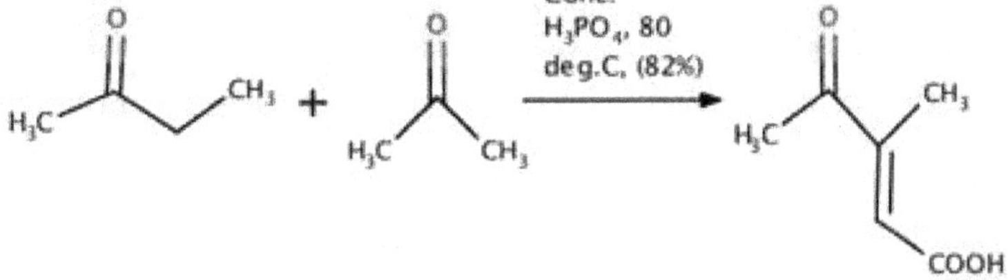

This first reaction of the B12 reaction scheme involves an ethyl methyl ketone (Compound 1A) reacting with acetaldehyde (Compound 1B) using reagents which are concentrated phosphoric acid (H_3PO_4) at 80°C, and the yield is 82%. The type of reaction that is occurring the Claisen-Schmidt condensation in which you have the formation of (2E)-3-methyl-4-oxopent-2-enoic acid. This reaction plays the role in producing the dienophile that will be used in the following reaction.

Overall if simplified this reaction is a type of condensation that results in the formation of an electron-poor molecule.

Mechanism:

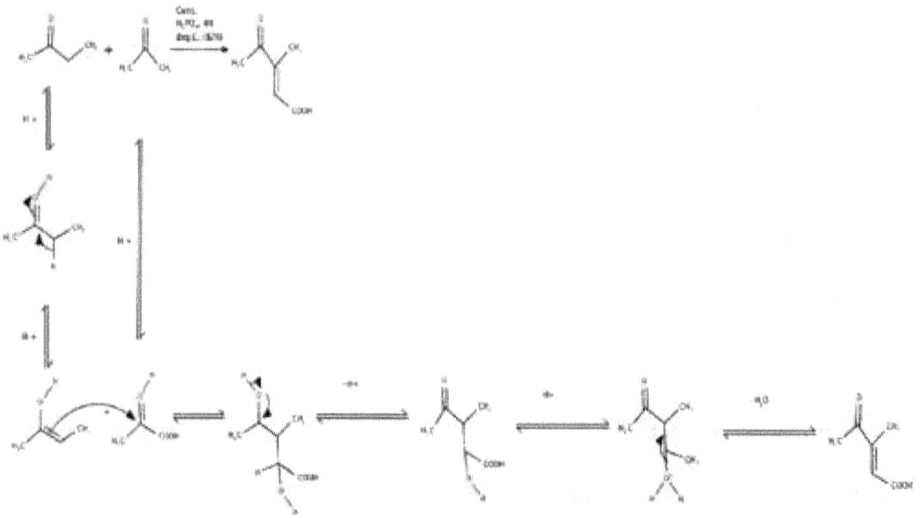

2. and 3. Diels-Alder

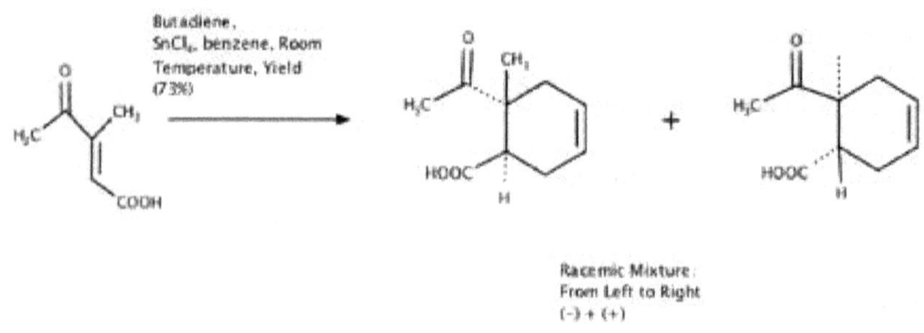

The second reaction of the B12 schema involves (2E)-3-methyl-4-oxopent-2-enoic acid (Compound 2) reacting with butadiene in tin (IV) chloride ($SnCl_4$) and benzene at conditions of room temperature. This results in a yield of 73%. The type of reaction that is occurring is the Diels-Alder reaction which involves the formation of a racemic mixture of two carboxylic acid-like molecules with ketone-like moieties attached to it. For the purposes of this discussion, the products will be labeled compounds 3A(−) and 3B(+).

Overall if simplified the type of reaction, Diels-Alder, is stereospecific and a type of concerted reaction in that all the bond breaking and bond forming occur at the same time. Moreover addition is syn. Also if this reaction follows the typical Diels-Alder format, it is a one-step cyclo-addition or conjugate addition. This reaction, which resulted in enantiomers which were resolved using phenylethylamine in chloroform and hexane, followed by the use of diluted HCl.

Mechanism:

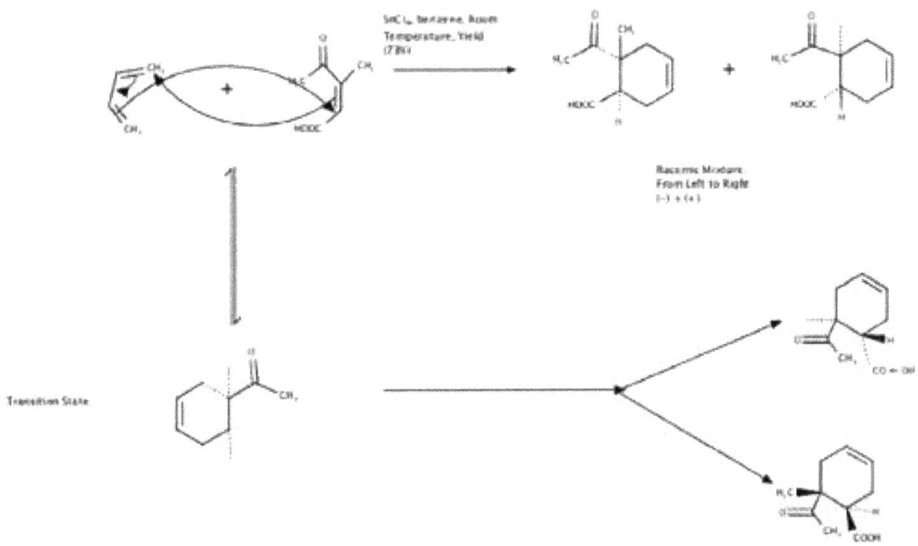

4. Oxidation

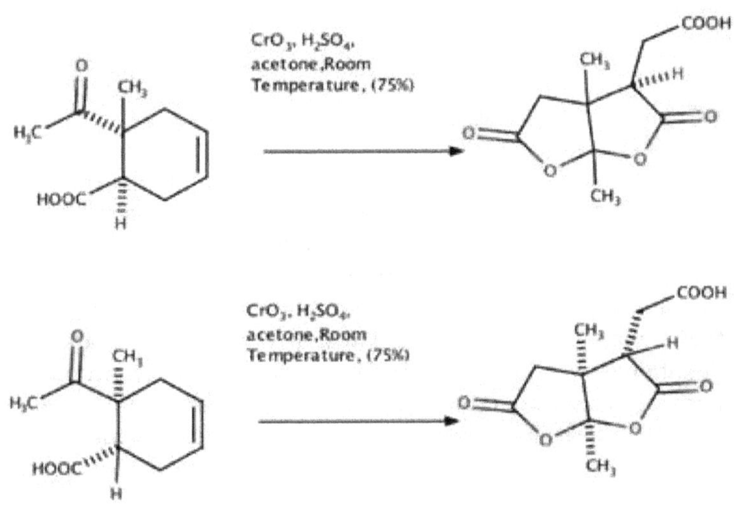

The fourth reaction in the B12 reaction schema involves compounds 3A(−) and 3B(+) reacting with chromate and sulfuric acid in acetone at room temperature to form dilactone carboxylic acids which will be labeled compounds 4B(−) and 4A(+) (from top to bottom), both of which are starting products for B12 ring reactors. This fourth reaction has a predicted yield of 75%. From the reagents and the reactants, this appears to be an organic redox reaction, possibly a Jones oxidation, in which we have a molecule being oxidized and or gaining hydrogen deficiency in the form of another ring. Stated simply, these reactions involve the oxidation of compounds 3A (−) and 3B(+) into ten carbon dilactone-carboxylic acids using reagents that normally are used in a type of organic redox reaction.

Mechanism:

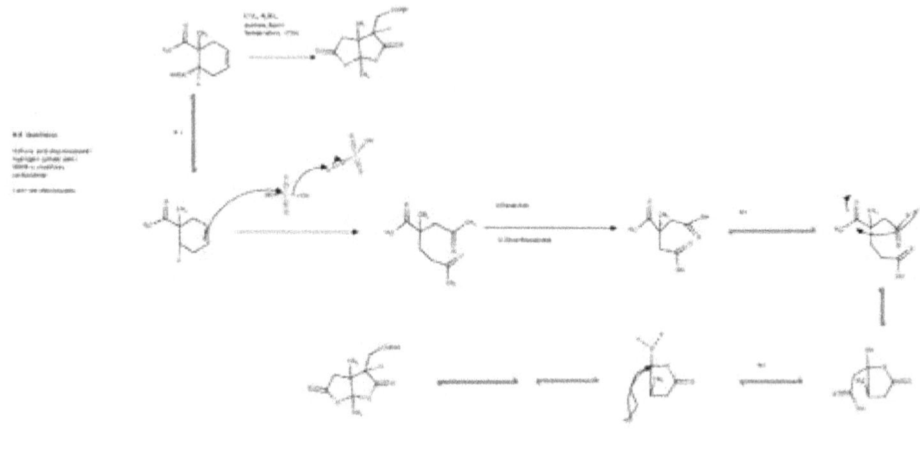

5. Arndt-Eistert

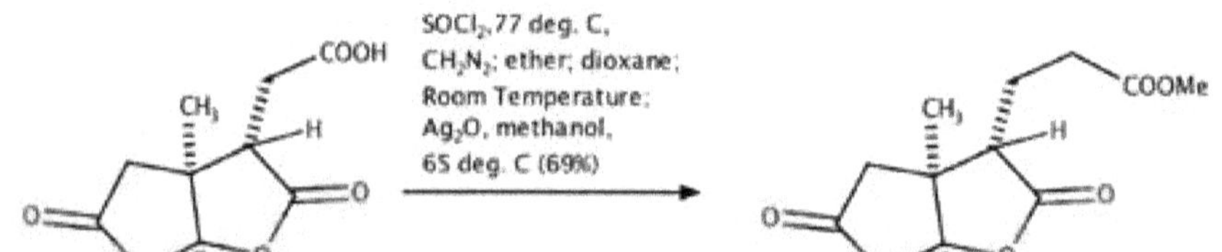

The fifth reaction in the B12 reaction schema involves compound 4A(+) reacting with thionyl chloride at 77°C. This was followed by reacting the acid chloride with diazomethane in ether at room temperature. After which it was reacted with silver dioxide in methanol at 65°C. This fifth reaction had a predicted yield of 69%. However the overall name of the reaction that is occurring is an Arndt-Eistert synthesis. Additionally an important step in the Arndt-Eistert reaction is the Wolff rearrangement of diazoketones to ketenes. The overall Arndt-Eistert reaction, excluding the Wolff rearrangement, can be seen from the reaction drawn below; this sequence involves several steps that result in a higher-order or homologated carboxylic acid.

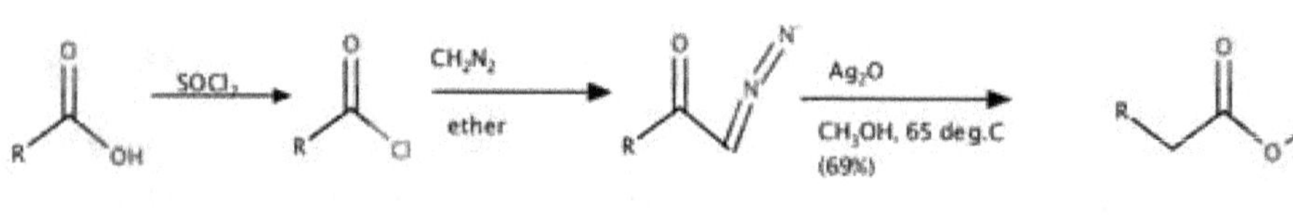

Stated simply this is a multistep reaction that involves the conversion of a carboxylic acid to an acid chloride, then to a diazo-ketone type molecule, and then the ester.

Some key points to note on the reaction, from Eschenmoser's notes for his 1973 German lecture at ETH Zurich, are as follows: "the treatment of the acid chloride with methanol/ pyridine at room temperature gives the same methyl ester as obtained by esterification with diazomethane; in the preparation of the acid chloride, there is no other structural change."

Mechanism:

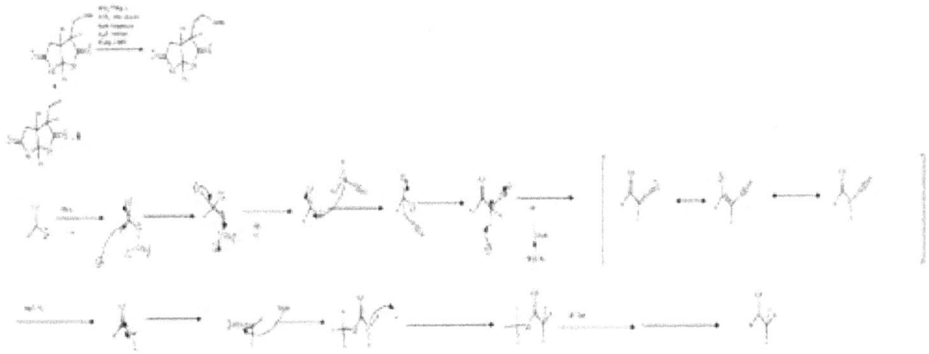

6. Ammonolysis

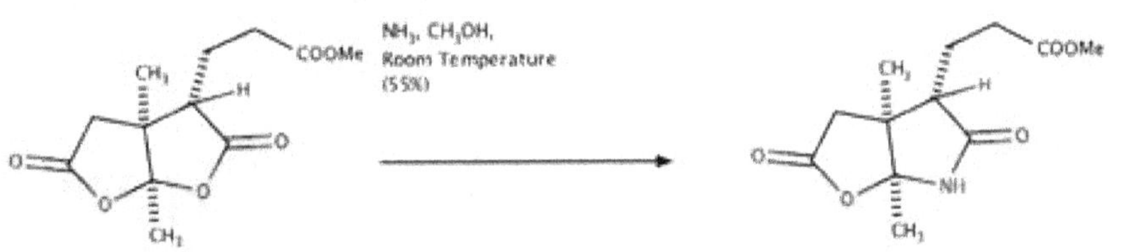

The sixth reaction in the B12 reaction schema involves compound 5A reacting with ammonia in methanol at room temperature. The sixth reaction had a yield of 55%. From this reaction it appears to be an ammonolysis reaction in the presence of a methanolic solvent. Also "the carbonyl groups of the dilactone moiety are much more nucleophilic towards ammonia than normal lactone or ester groups. 'Ammonolysis' of this type are much faster in methanol than in non-hydroxyl containing solvents. The constitution assignment for the isomeric lactone-lactams resulted from the identity of compound 6A with the main product of intramolecular NH transfer." Stated simply this reaction involved an intramolecular NH transfer using ammonia in methanol at room temperature [6].

Mechanism:

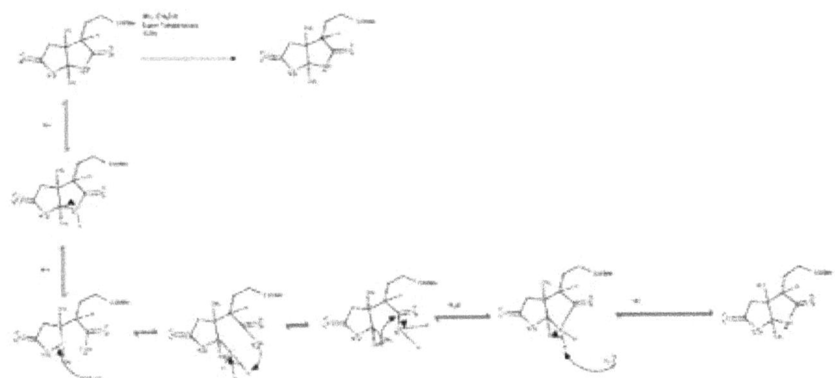

7. Ring opening (step 11 in Eschenmoser's overall schema)

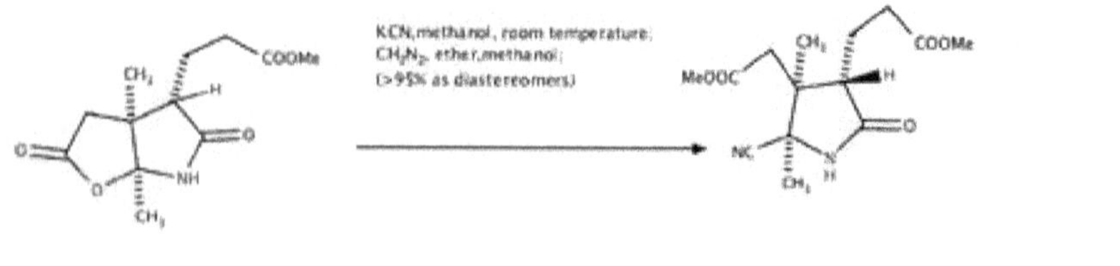

The eleventh reaction in the B12 reaction schema involves compound 6A reacting with potassium cyanide in methanol at room temperature, followed by a reaction with diazomethane in ether and methanol. This resulted in 95% being diastereomers. From this reaction we can see that a lactone ring is opened and a respective ester and cyano group are on the ends. Based on observation this appears to have gone through acid-catalyzed (methanol) ring opening, followed by nucleophilic attack by the cyanide anion from the potassium cyanide. Stated simply this involves the conversion of a 12 -carbon-dicarbonyl-bicyclic compound to a cyclic compound with the other ring being cleaved to form an ester and a cyanide at the ends where the ring broke.

Mechanism:

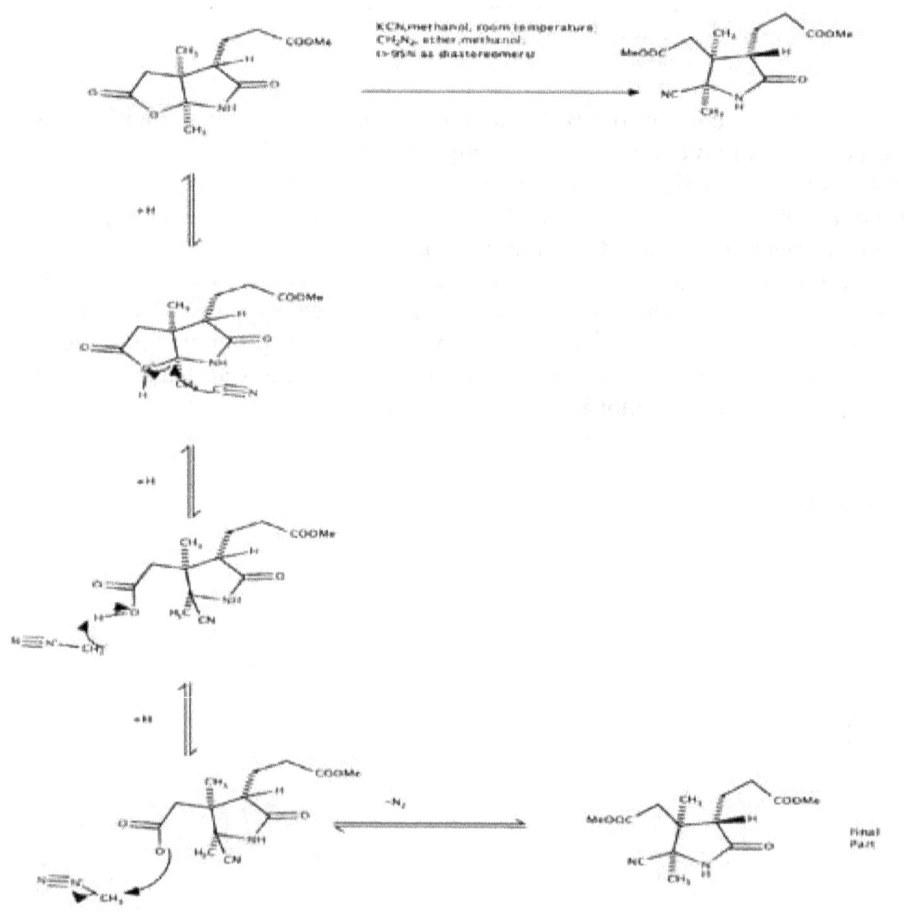

8. Thionation (step 12 in Eschenmoser's overall schema)

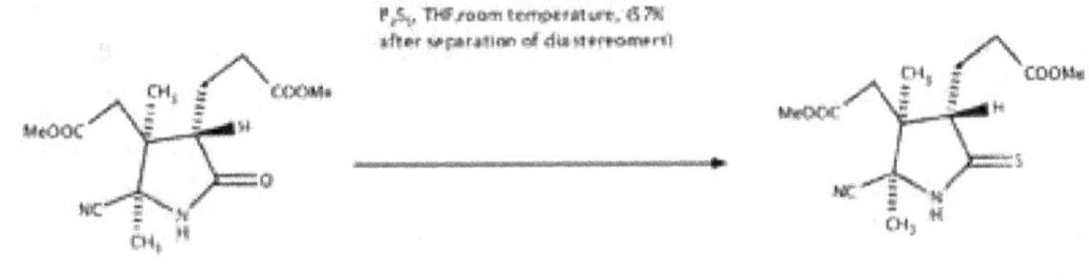

The twelfth reaction in the B12 reaction schema involves compound 11A reacting with diphosphorus pentasulfide and tetrahydrofuran at room temperature to form compound 12 A. Based on the observation of compound 11A, a 14 carbon-monocyclic compound going through a thionation of the carbonyl to form compound 12A, a 14 carbon-monocyclic compound. Stated simply this involves the conversion of a carbonyl to a thio-carbonyl on a 14-carbon monocyclic compound.

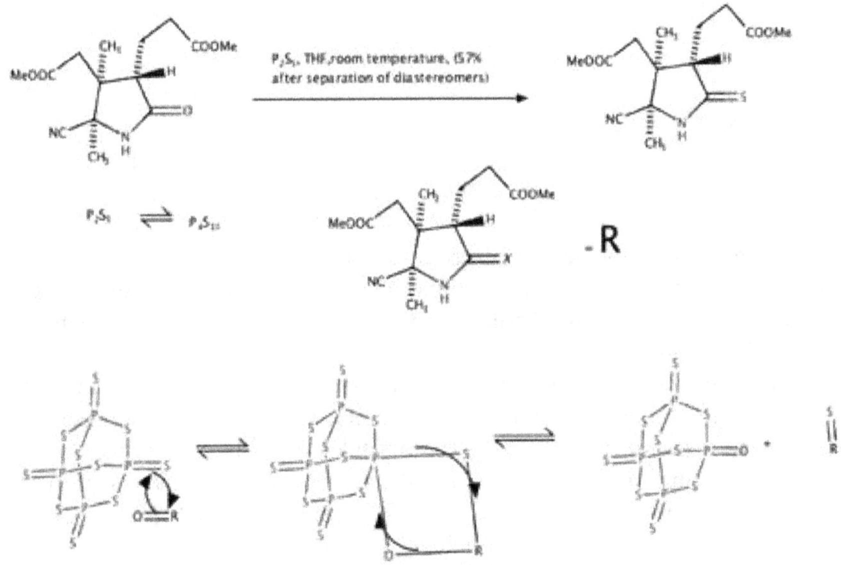

Ring B:

7. Thionation

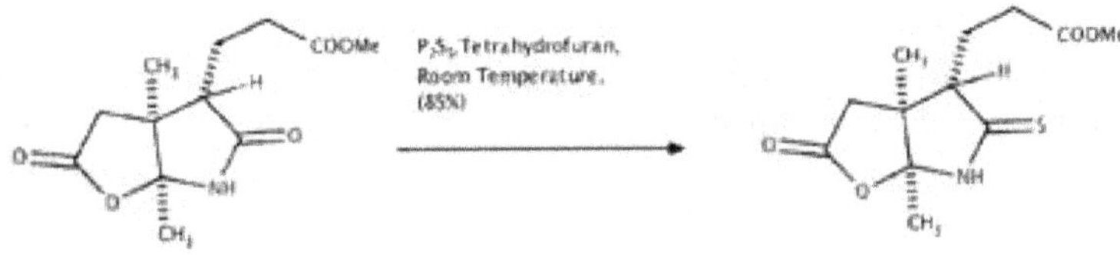

The seventh reaction in the B12 reaction schema involves compound 6A reacting with diphosphorus pentasulfide in tetrahydrofuran at room temperature. The seventh reaction had a yield of 85%. From the reaction compound, 6A is converted to compound 7A with a subsequent thionation in which the carbonyl is converted to a thiocarbonyl. Thionation is the conversion of the carbonyl group to thiocarbonyl, which is a commonly used procedure for the preparation of organosulfur compounds. In many instances with thionations, both the ketone and ester carbonyl groups of the oxoester can be affected by P_4S_{10} but typically in rather low yield. This thionation was specific in that both carbonyl groups were not thionated to thiocarbonyls. Simply put the seventh reaction involved the conversion of one of the carbonyls in the C10-dilactone-ester to a thiocarbonyl using thionating reagents at room temperature. An interesting fact to note is that compound 7A was a precursor for ring B of the macrocyclic corrin that composes the cyanocobalamin.

Mechanism:
(See thionation mechanism above)
Ring C:

7. Esterification and methanolysis (step 8 in the reference)

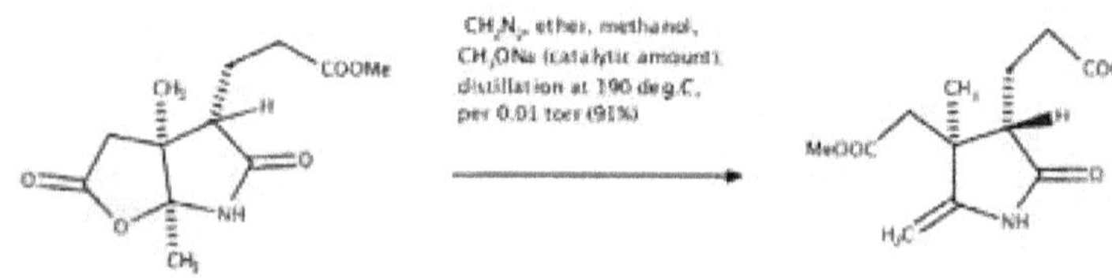

The eighth reaction in the B12 reaction schema involves compound 6A reacting with diazomethane in ether with methanol and a catalytic amount of sodium methoxide, after which is the distillation at 190°C at a pressure of 0.01 torr. This reaction had a yield of 91%. From the reaction compound, 6A is converted to compound 8A in which an esterification occurs, resulting in the formation of a methoxy-ester and the formation of a double bond with a methene. Through the use of Dr. Albert Eschenmoser's 1973 ETH Zurich German lecture notes, we gain a better understanding. It states that "Normally when diazomethane is esterified, the free carboxylic acids are transformed with an ethereal solution of CH_2N_2" and the hypothetical mechanism can be seen below:

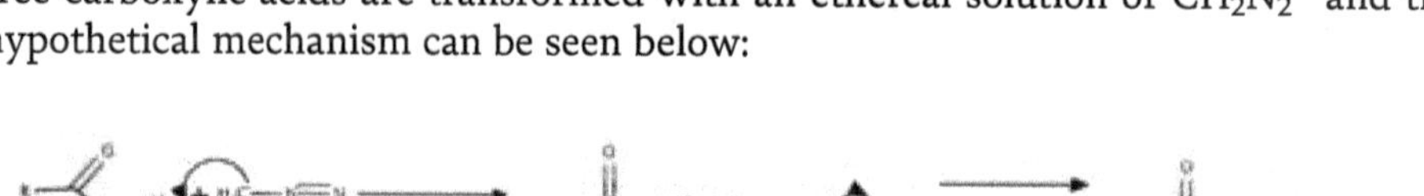

Conversion of compound 6A to compound 8A is "one of the rare examples of esterification in a basic mechanism."

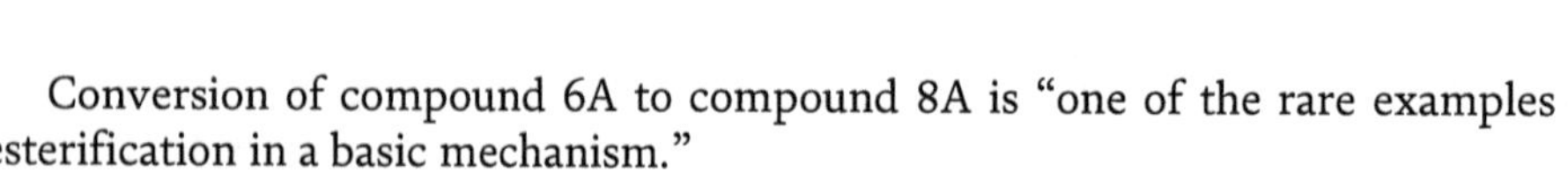

The catalytic amount of sodium methoxide serves to adjust the following equilibrium.

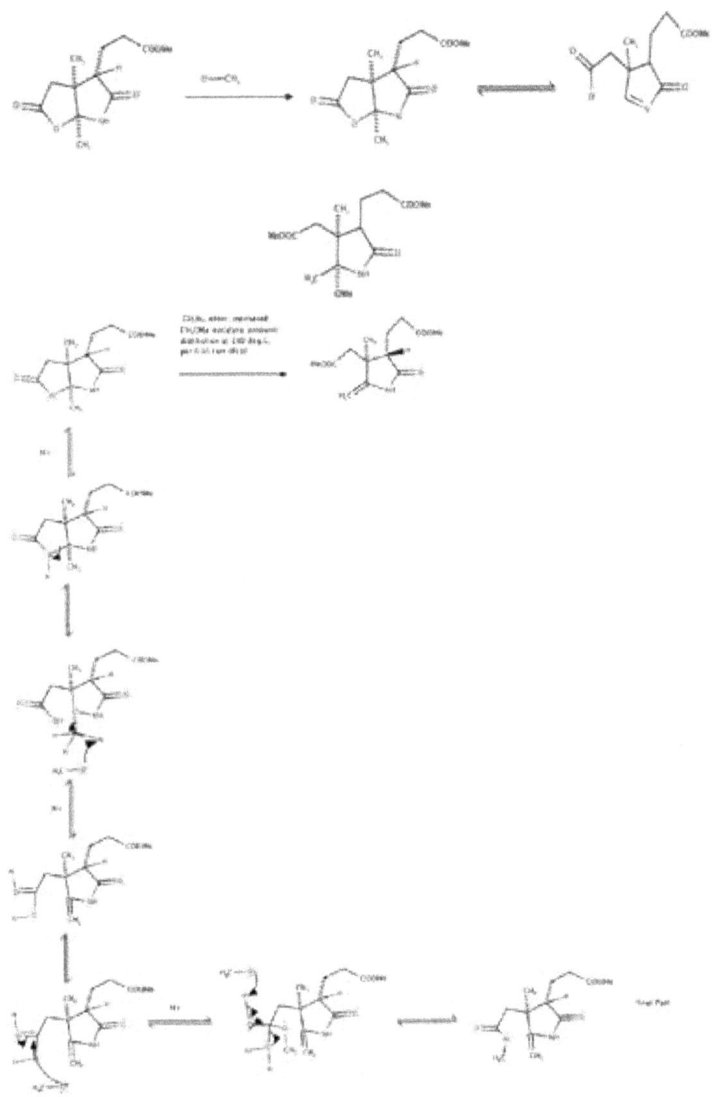

9. Reductive decarbonylation (step 10 in Eschenmoser's overall schema)

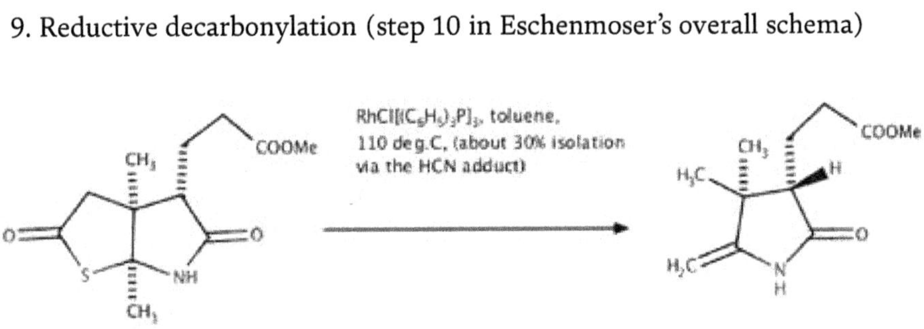

The tenth reaction in the B12 reaction schema involves compound 9A reacting with a rhodium-based catalyst in toluene at 110°C, which resulted in about 30%

isolation through the use of an HCN adduct. From the reaction a thiolactam ring is opened, resulting in a separate methyl and ethylene. As seen the remainder of the bicyclic reactant structure remains the same. However it is worth noting that in Eschenmoser's 1973 lecture notes, it includes that there are several products including the two cyclic structures, a phosphor-sulfuryl and a rhodium-based compound, all of which are reflective of the reagents, the reactants react with. Added to that, one of the groups of products is reacted again with silver ions in the presence of methanol (Ag^{+}/CH_3OH) to form the final pyrrolidine-like product, which is a precursor to ring C for the vitamin B12 synthesis. Additionally the ring precursor can be converted back to the reactant by the use of potassium cyanide in methanol (KCN, methanol). Both the conversion of ring C from the intermediate group of products to the final product and the reversed conversion back to the reagent in the group of products have yields of 90%. Stated simply this reaction involves the conversion of a bicyclic dicarbonyl-12-carbon ester to a cyclic 8 carbon pyrrolidine-like molecule using a catalyst in organic solvent. In other words, the "corresponding thiolactone is ran through reductive decarbonylation brought about by the chloro-tris-trisphenyl-phosphine complex of rhodium (I)" [6]. Then through the use of HCN, there was about 30% isolation. The entire reaction scheme for this step can be seen below:

Steps:

Note: In ten some insight was gained from Eschenmoser's German lecture.

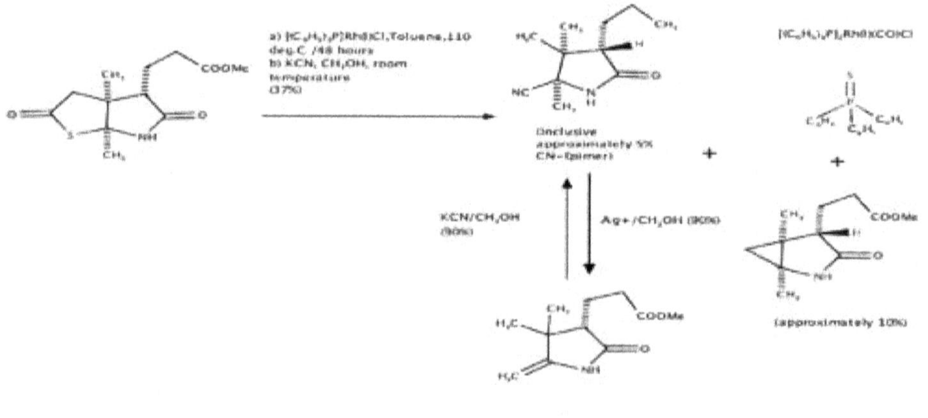

Ring D:

8. Hydrolysis, decarboxylation, and esterification (step 16 in the reference)

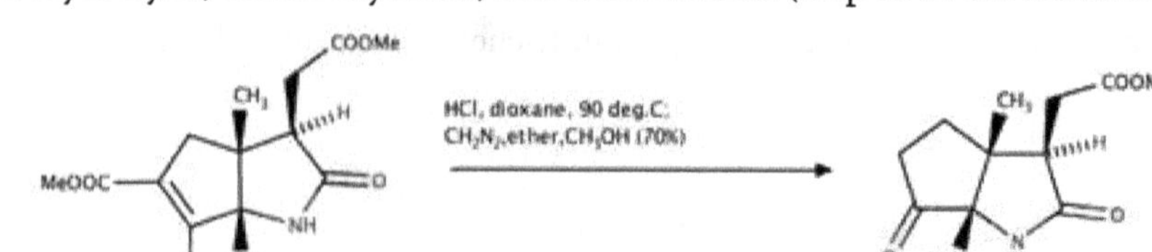

The general reaction involves compound 15B reacting with hydrochloric acid in dioxane at 90°C. This is followed by the reaction with diazomethane in ether and methanol. Compound 15B is a 12-carbon-bicyclic system with one of the cycles having a unit of unsaturation, i.e., a double bond, and an ester moiety and an amine moiety are attached to the cycle with the unit of unsaturation. Based on observation the two esters on compound 15B are hydrolyzed to acids, followed by the hydrolysis of anexamine to a ketone and then the decarboxylation of the beta-ketoacid, and finally the diazomethane is used to convert the remaining acid to an ester. Stated simply, this involves the conversion of a 12-carbon-cyclic compound to an 11-carbon dicarbonyl-bicyclic compound, by hydrolysis, decarboxylation, and then esterification.

Note: Some insight was gained from Dr. S. S.

4. Section 4

Final steps in the synthesis of cyanocobalamin.

1. Iminoester condensation and sulfide contraction (step 24 in Eschenmoser's overall schema).
2. Thionation (step 25 in Eschenmoser's overall schema).
3. Sulfide contraction via alkylative coupling (step 26 in Eschenmoser's overall schema).
4. Ammonolysis (step 27 in Eschenmoser's overall schema).
5. Iodination (step 28 in Eschenmoser's overall schema).
6. Elimination (step 29 in Eschenmoser's overall schema).
7. Photochemical A/D cycloisomerization (step 30 in Eschenmoser's overall schema).
8. Metal complexation (step 31 in Eschenmoser's overall schema).
9. Lactonization (iodolactonization) (step 32 in Eschenmoser's overall schema).
10. Alkylation (step 33 in Eschenmoser's overall schema).
11. Reduction and esterification (step 34 in Eschenmoser's overall schema).
12. Reduction (step 35 in Eschenmoser's overall schema).
13. Hydrolysis and ammonolysis (steps 36 and 37 in Eschenmoser's overall schema).
14. Final step: cobyric acid to cyanocobalamin (step 38 from K. Bernhauer and Eschenmoser's lecture).

5. Section 5

Characteristics of the cobyric acid molecule complex

Altogether the entire molecule "contains all peripheral carboxy functions in the primary amide form, except that of the propionic side chain in ring D."

Common steps in the entire synthesis:

The first four steps in the synthesis

- The Claisen-Schmidt condensation
- Diels-Alder (steps 2 and 3)
- Oxidation

Common problems in the synthesis of cobyric acid:

- Introduction of cobalt
- Closure of the macrocyclic ring
- Ester differentiation
- Introduction of methyl groups at bridges
- Restoration of lost stereochemistry [13]

Problems that had to be solved

- For rings A and B they are:
- Elongation of the free acetic acid chain by one methylene unit
- Specific replacement of one lactone oxygen by NH
- Conversion of the potential methylketone group into the enamide form

General approaches to problems

The sources especially Eschenmoser's and Woodward's lecture notes listed the general approaches involved:

1. Collaboration with other scientists.
2. Exhaustive study of the relationships between thioethers.
3. Purifications using analytic instrumentations such as high-performance liquid chromatography.
4. Use of pure reagents, exclusion of oxygen and moisture [14].

Possible future studies

Some possible future studies may involve the role of sulfur-aromatic interactions in certain mechanistic steps as well as carbocation-conjugate base interactions or stabilization. Added to this are variations of Markovnikov rules in the context of heterocycles. Additionally, whether through computation chemistry and/or experimental evidence, hypothesized organic chemistry mechanism testing can be done, considering how the plausibility of the mechanism is tied to the reality of the reaction.

6. Conclusion

Indeed "the emergence of the Woodward-Hoffman rules out of such a situation is an extreme example and its impact on chemistry" is significant, albeit "the very existence of these rules had stimulated, encouraged and assisted experimental involvement in a research project which eventually led to a new type of corrin synthesis" [14].

Added to that persistence in scientific research is very important given that there were many obstacles notably with the photochemical cyclo-isomerization. “In short, those transition-metal ions that quench luminescence of the excited corrin chromophore by virtue of their unfilled d-shells, also seem to thwart photochemical cycloisomerization of the corresponding A/D-seco-corrinoid complexes.” However amidst the new challenge, new approaches and ideas developed in that it became “increasingly clear that the A/D-seco-corrin to corrin system offers an optimal opportunity to study relationships between the nature of the metal ion complexation centers and the photochemical behavior of excited porphinoid ligand chromophores.”

Also there was “an essentially analogous reaction sequence starting from the enantiomeric form of the C10-dilactone acid leads to the skeleton of the ring D precursor, provided that not the free, but the lactonized (—CH_2—COO)— chain is lengthened by one methylene unit.” Additionally, the conversion of ring B to the precursor of ring C requires a method for specific removal of the carbomethoxy group of acetic acid side chain and its replacement by hydrogen.

As Eschenmoser [6] wrote in his lecture, “but I should perhaps propose that we enjoy the figure just from an aesthetic point of view, by watching the corrinoid chromophore system evolve, like a bud blooming into a flower.”

Author details

David Joshua Ferguson
Taylor University, Upland, Indiana, United States of America

*Address all correspondence to: davidfergusonwarrior@gmail.com

References

[1] Peters RA. The vitamin B complex. British Medical Journal. 1936;**2**(3957): 903-905. DOI: 10.1136/bmj.2.3957.903

[2] Homologation Reactions. Available form: https://en.wikipedia.org/wiki/Homologation_reactions [Accessed: 23 November 2019]

[3] Corrin Explanations. Available from: https://en.wikipedia.org/wiki/Corrin[Accessed: 23 November 2019]

[4] Stephenson NA, Gellman SH, Stahl SS. Ammonolysis of anilides promoted by ethylene glycol and phosphoric acid. RSC Advances. 2014; **4**(87):46840-46843. DOI: 10.1039/C4RA09065A

[5] Ranganathan S. A tale of two topologies: Woodward-Hoffmann rules at your fingertips! Resonance. 2011; **16**(12):1211-1222. DOI: 10.1007/s12045-011-0136-7

[6] Becker A. Research Collection: Theses. 2017. pp. 12-19. DOI: 10.3929/ethz-a-010782581

[7] Vitamin B12 Total Synthesis. Available from: https://en.wikipedia.org/wiki/Vitamin_B12_total_synthesis [Accessed: 23 November 2019]

[8] Adapted from: Lecture material from Dr. RB Woodward by Pure and Applied Chemistry; Lecture material from Dr. A. Eschenmoser from ETH Zurich Research Collection, The numbering of the reactions was adapted from Dr. Albert Eschenmoser's german lecture paper

[9] Organic Synthesis Search. Available from: https://www.organic-chemistry.org/synthesis/ [Accessed: 23 November 2019]

[10] Maurya CK, Mazumder A, Gupta PK. Phosphorus Pentasulfide Mediated Conversion of Organic Thiocyanates to Thiols. Available from: https://www.ncbi.nlm.nih.gov/pmc/articles/PMC5496577/

[11] File:Arndt-Eistert-Homologation mechanism V2.svg. Available from: https://commons.wikimedia.org/wiki/File:Arndt-Eistert-Homologation_mechanism_V2.svg [Accessed: 23 November 2019]

[12] Sudalai A, Kanagasabapathy S, Benicewicz BC. Phosphorus pentasulfide: A mild and versatile catalyst/reagent for the preparation of dithiocarboxylic esters. Organic Letters. 2000;**2**(20):3213-3216. DOI: 10.1021/ol006407q

[13] Protection of Carbonyl Groups. Available from: https://en.chem-station.com/reactions-2/2014/04/protection-of-carbonyl-groups.html [Accessed: 23 November 2019]

[14] Marini-Bettolo GB. Recent advances in the chemistry of natural products from Latin American Flora. Svensk Farmaceutisk Tidskrift. 1955;**59**(5): 130-136

4

Therapeutic Significance of 1,4-Dihydropyridine Compounds as Potential Anticancer Agents

Tangali Ramanaik Ravikumar Naik

Abstract

A series of 1,4-dihydropyridines have been prepared from a three-component one-pot condensation reaction of β-diketonates, an aromatic aldehyde, and ammonium acetate under microwave irradiation. The reaction is performed using crystalline nano-ZnO in ethanol under microwave irradiation (CEM discover). A wide range of functional groups was tolerated in the developed protocol. The present methodology offers several advantages such as simple procedure, greener condition, excellent yields and short reaction time. The synthesized compounds were evaluated for DNA photocleavage, SAR analysis and molecular docking studies. The compound (**4b, 4c, 4 h, 4i, 4n** and **4o**) showed potent DNA cleavage activities compared to other derivatives. The molecular interactions of the active compounds within the binding site of B-DNA were studied through molecular docking simulations; the compound (**4b, 4c, 4 h, 4i, 4n** and **4o**) showed good docking interaction with minimum binding energies. All synthetic compounds were characterized by different spectroscopic techniques.

Keywords: 1,4-Dihydropyridines, DNA photocleavage, molecular docking, SAR analysis, ZnO nanoparticle

1. Introduction

Facile and efficient synthesis of biological active molecules is one of the main objectives of organic and medicinal chemistry. In recent years, multicomponent reactions have become one of the important tools in the synthesis of structurally diverse chemical libraries of drug-like polyfunctional organic molecules [1–4]. Furthermore, MCRs offer the advantage of simplicity and synthetic efficiency over conventional chemical reactions in several aspects. MCRs allow the construction of combinatorial libraries of complex organic molecules for an efficient lead structure identification and optimization in drug discovery [5–10].

In continuation of our ongoing research work on microwave assisted synthesis of nano materials [11, 12] we have found that, nano-crystalline metal oxides have attracted considerable attention of synthetic and medicinal chemists because of their high catalytic activity and reusability [13–25]. Zinc oxide is an inexpensive, moisture stable, reusable, commercially available and is non-toxic, insoluble in polar as well as non-polar solvents [26–31]. A wide range of organic reactions that include Beckmann rearrangements [32], N-benzylation [33], acylation [34], dehydration of oximes [35], nucleophilic ring opening reactions of epoxides [36], synthesis of cyclic urea [37], N-formylation of amines [38]. In particular crystalline nano-ZnO oxide exhibit better catalytic activity compared to their bulk sized counterparts [29, 39–42].

In recent years, much attention has been directed toward the synthesis of dihydropyridine compounds owing to their tremendous application in various research fields including biological science and medicinal chemistry [43, 44]. Many DHPs are already commercial products such as: amlodipine, felodipine, isradipine, lacidipine, nicardipine, nitrendipine, nifedipine and nimodipine B, of which nitrendipine and nemadipine B exhibit potent calcium channel blocking activities [45–49] (**Figure 1**) and have emerged as one of the most important classes of drugs for the treatment of cardiovascular diseases [50, 51]. Moreover dihydropyridine derivatives possess a variety of biological activities like, geroprotective, hepatoprotective, anti-atherosclerotic, antitumor, and antidiabetic activities [46, 52, 53]. Widespread studies have uncovered that dihydropyridine unit containing compounds exhibit various medicinal functions such as neuroprotectant, platelet anti-aggregatory activity, cerebral anti ischemic activity in the treatment of Alzheimer's disease, chemosensitizer in tumor therapy [54–56]. Drug-resistance modifiers [57], antioxidants [58] and a drug for the treatment of urinary urge incontinence [59].

In order to model and understand these biological properties and to develop new chemotherapeutic agents based upon the 1,4-DHP compounds, significant effort has been devoted to establish effective methods for their synthesis. Generally, 1,4-DHPs were synthesized by Hantzsch method [60], which involves cyclocon-densation of an aldehyde, a β-ketoester and ammonia either in acetic acid or under reflux in alcohols for long reaction times which typically leads to low yields [46, 61, 62]. Other methods comprise the use of microwaves [63–65], high temperatures at reflux [66–69], organocatalysts [70] and metal triflates [71].

Recently, DNA is an important drug target and it regulates many biochemical processes that occur in the cellular system. Small-molecule interactions with DNA continue to be intensely and widely studied for their usefulness as probes of cellular replication and transcriptional regulation and for their potential as pharmaceuticals [72–75]. In particular, designing of the compound based on their ability to cleave DNA is of great importance not only from the primary biological point of view

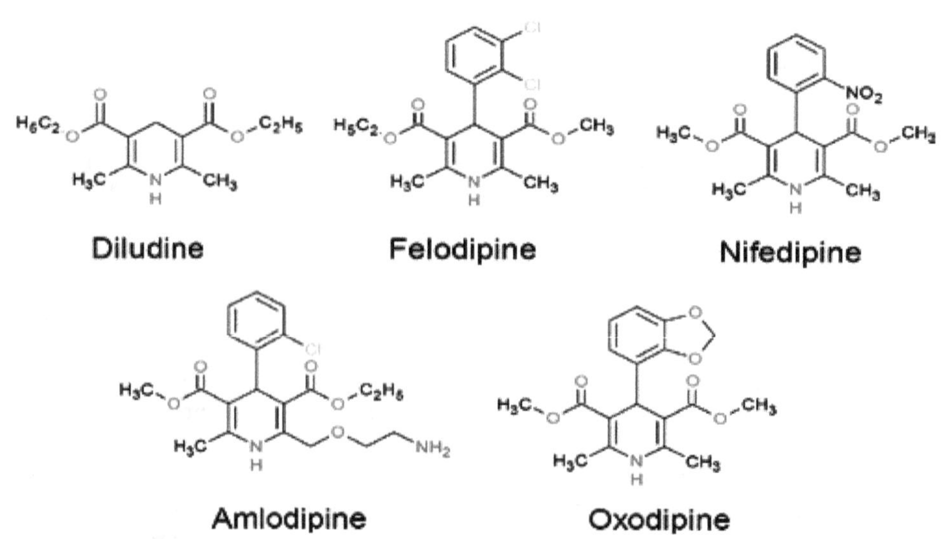

Figure 1.
Drugs containing 1,4-DHP moieties.

but also in terms of photodynamic therapeutic approach to develop potent drugs [72–75]. 1,4-Dihydropyridine derivatives have attracted the attention of the chemists because of their diverse biological applications [76]. The biological significance of this class of compounds impelled us to extend this series by working on the synthesis and DNA photocleavage studies of 1,4-dihydropyridine derivatives. In this communication, synthesis of 1,4-dihydropyridine derivatives and their DNA photocleavage studies and molecular docking have been reported.

In literature, there are several methods known for the synthesis of 1,4-dihydropyridine derivatives. In continuation of our program on the chemistry of nano material, herein we report an efficient microwave method for the synthesis of crystalline ZnO-NPs. The ZnO used in this work was synthesized according to a modified method. The prepared crystalline ZnO-nano-particle was characterized using powder XRD, SEM, EDX (**Figure 2**). Our synthetic approach started with the condensation of 1 equiv. of benzaldehyde **1a** with 2 equiv. of ethyl acetoacetate **2a** and 2 equiv. of NH_4OAc **3a** in the presence of ZnO-Nps resulted in the formation of Hantzsch 1,4-dihydropyridine **4a** (**Figure 3**). The reaction was complete in 5 min under microwave irradiation and the product was isolated by the usual work-up, in 90% yield and high purity. Under similar conditions, various substituted aromatic aldehydes carrying either electron-donating or -withdrawing substituents reacted with 1,3-diketones to form 1,4-DHPs in good to excellent yields, and the results are summarized in **Table 1**.

A microwave irradiation-assisted process very often minimizes the formation of byproducts and requires much less time than thermal methods. The main benefits of performing reactions under controlled conditions in sealed vessels are the significant rate enhancements and the higher product yields that can frequently be achieved. Therefore, in continuation of our studies on microwave synthesis of nano-materials [77–81], we have attempted to develop a rapid, microwave-assisted protocol for the synthesis of 1,4-DHPs using crystalline ZnO-nano catalyst (**Figure 3**).

The DNA cleavage of 1,4-DHP derivatives were studied by agarose gel electrophoresis. When circular plasmid DNA was subjected to electrophoresis, relatively fast migration was observed for the intact supercoiled DNA (type I). If scission occurs on one strand (nicking), the supercoiled DNA will relax to generate a slower moving open circular form (type II). If both strands are cleaved, a linear form

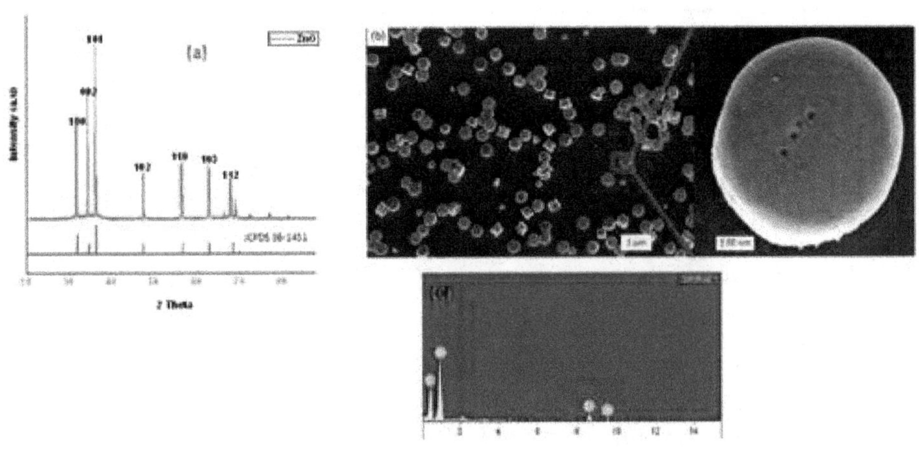

Figure 2.
(a) Powder XRD of obtained ZnO nano particles by microwave method; (b) SEM images of ZnO-NPs; (c) EDX analysis spectrum of obtained ZnO nano particles by microwave method.

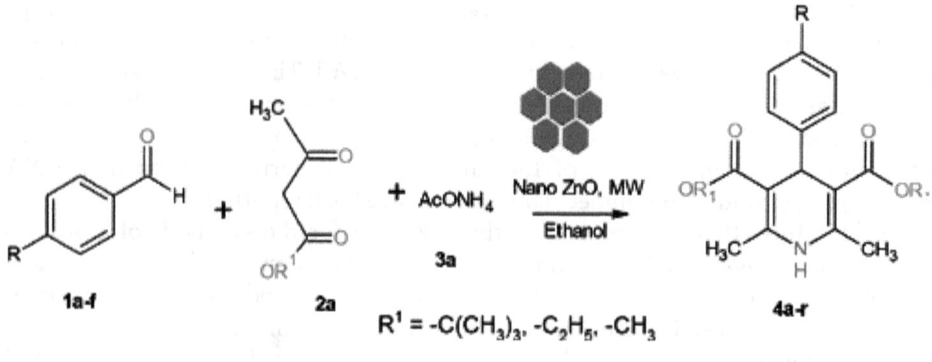

Figure 3.
Synthesis of 1,4-dihydropyridines.

Entry[a]	R	R^1	Products	Entry[a]	Yield (%)[b]
1	C_6H_5	*t*-Bu	**4a**	1	90
2	4-MeO-C_6H_5	*t*-Bu	**4b**	2	95
3	4-OH-C_6H_5	*t*-Bu	**4c**	3	95
4	4-F-C_6H_5	*t*-Bu	**4d**	4	95
5	4-Cl-C_6H_5	*t*-Bu	**4e**	5	90
6	4-NO_2-C_6H_5	*t*-Bu	**4f**	6	95
7	C_6H_5	Et	**4g**	7	90
8	4-MeO-C_6H_5	Et	**4h**	8	95
9	4-OH-C_6H_5	Et	**4i**	9	92
10	4-F-C_6H_5	Et	**4j**	10	92
11	4-Cl-C_6H_5	Et	**4k**	11	90
12	4-NO_2-C_6H_5	Et	**4l**	12	90
13	C_6H_5	Me	**4m**	13	90
14	4-MeO-C_6H_5	Me	**4n**	14	87
15	4-OH-C_6H_5	Me	**4o**	15	90
16	4-F-C_6H_5	Me	**4p**	16	90
17	4-Cl-C_6H_5	Me	**4q**	17	90
18	4-NO_2-C_6H_5	Me	**4r**	18	90

[a]*All the products were characterized by 1H NMR and 13C NMR studies and compared with the literature mps.*
[b]*Yields of isolated products*

Table 1.
Synthesis of 1,4-dihydropyridines.

(type III) that migrates between type I and type II will be generated [82–85]. The conversion of type I (supercoiled) to type II (nicked circular) was observed with different concentration of 1,4-DHP and irradiated for 2 h, in 1:9 DMSO/trisbuffer (20 μM, pH- 7.2) at 365 nm. No DNA cleavage was observed for the control in which 1,4-DHP was absent (lane 1) (**Figure 4**). With increasing concentration of these 1,4-DHP the amount of type I of pUC 19 DNA diminished gradually, whereas type II increased (**Figure 4**).

At 40 μM concentration, the Compound (**4c**) can promote only 30% conversion of DNA from type I to II (**Figure 5**). At the concentration of 80 μM, compound (**4c**)

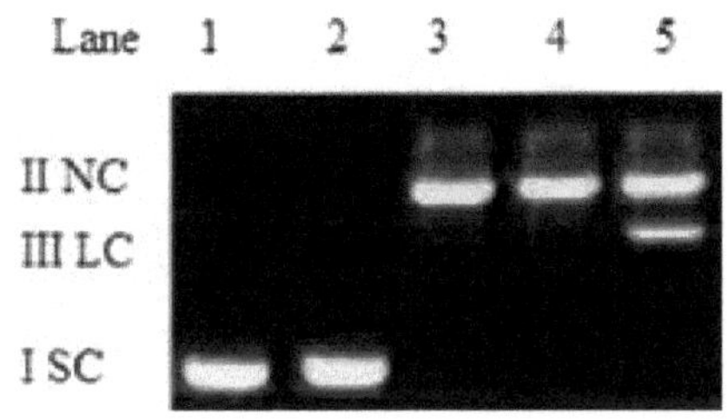

Figure 4.
*Light-induced DNA cleavage by 1,4-DHP. The 1,4-DHP was irradiated with UV light at 365 nm. Lane; 1: Control DNA (with out compound), lane; 2: 20 μM (**4c**), lane; 3: 40 μM (**4c**), lane; 4: 60 μM (**4c**), lane; 5: 80 μM (**4c**).*

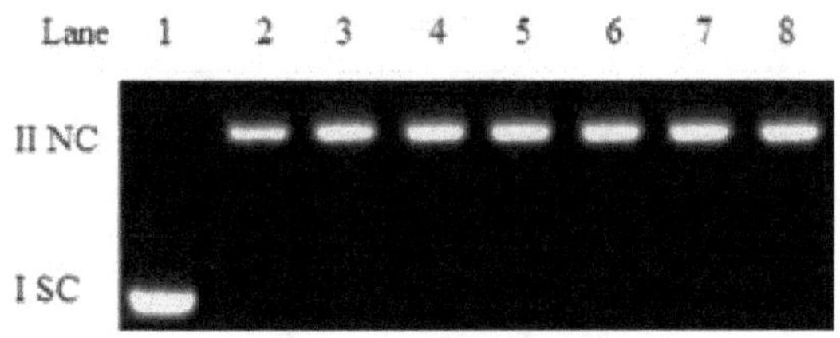

Figure 5.
*Light-induced DNA cleavage by 1,4-DHP. The 1,4-DHP was irradiated with UV light at 365 nm. Lane; 1: Control DNA (with out compound), lane; 2: 40 μM (**4a**), lane; 3: 40 μM (**4b**), lane; 4: 40 μM (**4c**), lane; 5: 40 μM (**4d**), lane; 5: 40 μM (**4e**), lane; 5: 40 μM (**4f**), lane; 5: 40 μM (**4 g**).*

can almost promote the about 80% conversion of DNA from type I to II (**Figure 5**). The cleavage potential of the test compounds were assessed by comparing the bands appeared in control and test compounds at 80 μM concentration. However, other derivatives exhibits much lower cleaving efficiency for pUC 19 DNA. Even at the concentration of 80 μM, it can promote only 40% conversion of DNA from type I to II (**Figure 5**).

But at higher concentrations around 130 μM, the compounds get precipitated and there is no moment in the DNA. The image (**Figure 6**) clearly demonstrates that compounds (**4b, 4c, 4d, 4e, 4f** and **4 g**) shows DNA cleavage of pUC19 DNA at 80 μM concentration. The results indicated that compounds bearing $-OCH_3$ and $-OH$ at *-para* position of phenyl ring (C-6) did cleave the DNA completely, other compounds have displayed nearly complete cleavage of DNA. Overall, it indicates that, the alkoxy groups are highly reactive radicals, which abstracts hydrogen atoms efficiently at C-4′ of 2-deoxyribose. It is of interest to note that hydroxyl group has been reported to bring about oxygen radical mediated DNA damage in the presence of photoirradiation [86].

The structure–activity relationship studies of 1,4-DHPs with regard to DNA photocleavage studies shows that, the changes in the substitution pattern at C-3, C-4, and C-5 positions alter the 1,4-DHP ring. Osiris Property Explorer is one such knowledge based activity prediction tool which predicts drug likeliness, drug score and undesired properties such as mutagenic, tumorigenic, irritant and reproductive effect of novel compounds based on chemical fragment data of available drugs and non-drugs as reported (**Table 2**) [87]. It was observed that, the compounds having aliphatic groups such as $-CH_3$, $-COOCH_3$, $-COOC_2H_5$ and $-COOC(CH_3)_3$, attached to C-2 and C-3 of 1,4-DHP exhibited good activity. Other derivatives possessing, an electron-donating substituent, such as hydroxy and methoxy group on the phenyl ring (C-6) increases DNA photocleavage activity. A lone pair of electrons on oxygen atom of methoxy group delocalizes into the π space of benzene ring,

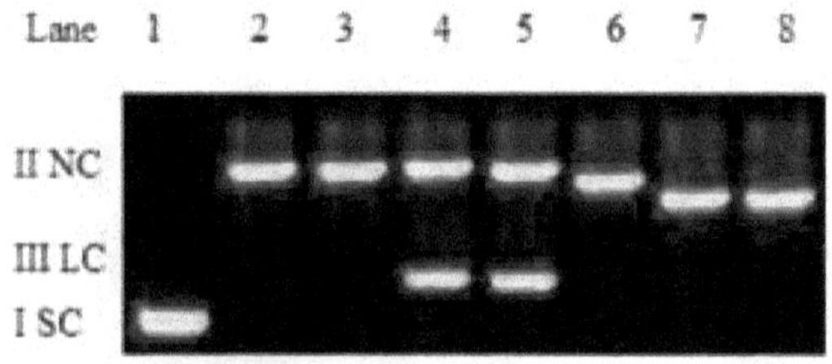

Figure 6.
*Light-induced DNA cleavage by 1,4-DHP. The 1,4-DHP was irradiated with UV light at 365 nm. Lane; 1: Control DNA (with out compound), lane; 2: 80 µM (**4a**), lane; 3: 80 µM (**4b**), lane; 4: 80 µM (**4c**), lane; 5: 80 µM (**4d**), lane; 5: 80 µM (**4e**), lane; 5: 80 µM (**4f**), lane; 5: 80 µM (**4 g**).*

Compounds	Mol. wt	C log P	Drug-likeness	Drug-score	Toxicity risks[a]			
					M[b]	T[c]	I[d]	R[e]
4a	329	3.29	2.41	0.77	(+)	(+)	(+)	(+)
4b	359	3.22	2.34	0.75	(+)	(+)	(+)	(+)
4c	345	2.94	2.48	0.79	(+)	(+)	(+)	(+)
4d	347	3.39	1.65	0.70	(+)	(+)	(+)	(+)
4e	419	5.37	−17.92	0.22	(+)	(+)	(+)	(+)
4f	430	4.17	−19.36	0.10	(−)	(+)	(+)	(−)
4 g	301	2.48	4.04	0.87	(+)	(+)	(+)	(+)
4 h	331	2.41	3.87	0.51	(+)	(+)	(+)	(+)
4i	317	2.13	4.08	0.53	(+)	(+)	(+)	(+)
4j	319	2.58	3.29	0.50	(+)	(+)	(+)	(+)
4 k	363	3.89	3.33	0.68	(+)	(+)	(+)	(+)
4 l	374	2.69	1.92	0.25	(−)	(+)	(+)	(−)
4 m	269	2.98	4.09	0.50	(+)	(+)	(+)	(+)
4n	299	2.91	3.94	0.49	(+)	(+)	(+)	(−)
4o	285	2.63	4.14	0.51	(+)	(+)	(+)	(+)
4p	287	3.08	3.42	0.47	(+)	(+)	(+)	(−)
4q	335	3.08	4.97	0.48	(+)	(+)	(+)	(−)
4r	346	1.88	3.50	0.30	(−)	(+)	(+)	(−)

[a]*Ranking as (+) no bad effect, (+/−) medium bad effect, (−) bad effect.*
[b]*M (mutagenic effect);*
[c]*T (tumorigenic effect);*
[d]*I (irritant effect);*
[e]*R (reproductive effect).*

Table 2.
Drug likeliness properties of 1,4-dihydro pyridines according to Osiris property explorer tool.

thereby increasing the activity. Similarly, electron-withdrawing substituent's, such as 4-fluorophenyl, 4-chloro phenyl of 1,4-DHP lower the activity. These results indicate that, the alkoxy substituent's and nitrogen of pyridine ring in the 1,4-DHP structure are the responsible for DNA cleavage.

In order to rationalize the observed spectroscopic results and to get more insight into the intercalation modality, the 1,4-DHP (**4a–r**) were successively docked [88–90] within the DNA duplex of sequence d(CGCGAATTCGCG)$_2$ dodecamer

(PDB ID: 1BNA) in order to predict the chosen binding site along with preferred orientation of the ligand inside the DNA minor groove. All synthesized 1,4-DHP derivatives were drawn in ChemSketch and structures were saved in .mol format. Afterwards the .mol format was used in Hyperchem-7, to adjust their fragments, followed by total energy minimization of ligands so that they can attain a stable conformation and the file was saved in .pdb format.

Protein 3D structure of B-DNA was obtained from RCSB PDB (an information portal to biological macromolecular structures). The water molecules were removed from the file, and the protein was protonated in 3D to add polar hydrogen's. Binding pocket was identified using site finder, and the respective residues were selected. Docking parameters were set to default values and scoring algorithm, the docking runs were retained to 30 conformations per ligand. The docked protein structures were saved in .pdb format, and ligand's conformations were investigated one by one. Complexes with best conformations were selected on the basis of highest score, lowest binding energy and minimum RMSD values [91].

The synthesized organic compounds perform their biological activity more efficiently by binding respective protein or DNA at their specific binding site. Identification of interacting residues with ligands is a necessary step toward rational drug designing, understanding of molecular pathway and mechanistic action of protein.

Molecular docking was carried out between rigid receptor protein and the flexible ligands. **Table 3** shows the details of the docking results including RMSD and binding energy values of protein–ligand complexes. The ligands (**4b, 4c, 4 h, 4i, 4n** and **4o**) bind strongly to B-DNA as inferred by their minimum binding energy values, that is, −13.8, −12.9 and – 12.3 kcal/mol, respectively (**Figure 7**).

Figure 8 shows the position of active site in the helical structure of DNA and it also shows that all docked ligands clustered inside the pocket. **Figure 8** exhibited

Products	Docking energy (Kcal/mol)	Inhibition constant (M)	RMSD
4a	−6.23	4.35×10^{-7}	2.5
4b	−24.12	1.81×10^{-16}	1.1
4c	−21.74	1.96×10^{-16}	1.5
4d	−5.72	5.96×10^{-7}	3.4
4e	−7.24	6.31×10^{-7}	3.4
4f	−6.85	4.88×10^{-7}	3.8
4g	−7.41	4.51×10^{-7}	2.0
4h	−22.35	1.92×10^{-16}	1.0
4i	−19.81	2.32×10^{-16}	1.0
4j	−6.34	5.88×10^{-7}	2.1
4k	−6.68	6.76×10^{-7}	2.1
4l	−8.22	5.18×10^{-7}	2.4
4m	−7.55	4.68×10^{-7}	2.3
4n	−22.64	1.96×10^{-16}	1.1
4o	−20.36	2.18×10^{-16}	1.0
4p	−6.78	6.20×10^{-7}	1.5
4q	−6.52	7.15×10^{-7}	1.8
4r	−7.89	6.32×10^{-7}	1.5

Table 3.
Molecular docking studies of 1,4-dihydropyridines.

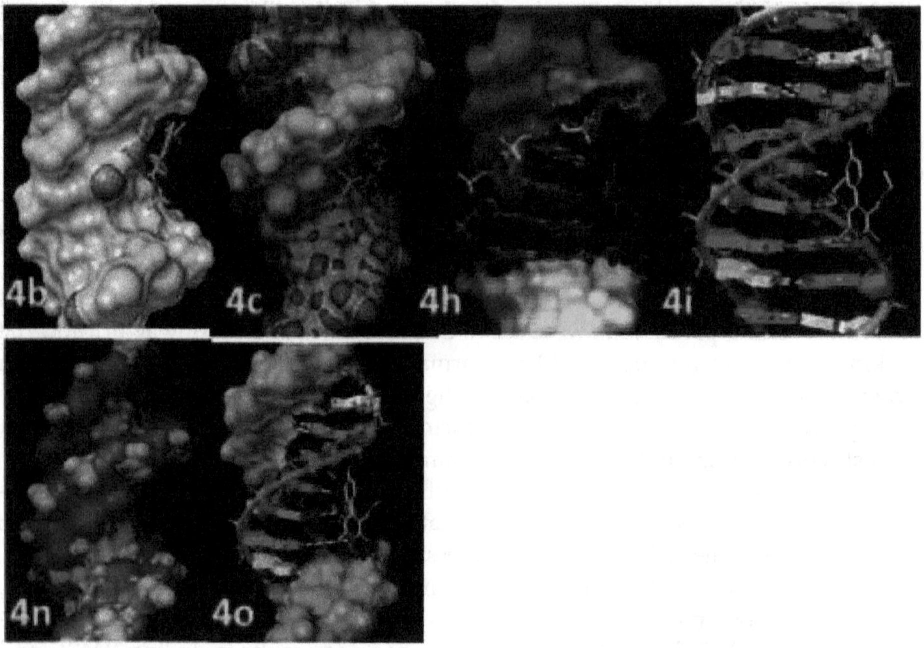

Figure 7.
1,4-DHP was successively docked within the DNA duplex of sequence d(CGCGAATTCGCG)$_2$ dodecamer (PDB ID: 1BNA).

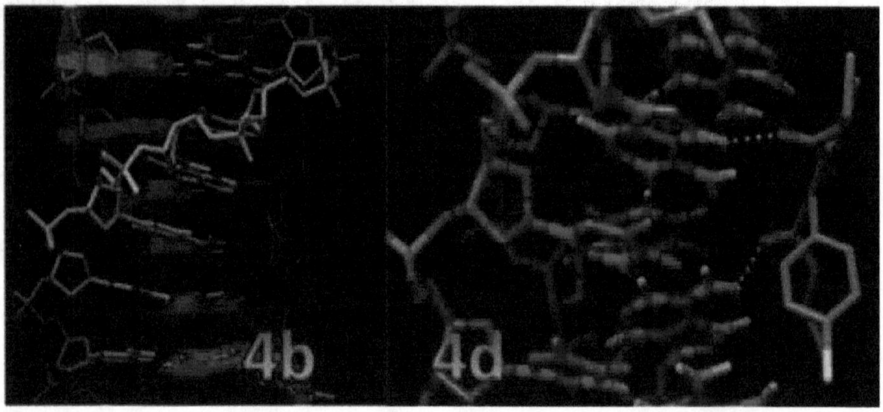

Figure 8.
Interaction of 1,4-DHP with DNA duplex of sequence d(CGCGAATTCGCG)$_2$ dodecamer (PDB ID: 1BNA).

the hydrogen bond interaction of **4c** and **4d** with key residues in active site inside the helical structure of DNA. In this model, it is clearly indicated that the compound **4c** formed hydrogen bonded between the –OH and N1 of thymine, which is DT7 and DT19 with the bond length of 2.02 and 2.05 Å respectively. Moreover, the other derivatives of 1,4-DHP formed less H-bond interaction with the DNA due to the orientation of aromatic ring involved in van der Waals interactions (Wireframe model) and flat hydrophobic regions of the binding sites of DNA (**Table 3**). These results demonstrated the in silico molecular docking studies of 1,4-DHPs with B-DNA suggested that 1,4-DHPs possess the potential to disturb hydrophobic and H-bond interactions thereby affecting the stability of attachment of B-DNA, and may be effective for cancer cell lines.

2. Experimental

2.1 Materials and method

All the chemicals used in the present study are of AR grade. Whenever analytical grade chemicals were not available, laboratory grade chemicals were purified and used. $AlCl_3$, $ZnCl_2$, $Yb(OTF)_3$, $FeCl_3$ and Zinc acetate obtained from Merck chemicals and are directly used without further purification. Melting points were recorded on an open capillary tube with a Buchi melting point apparatus and are uncorrected. ^1H- NMR spectra were obtained using a 400 MHz on a Bruker spectrometer (chemical shifts in δ ppm).

2.1.1 General procedure for the preparation of ZnO-Nps

In a typical synthesis process, zinc acetate dihydrate (1.1 g, 0.01 M) was dissolved in 20 mL of ethanol with constant stirring for 20 min. Then KOH (0.178 M) was added into the above mixed solution. After further stirring for 5 min, the reaction mixture was put into a CEM microwave synthesizer to irradiate for 10 min with the power set at 150 W, Temperature at 150°C and Pressure 150 C^0. After completion of reaction, the white precipitate was collected by centrifugation, washed twice with deionized water, ethanol and dried in vacuum oven at 60°C for 5 h.

Crystalline structure of the prepared ZnO-Nps was determined by powder X-ray diffraction (XRD). The strong intensity and narrow width of diffraction peaks indicate the high crystallinity of the prepared ZnO-Nps (**Figure 2a**). The peaks are indexed as 31.82° (100), 34.54° (002), 36.42° (101), 47.46° (102), 56.74° (110), 62.92° (103), 66.06° (200), 68.42° (112), 69.06° (201) and 78.82° (202) respectively. This revealed that the resultant nanoparticles were pure ZnO with a hexagonal structure (JCPDS 36-1451). No impurities could be detected in this pattern, which implies hexagonal phase ZnO nanoparticles could be obtained under the current microwave method. X-ray diffraction shows that metal oxide is pure ZnO having hexagonal structure. Sharpness of the peaks shows good crystal growth of the oxide particles. Average particle sizes of the ZnO have been calculated using from high intensity peak using Image J.

2.1.2 General procedure for the synthesis of 1,4-DHP by microwave method

A mixture of aromatic aldehydes **1a** (5 mmol), ethyl acetoacetate **2** (10 mmol), and ammonium acetate **3** (10 mmol) and ZnO (10 mol %) was taken in ethanol (20 mL) and the mixture was heated at microwave irradiation for 5 min (monitored by TLC after 5 min. interval). After 5 min, the reaction mixture was cooled to room temperature and then it was poured into cold water. The product was extracted with ethyl acetate. The organic layer was washed with brine, water and dried over anhydrous Na_2SO_4. The crude product thus obtained was recrystallized from EtOH to obtain desired product (**Figure 3, Table 1**).

4a. Di-tert-Butyl – 1,4-dihydro-2,6-dimethyl-4-phenylpyridine-3,5-dicarboxylate

Solid: MP 180–182°C; ^1H NMR (500 MHz, $CDCl_3$) δ 1.43 (s, 18H), 2.30 (s, 6H), 4.83 (s, 1H), 5.58 (brs, 1H), 7.05-7.10 (m, 1H), 7.10-7.20 (m, 2H), 7.23-7.30 (m, 2H); ^{13}C NMR (125 MHz, $CDCl_3$) δ 20.0, 28.4, 40.0, 80.0, 105.5, 125.6, 127.5, 128.5, 129.2, 143.0, 147.5, 167.3.

4b. Di-tert-butyl 4-(4-methoxyphenyl)-1,4-dihydro-2,6-dimethylpyridine-3,5-dicarboxylate

Solid: MP 168–170°C; ^{1}H NMR (500 MHz, $CDCl_3$) δ 1.40 (s, 18H), 2.25 (s, 6H), 3.86 (s, 3H), 4.81 (s, 1H), 5.51 (brs, 1H), 7.10-7.20 (d, 2H), 7.40-7.50 (d, 2H); ^{13}C NMR (125 MHz, $CDCl_3$) δ 19.8, 30.0, 41.0, 56.0, 81.0, 106.1, 125.6, 127.8, 135.0, 146.4, 153.2, 160.0, 167.5.

4c. Di-tert-butyl 4-(4-hydroxy-phenyl)-1,4-dihydro-2,6-dimethylpyridine-3,5-dicarboxylate

Solid: MP 230–232°C; ^{1}H NMR (500 MHz, $CDCl_3$) δ 1.36 (s, 18H), 2.28 (s, 6H), 4.90 (s, 1H), 5.56 (brs, 1H), 6.86-6.90 (d, 2H), 7.10-7.20 (d, 2H), 10.10 (s, 1H); ^{13}C NMR (125 MHz, $CDCl_3$) δ 24.5, 32.8, 45.3, 88.0, 108.4, 128.3, 131.0, 134.2, 134.6, 136.8, 148.4, 154.6, 172.6.

4d. Di-tert-butyl - 4-(4-fluorophenyl)-1,4-dihydro-2,6-dimethylpyridine-3,5-dicarboxylate

Solid: MP 150–152°C; ^{1}H NMR (500 MHz, $CDCl_3$) δ 1.43 (s, 18H), 2.30 (s, 6H), 4.81 (s, 1H), 5.50 (brs, 1H), 6.90-6.96 (d, 2H), 7.15-7.20 (d, 2H); ^{13}C NMR (125 MHz, $CDCl_3$) δ 20.0, 21.3, 38.9, 40.0, 79.8, 106.0, 114.2, 113.7, 125.4, 126.8, 129.2, 142.5, 143.2, 160.0, 162.5, 167.1.

4e. Di-tert-butyl 4-(4-chlorophenyl)-1,4-dihydro-2,6-dimethylpyridine-3,5-dicarboxylate

Solid: MP 188–190°C; ^{1}U NMR (500 MHz, $CDCl_3$) δ 1.38 (s, 18H), 2.25 (s, 6H), 4.85 (s, 1H), 5.50 (brs, 1H), 6.80-6.85 (d, 2H), 7.00-7.08 (d, 2H); ^{13}C NMR (125 MHz, $CDCl_3$) δ 24.3, 33.4, 45.1, 86.2, 108.8, 128.9, 130.4, 133.5, 134.3, 136.1, 148.6, 151.6, 172.4.

4 f. Di-tert-butyl – 4-(4-nitrophenyl)-1,4-dihydro-2,6-dimethylpyridine-3,5-dicarboxylate

Solid: MP 176–178°C; ^{1}H NMR (500 MHz, $CDCl_3$) δ 1.38 (s, 18H), 2.30 (s, 6H), 4.86 (s, 1H), 5.55 (brs, 1H), 7.00–7.10 (d, 2H), 7.15–7.25 (d, 2H); ^{13}C NMR (125 MHz, $CDCl_3$) δ 20.5, 22.4, 38.6, 40.1, 79.6, 107.0, 114.5, 114.6, 126.2, 126.8, 129.6, 142.6, 144.6, 161.0, 167.1.

4 g. 2,6-Dimethyl-4-phenyl-1,4-dihydro-pyridine-3,5-dicarboxylic acid diethyl ester

Solid: MP 158–160°C; ^{1}H NMR ($CDCl_3$, 400 MHz): δ 1.20 (t, J = 9.7 Hz, 6H, $2CH_3CH_2$), 2.28 (s, 6H, $2CH_3$), 4.10 (q, J = 6 Hz, 4H, $2CH_3CH_2$), 5.00 (s, 1H, CH), 5.75 (s, 1H, NH), 7.10–7.50 (m, 5H); ^{13}C NMR ($CDCl_3$, 75 MHz): δ = 14.20 (C-3″), 19.5 (C-1″), 39.6 (C-4), 59.5 (C-2″), 104.1 (C-3 and C-5), 126.0 (C-4′), 127.8 (C-3′ and C-5′), 130.0 (C-2′ and C-6′), 143.8 (C-2 and C-6), 148.0 (C-1′), 168.0 (C-4″).

4 h. 2,6-Dimethyl-4-(4-methoxy-phenyl)-1,4-dihydro-pyridine-3,5-dicarboxylic acid diethyl ester

Solid: MP 160–162°C; ^{1}H NMR ($CDCl_3$, 400 MHz): δ 1.21 (t, J = 7.0 Hz, 6H), 2.30 (s, 6H), 3.78 (s, 3H), 4.10 (q, J = 6.3 Hz, 4H), 4.95 (s, 1H), 5.60 (s, 1H), 6.80

(d, J = 8.4 Hz, 2H), 7.18 (d, J = 8.7 Hz, 2H); ^{13}C NMR ($CDCl_3$, 75 MHz): δ 14.2, 19.6, 38.8, 55.2, 59.8, 104.0, 115.0, 128.8, 140.0, 145.3, 156.7, 168.0.

4i. 2,6-Dimethyl-4-(4-hydroxy-phenyl)-1,4-dihydro-pyridine-3,5-dicarboxylic acid diethyl ester

Solid: MP 238–240°C; ^{1}H NMR ($CDCl_3$, 400 MHz): δ 1.18 (t, J = 7.2 Hz, 6H), 2.28 (s, 6H), 4.05 (q, J = 6.6 Hz, 4H), 4.90 (s, 1H), 5.61 (s, 1H), 6.70 (d, J = 8.7 Hz, 2H), 7.15 (d, J = 8.4 Hz, 2H), 9.90 (s, 1H); ^{13}C NMR ($CDCl_3$, 75 MHz): δ 14.0, 18.9, 39.0, 59.0, 103.0, 114.2, 128.3, 139.4, 144.2, 154.1, 167.6.

4j. 2,6-Dimethyl-4-(4-fluoro-phenyl)-1,4-dihydro-pyridine-3,5-dicarboxylic acid diethyl ester

Solid: MP 152–154°C; 1H NMR ($CDCl_3$, 400 MHz): δ 1.10 (t, J = 7.2 Hz, 6H), 2.25 (s. 6H), 4.00 (q, J = 5.7 Hz, 4H), 4.88 (s, 1H), 5.68 (s, 1H), 6.80 (m, 2H), 7.15(m, 2H); ^{13}C NMR ($CDCl_3$, 75 MHz): δ 14.3, 19.7, 39.6, 60.1, 104.2, 114.4, 129.4, 129.7, 130.0, 143.5, 147.0, 167.5.

4 k. 2,6-Dimethyl-4-(4-chloro-phenyl)-1,4-dihydro-pyridine-3,5-dicarboxylic acid diethyl ester

Solid: MP 153–155°C; 1H NMR ($CDCl_3$, 400 MHz): δ 1.12 (t, J = 7.2 Hz, 6H), 2.35 (s. 6H), 4.12 (q, J = 5.7 Hz, 4H), 5.10 (s, 1H), 5.82 (s, 1H), 7.50 (d, 2H), 8.16 (d, 2H); ^{13}C NMR ($CDCl_3$, 75 MHz): δ 14.2, 18.6, 39.6, 60.0, 101.6, 116.8, 127.8, 129.3, 130.2, 144.8, 147.2, 166.8.

4 l. 2,6-Dimethyl-4-(4-nitro-phenyl)-1,4-dihydro-pyridine-3,5-dicarboxylic acid diethyl ester

Solid: MP 178–180°C; 1H NMR ($CDCl_3$, 400 MHz): δ 1.26 (t, J = 7.2 Hz, 6H), 2.35 (s. 6H), 4.06 (q, J = 5.7 Hz, 4H), 5.08 (s, 1H), 5.76 (s, 1H), 7.48 (m, 2H), 8.02 (m, 2H); ^{13}C NMR ($CDCl_3$, 75 MHz): δ 14.2, 19.5, 39.6, 59.6, 104.2, 121.3, 1234.0, 128.4, 136.8, 144.5, 147.8, 148.8, 167.5.

4 m. 2,6-Dimethyl-4-phenyl-1,4-dihydro-pyridine-3,5-dicarboxylic acid dimethyl ester

Solid: MP 194–196°C; ^{1}H NMR ($CDCl_3$, 400 MHz): δ 2.30 (s, 6H, $2CH_3$), 3.66 (s, 6H, $2CH_3$), 5.00 (s, 1H, CH), 5.80 (b, 1H), 7.20-7.56 (m, 5H); ^{13}C NMR ($CDCl_3$, 75 MHz): δ = 19.7, 38.7, 50.5, 105.5, 126.2, 127.0, 128.0, 144.1, 147.1, 168.2.

4n. 2,6-Dimethyl-4-(4-methoxy-phenyl)-1,4-dihydro-pyridine-3,5-dicarboxylic acid dimethyl ester

Solid: MP 185–187°C; ^{1}H NMR ($CDCl_3$, 400 MHz): δ 2.28 (s, 6H, $2CH_3$), 3.60 (s, 6H, $2CH_3$), 3.78 (s, 3H), 4.89 (s, 1H, CH), 5.30 (b, 1H), 6.80–7.10 (m, 4H); ^{13}C NMR ($CDCl_3$, 75 MHz): δ 19.5, 38.7, 55.1, 51.8, 104.4, 113.2, 128.9, 140.4, 143.4, 158.0, 167.7.

4o. 2,6-Dimethyl-4-(4-hydroxy-phenyl)-1,4-dihydro-pyridine-3,5-dicarboxylic acid dimethyl ester

Solid: MP 228–230°C; ^{1}H NMR ($CDCl_3$, 400 MHz): δ 2.26 (s, 6H, 2CH3), 3.63 (s, 6H, $2CH_3$), 5.00 (s, 1H, CH), 5.40 (b, 1H), 6.95–7.20 (m, 4H); ^{13}C NMR ($CDCl_3$, 75 MHz): δ 18.4, 38.4, 51.8, 103.1, 114.2, 128.4, 139.0, 144.2, 155.0, 167.6.

4p. 2,6-Dimethyl-4-(4-fluoro-phenyl)-1,4-dihydro-pyridine-3,5-dicarboxylic acid dimethyl ester.

Solid: MP 170–172°C; 1H NMR ($CDCl_3$, 400 MHz): δ 2.32 (s, 6H, $2CH_3$), 3.64 (s, 6H, $2CH_3$), 4.98 (s, 1H, CH), 5.78 (b, 1H), 7.10 (t, 2H), 7.32 (t, 2H); ^{13}C NMR ($CDCl_3$, 75 MHz): δ 19.5, 40.0, 51.0, 104.1, 114.4, 129.3, 130.0, 144.1, 145.3, 160.5, 162.3, 167.6.

4q. 2,6-Dimethyl-4-(4-chloro-phenyl)-1,4-dihydro-pyridine-3,5-dicarboxylic acid dimethyl ester

Solid: MP 194–196°C; 1H NMR ($CDCl_3$, 400 MHz): δ 2.30 (s, 6H, $2CH_3$), 3.66 (s, 6H, $2CH_3$), 4.95 (s, 1H, CH), 5.76 (b, 1H), 7.15 (m, 2H), 7.36 (m, 2H); ^{13}C NMR ($CDCl_3$, 75 MHz): δ 19.5, 39.6, 51.1, 103.6, 113.8, 128.2, 130.0, 144.4, 146.2, 160.4, 167.8.

4r. 2,6-Dimethyl-4-(4-nitro-phenyl)-1,4-dihydro-pyridine-3,5-dicarboxylic acid dimethyl ester

Solid: MP 210–212°C; 1H NMR ($CDCl_3$, 400 MHz): δ 3.00 (s, 6H, $2CH_3$), 3.61 (s, 6H, $2CH_3$), 5.08 (s, 1H, CH), 5.86 (b, 1H), 7.30 (m, 2H), 7.62 (m, 2H); ^{13}C NMR ($CDCl_3$, 75 MHz): δ 19.7, 40.1, 51.2, 103.2, 114.4, 128.7, 145.0, 146.1, 156.2, 167.6.

3. Conclusion

In conclusion, the present study describes the ZnO-NPs catalyzed synthesis of 1,4-dihydropyridines (**4a–r**) under microwave irradiation, giving excellent yields in shorter reaction time as compared to conventional method. All the synthesized compounds were evaluated for DNA photocleavage, SAR and DNA docking studies. DNA cleavage by gel electrophoresis method revealed that compounds (**4b** and **4c**) were found to cleave the DNA completely. The preliminary SAR study revealed that the $-OCH_3$ and –OH substituted compounds, were more favorable for activity, particularly at *-para* position of the phenyl ring. Docking studies indicated that one of the ester moieties of these compounds played a key role in their interactions with the DNA. However, the nature of reactive intermediates involved in the DNA cleavage by the 1,4-dihydropyridines has not been clear. Needless to say, further understanding the mechanism of biological action are still required in order to fully develop these compounds as potent anticancer drugs.

Acknowledgements

We are grateful to Prof. H. S. Bhojya Naik, Department of Industrial chemistry, Kuvempu University, for his suggestions, and CeNSE, Indian Institute of science, Bangalore, for providing all necessary facility to carry-out this result.

Author details

Tangali Ramanaik Ravikumar Naik
Vijayanagara Srikrishna Devaraya University (VSKU), Ballari, Karnataka

*Address all correspondence to: naikravi7@gmail.com

References

[1] Raman DJ, Yus M. Asymmetric multicomponent reactions (AMCRs): The new frontier. Angewandte Chemie, International Edition. 2005;**44**:1602

[2] Domling A. Recent developments in isocyanide based multicomponent reactions in applied chemistry. Chemical Reviews. 2006;**106**:17

[3] Domling A, Ugi I. Multicomponent Reactions with Isocyanides. Angewandte Chemie, International Edition. 2000;**39**:3168

[4] Zhu J, Bienayme H, editors. Multicomponent Reaction. Weinheim: Wiley-VCH; 2005

[5] Tanaka K, Toda F. Solvent-free organic synthesis. Chemical Reviews. 2000;**100**:1025

[6] Li C-J. Organic reactions in aqueous media with a focus on carbon-carbon bond formations: A decade update. Chemical Reviews. 2005;**105**:3095

[7] Paul S, Bhattacharyya P, Das AR. One-pot synthesis of dihydropyrano[2,3-*c*] chromenes via a three component coupling of aromatic aldehydes, malononitrile, and 3-hydroxycoumarin catalyzed by nano-structured ZnO in water: A green protocol. Tetrahedron Letters. 2011;**52**:4636-4641

[8] Ghosh PP, Das AR. Nano crystalline ZnO: A competent and reusable catalyst for one pot synthesis of novel benzylamino coumarin derivatives in aqueous media. Tetrahedron Letters. 2012;**53**:3140

[9] Bhattacharyya P, Pradhan K, Paul S, Das AR. Nano crystalline ZnO catalyzed one pot multicomponent reaction for an easy access of fully decorated 4*H*-pyran scaffolds and its rearrangement to 2-pyridone nucleus in aqueous media. Tetrahedron Letters. 2012;**53**:4687

[10] Ghosh PP, Pal G, Paul S, Das AR. Design and synthesis of benzylpyrazolyl coumarin derivatives via a four-component reaction in water: Investigation of the weak interactions accumulating in the crystal structure of a signified compound. Green Chemistry. 2012;**14**:2691

[11] Jena A, Vinu R, Shivashankar SA, Giridhar M. Microwave assisted synthesis of nanostructured titanium dioxide with high photocatalytic activity. Industrial and Engineering Chemistry Research. 2010;**49**(20):9636-9643

[12] Sai R, Kulkarni SD, Vinoy KJ, Bhat N, Shivashankar SA. $ZnFe_2O_4$: Rapid and sub-100°C synthesis and anneal-tuned magnetic properties. Journal of Materials Chemistry. 2012;**22**:2149-2156

[13] Reddy KH, Reddy VVP, Shankar J, Madhav B, Kumar BSPA, Nageswar YVD. Copper oxide nanoparticles catalyzed synthesis of aryl sulfides via cascade reaction of aryl halides with thiourea. Tetrahedron Letters. 2011;**52**:2679-2682

[14] Cristau H-J, Cellier PP, Spindler J-F, Taillefer M. Highly efficient and mild coppercatalyzed N- and C-arylations with aryl bromides and iodides. Chemistry. 2004;**10**(22):5607-5622

[15] Mittapelly N, Reguri BR, Mukkanti K. Copper oxide nanoparticles-catalyzed direct Nalkylation of amines with alcohols. Der Pharma Chemica. 2011;**3**:180-189

[16] Chassaing S, Kumarraja M, Sido ASS, Pale P, Sommer J. Click chemistry in CuI-zeolites: The Huisgen [3 + 2]-cycloaddition. Organic Letters. 2007;**9**:883-886

[17] Hudson R, Feng Y, Varma RS, Moores A. Bare magnetic nanoparticles:

Sustainable synthesis and applications in catalytic organic transformations. Green Chemistry. 2014;**16**:4493-4505

[18] Meldal M, Tornoe CW. Cu-catalyzed azide-alkyne cycloaddition. Chemical Reviews. 2008;**108**:2952-3015

[19] Hein JE, Fokin VV. Copper-catalyzed azide-alkyne cycloaddition (CuAAC) and beyond: New reactivity of copper(I) acetylides. Chemical Society Reviews. 2010;**39**:1302-1315

[20] Jin T, Yan M, Yamamoto Y. Click chemistry of alkyne-azide cycloaddition using nano-structured copper catalysts. ChemCatChem. 2012;**4**:1217-1229

[21] Zhou Y, He T, Wang Z. Nanoparticles of silver oxide immobilized on different templates: Highly efficient catalyst for three-component coupling of aldehydeamine-alkyne. ARKIVOC. 2008;**xiii**:80-90

[22] Zhou X, Lu Y, Zhai L-L, Zhao Y, Liu Q , Sun W-Y. Propargylamines formed from three-component coupling reactions catalyzed by silver oxide nanoparticles. RSC Advances. 2013;**3**:1732-1734

[23] Kwon SG, Hyeon T. Colloidal chemical synthesis and formation kinetics of uniformly sized nanocrystals of metals, oxides, and chalcogenides. Accounts of Chemical Research. 2008;**41**:1696

[24] Hu A, Yee GT, Lin W. Magnetically recoverable chiral catalysts immobilized on magnetite nanoparticles for asymmetric hydrogenation of aromatic ketones. Journal of the American Chemical Society. 2005;**127**:12486

[25] Kawamura M, Sato K. Magnetically separable phase-transfer catalysts. Chemical Communications. 2006;**45**:4718

[26] Xia YN, Yang PD, Sun YG, Wu YY, Mayers B, Gates B, et al. One-dimensional nanostructures: Synthesis, characterization, and applications. Advanced Materials. 2003;**15**:353-389

[27] Comparelli R, Fanizza E, Curri ML, Cozzoli PD, Mascolo G, Agostiano A. UV-induced photocatalytic degradation of azo dyes by organic-capped ZnO nanocrystals immobilized onto substrates. Applied Catalysis B: Environmental. 2005;**60**:1-11

[28] Moghaddam FM, Saeidian H. Controlled microwave-assisted synthesis of ZnO nanopowder and its catalytic activity for *O*-acylation of alcohol and phenol. Materials Science and Engineering B. 2007;**139**:265-269

[29] Mirjafary Z, Saeidian H, Sadeghi A, Moghaddam FM. ZnO nanoparticles: An efficient nanocatalyst for the synthesis of β-acetamido ketones/esters via a multi-component reaction. Catalysis Communications. 2008;**9**:299-306

[30] Gupta M, Paul S, Gupta R, Loupy A. ZnO: A versatile agent for benzylic oxidations. Tetrahedron Letters. 2005;**46**:4957-4960

[31] Lietti L, Tronconi E, Forzatti P. Surface properties of zno-based catalysts and related mechanistic features of the higher alcohol synthesis by FT-IR spectroscopy and TPSR. Journal of Molecular Catalysis. 1989;**55**:43-54

[32] Sharghi H, Hosseini M. Solvent-free and one-step beckmann rearrangement of ketones and aldehydes by zinc oxide. Synthesis. 2002:1057

[33] Dhakshinamoorthy A, Visuvamithiran P, Tharmaraj V, Pitchumani K. Clay encapsulated ZnO nanoparticles as efficient catalysts for N-benzylation of amines. Catalysis Communications. 2011;**16**:15-19

[34] Sarvari MH, Sharghi H. Reactions on a solid surface. A simple, economical

and efficient Friedel-Crafts acylation reaction over zinc oxide (ZnO) as a new catalyst. The Journal of Organic Chemistry. 2004;**69**:6953

[35] Sarvari MH. ZnO/CH3COCl: A new and highly efficient catalyst for dehydration of aldoximes into nitriles under solvent-free condition. Synthesis. 2005;**5**:787

[36] Sarvari MH. Synthesis of β-aminoalcohols catalyzed by ZnO. Acta Chimica Slovenica. 2008;**55**:440

[37] Kim YJ, Varma RS. Microwave assisted preparation of cyclic ureas from diamines in the presence of ZnO. Tetrahedron Letters. 2004;**45**:7205

[38] Sarvari MH, Sharaghi H. ZnO as a new catalyst for N-formylation of amines under solvent-free conditions. The Journal of Organic Chemistry. 2006;**71**:6652

[39] Zhang M, Wang L, Ji H, Wu B, Zenge X. Cumene liquid oxidation to cumene hydroperoxide over CuO nanoparticle with molecular oxygen under mild condition. Journal of Natural Gas Chemistry. 2007;**16**:393-417

[40] Beydoun D, Amal R, Low G, McEvoy S. Role of nanoparticles in photocatalysis. Journal of Nanoparticle Research. 1999;**1**:439-458

[41] Kassaee MZ, Masrouri H, Movahedi F. ZnO-nanoparticle-promoted synthesis of polyhydroquinoline derivatives via multicomponent Hantzsch reaction. Monatshefte für Chemie. 2010;**141**:317-322

[42] Prasad GK, Ramacharyulu PVRK, Singh B, Batra K, Srivastava AR, Ganesan K, et al. Sun light assisted photocatalytic decontamination of sulfur mustard using ZnO nanoparticles. Journal of Molecular Catalysis A: Chemical. 2011;**349**:55

[43] Evans BE, Rittle KE, Bock MG, Dipardo RM, Freidinger RM, Whitter WL, et al. Methods for drug discovery: Development of potent, selective, orally effective cholecystokinin antagonists. Journal of Medicinal Chemistry. 1998;**31**:2235

[44] Muller G. Medicinal chemistry of target family-directed masterkeys. Drug Discovery Today. 2003;**8**:681

[45] Bocker H, Guengerich FP. Oxidation of 4-aryl- and 4-alkyl-substituted 2,6-dimethyl- 3,5-bis(alkoxycarbonyl)-1,4-dihydropyridines by human liver microsomes and immunochemical evidence for the involvement of a form of cytochrome P-450. Journal of Medicinal Chemistry. 1986;**29**(9):1596-1603

[46] Sausins A, Duburs G. Synthesis of 1,4-dihydropyridines by cyclocondensation reactions. Heterocycles. 1988;**27**:269-289

[47] Goldman S, Stoltefuss J. 1,4-dihydropyridines: Effects of chirality and conformation on the calcium antagonist and calcium agonist activities. Angewandte Chemie International Edition in English. 1991;**30**:559-1578

[48] Bossert F, Meyer H, Wehinger E. 4-Aryldihydropyridines, a new class of highly active calcium antagonists. Angewandte Chemie International Edition in English. 1981;**20**:762-769

[49] Bossert F, Vater W. 1,4-dihydropyridines--a basis for developing new drugs. Medicinal Research Reviews. 1989;**9**(3):291-324

[50] Buhler FR, Kiowski W. Calcium antagonists in hypertension. Journal of Hypertension. 1987;**5**(3):S3-S10

[51] Reid JL, Meridith PA, Pasanisi F. Clinical pharmacological

aspects of calcium antagonists and their therapeutic role in hypertension. Journal of Cardiovascular Pharmacology. 1985;7:S18-S20

[52] Godfaid T, Miller R, Wibo M. Calcium antagonism and calcium entry blockade. Pharmacological Reviews. 1986;**38**:321-327

[53] Mannhold R, Jablonka B, Voigdt W, Schoenafinger K, Schravan K. Calcium- and calmodulin-antagonism of elnadipine derivatives: Comparative SAR. European Journal of Medicinal Chemistry. 1992;**27**:229-235

[54] Boer R, Gekeler V. Chemosensitizer in tumor therapy: New compounds promise better efficacy. Drugs of the Future. 1995;**20**:499-509

[55] Bretzel RG, Bollen CC, Maester E, Federlin KF. Nephroprotective effects of nitrendipine in hypertensive type I and type II diabetic patients. American Journal of Kidney Diseases. 1993;**21**:54-63

[56] Bretzel RG, Bollen CC, Maester E, Federlin KF. Trombodipine platelet aggregation inhibitor antithrombotic. Drugs of the Future. 1992;**17**:465-468

[57] Sridhar R, Perumal PT. A new protocol to synthesize 1,4-dihydropyridines by using 3,4,5-trifluorobenzeneboronic acid as a catalyst in ionic liquid: Synthesis of novel 4-(3-carboxyl-1*H*-pyrazol-4-yl)-1,4-dihydropyridines. Tetrahedron. 2005;**61**:2465

[58] Heravi MM, Behbahani FK, Oskooie HA, Shoar RH. Catalytic aromatization of Hantzsch 1,4-dihydropyridines by ferric perchlorate in acetic acid. Tetrahedron Letters. 2005;**46**:2775

[59] Moseley JD. Alternative esters in the synthesis of ZD0947. Tetrahedron Letters. 2005;**46**:3179

[60] Hantzsch A. Condensationprodukte aus aldehydammoniak und ketoniartigen verbindungen. Bernoulli. 1881;**14**:1637-1638

[61] Loev B, Snader KM. Oxidation dealkylation of certain dihydropyridines. The Journal of Organic Chemistry. 1965;**30**:1914

[62] Alajarin R, Vaquero JJ, Garcia JLN, Alvarez-Builla J. Synthesis of 1,4-dihydropyridines under microwave irradiation. Synlett. 1992:297

[63] Khadikar BM, Gaikar VG, Chitnavis AA. Aqueous hydrotrope solution as a safer medium for microwave enhanced Hantzsch dihydropyridine ester synthesis. Tetrahedron Letters. 1995;**36**:8083

[64] Ohberg L, Westman J. An efficient and fast procedure for the hantzsch dihydropyridine synthesis under microwave conditions. Synlett. 2001;**2001**(8):1296-1298

[65] Agarwal A, Chauhan PMS. Solid supported synthesis of structurally diverse dihydropyrido[2,3-*d*] pyrimidines using microwave irradiation. Tetrahedron Letters. 2005;**46**:1345

[66] Phillips AP. Hantzsch's pyridine synthesis. Journal of the American Chemical Society. 1949;**71**:4003

[67] Anderson GJR, Berkelhammer G. A study of the primary acid reaction on model compounds of reduced diphosphopyridine nucleotide. Journal of the American Chemical Society. 1958;**80**:992

[68] Dolly HS, Chimni SS, Kumar S. Acid catalysed enamine induced transformations of 1,3-dimethyl-5-formyluracil. A unique annulation reaction with enaminones. Tetrahedron. 1995;**51**:12775

[69] Breitenbucher JG, Figliozzi G. Solid-phase synthesis of 4-aryl-1,4-dihydropyridines via the Hantzsch three component condensation. Tetrahedron Letters. 2000;**41**:4311

[70] Kumar A, Maurya RA. Synthesis of polyhydroquinoline derivatives through unsymmetric Hantzsch reaction using organocatalysts. Tetrahedron. 2007;**63**:1946

[71] Wang L-M, Sheng J, Zhang L, Han J-W, Fan Z, Tian H, et al. Facile Yb(OTf)3 promoted one-pot synthesis of polyhydroquinoline derivatives through Hantzsch reaction. Tetrahedron. 2005;**61**:1539

[72] Ravikumar Naik TR, Bhojya Naik HS, Prakash Naik HR, Bindu PJ, Harish BG, Krishna V. Synthesis, DNA binding, docking and photoclevage studies of novel benzo[b][1,8] naphthyridines. Medicinal Chemistry. 2009;**5**(5):411

[73] Bindu PJ, Mahadevan KM, Ravikumar Naik TR. Sm(III)nitrate-catalyzed one-pot synthesis of furano[3,2c]-1,2,3,4 tetrahydroquinolines and DNA photocleavage studies. Journal of Molecular Structure. 2012;**1020**:142

[74] Bindu PJ, Mahadevan KM, Satyanarayan ND, Ravikumar Naik TR. Synthesis and DNA cleavage studies of novel quinoline oxime esters. Bioorganic & Medicinal Chemistry Letters. 2012;**22**(2):898

[75] Bindu PJ, Mahadevan KM, Naik TRR, Harish BG. Synthesis, DNA binding, docking and photocleavage studies of 2-chloro-3-quinolinyl-3-phenylpropen-2-ones. Medicinal Chemistry Communications. 2014;**5**:1708

[76] Ravikumar Naik TR, Shivashankar SA. Heterogeneous bimetallic $ZnFe_2O_4$ nanopowder catalyzed synthesis of Hantzsch 1,4-dihydropyridines in water. Tetrahedron Letters. 2016;**57**:4046-4049

[77] Janis RA, Silver PJ, Triggle DJ. Drug action and cellular calcium regulation. Aciv, Advances in Drug Research. 1987;**16**:309

[78] Lavilla R. Recent developments in the chemistry of dihydropyridines. Journal of the Chemical Society, Perkin Transactions 1. 2002:1141

[79] Kappe CO. Biologically active dihydropyrimidones of the Biginelli-type–a literature survey. European Journal of Medicinal Chemistry. 2000;**35**:1043

[80] Varache-Lemebge M, Nuhrich A, Zemb V, Devaux G, Vacher P, Vacher AM, et al. Synthesis and activities of a thienyl dihydropyridine series on intracellular calcium in a rat pituitary cell line (GH3/B6). European Journal of Medicinal Chemistry. 1996;**31**:547

[81] Alker D, Campbell SF, Cross PE, Burges RA, Carter AJ, Gardiner DG. Long-acting dihydropyridine calcium antagonists. 4. Synthesis and structure-activity relationships for a series of basic and nonbasic derivatives of 2-[(2-aminoethoxy)methyl]-1,4-dihydropyridine calcium antagonists. Journal of Medicinal Chemistry. 1990;**33**:585

[82] Reddy PR, Rao KS, Satyanarayana B. Synthesis and DNA cleavage properties of ternary Cu(II) complexes containing histamine and amino acids. Tetrahedron Letters. 2006;**47**(41):7311-7315

[83] Reddy DS, Hosamani KM, Devarajegowda HC. Design, synthesis of benzocoumarinpyrimidine hybrids as novel class of antitubercular agents, their DNA cleavage and X-ray studies. European Journal of Medicinal Chemistry. 2015;**101**:705-715

[84] Barton JK, Raphael AL. Photoactivated stereospecific cleavage of double-helical DNA by cobalt(III) complexes. Journal of the American Chemical Society. 1984;**106**:2466

[85] Sigman DS. Nuclease activity of 1,10-phenanthroline-copper ion. Accounts of Chemical Research. 1986;**19**:180

[86] Liu C, Zhou J, Xu H. Interaction of the copper(II) macrocyclic complexes with DNA studied by fluorescence quenching of ethidium. Journal of Inorganic Biochemistry. 1998;**71**:1-6

[87] Suvarna S, Krishna K, Kaushik SH, Rijesh K, Diwakar L, Reddy GC. Synthesis, anticancer and antioxidant activities of 2,4,5-trimethoxy chalcones and analogues from asaronaldehyde: Structureeactivity relationship. European Journal of Medicinal Chemistry. 2013;**62**:435-442

[88] Sun C, Aspland SE, Ballatore C, Castillo R, Smith AB, Castellino AJ. The design, synthesis, and evaluation of two universal doxorubicin-linkers: Preparation of conjugates that retain topoisomerase II activity. Bioorganic & Medicinal Chemistry Letters. 2006;**16**:104

[89] Zhang Y, Zheng W, Luo Q, Zhao Y, Zhang E, Liu S, et al. Dual-targeting organometallic ruthenium(ii) anticancer complexes bearing EGFR-inhibiting 4-anilinoquinazoline ligands. Dalton Transactions. 2015;**44**:13100-13111

[90] Tabassum S, Zaki M, Afzal M, Arjmand F. New modulated design and synthesis of quercetin-Cu(II)/Zn(II)-Sn2(IV) scaffold as anticancer agents: In vitro DNA binding profile, DNA cleavage pathway and Topo-I activity. Dalton Transactions. 2013;**42**(27):10029

[91] Taha M, Ismail NH, Khan A, Shah SAA, Anwar A, Halim SA, et al. Synthesis of novel derivatives of oxindole, their urease inhibition and molecular docking studies. Bioorganic & Medicinal Chemistry Letters. 2015;**25**(16):3285-3289

5

Total Synthesis of Macrolides

Chebolu Naga Sesha Sai Pavan Kumar

Abstract

Structurally complex macrolide natural products, isolated from a variety of marine and other sources, continue to provide a valuable source of targets for the synthetic chemist to embark. In this account, we provide the recent progress and pathways in the total synthesis of macrolides and discussed the synthesis of (+)-neopeltolide, aspergillide D, miyakolide and acutiphycin natural products.

Keywords: macrolide, aldolization, macrolactonization, ring closing metathesis, coupling reactions, total synthesis

1. Introduction

Macrolides are a class of antibiotics that consist of a large macrocyclic lactone ring attached to deoxy sugars. These antibiotics are bacteriostatic in nature and act by inhibiting protein synthesis of bacteria. These are obtained mainly from certain actinomycetes genus, such as *Streptomyces* and related species. The original macrolide complex, erythromycin A, was isolated in 1952 as a natural product of *Saccharopolyspora erythraea* (formerly *Streptomyces erythreus*). Other examples include clarithromycin, azithromycin, telithromycin, cethromycin, modithromycin, etc. Macrolides structurally contain three characteristic parts in every molecule, that is, a macrocyclic lactone ring, multiple ketone & hydroxyl group, and two deoxy sugars attached by glycosidic bond. According to the carbon number of lactone ring, macrolides are classified into several types. That is, 12-membered ring, 13-membered ring, 14-membered ring, 15-membered ring, 16-membered rings, etc. (**Figure 1**). Out of these, most of the antibiotic drugs comprised of 14-membered and 16-membered lactone rings.

The construction of macrocyclic structures is a recurrent and challenging problem in synthetic organic chemistry. Theoretically, macrocyclic systems can be generated by cyclization of open, long chain precursors or by cleavage of internal bonds in polycyclic systems. In the course of synthesis, numerous problems are encountered to achieve target molecules. Despite the several problems, however, recent interest in the chemistry of macrolide antibiotics and other biologically active macrolactones and macrolactams resulted in the discovery and development of several new synthetic methods for macrolide formation. In this chapter, total synthesis of some of the macrolides is discussed with scrupulous emphasis on the key macrolide ring forming reactions.

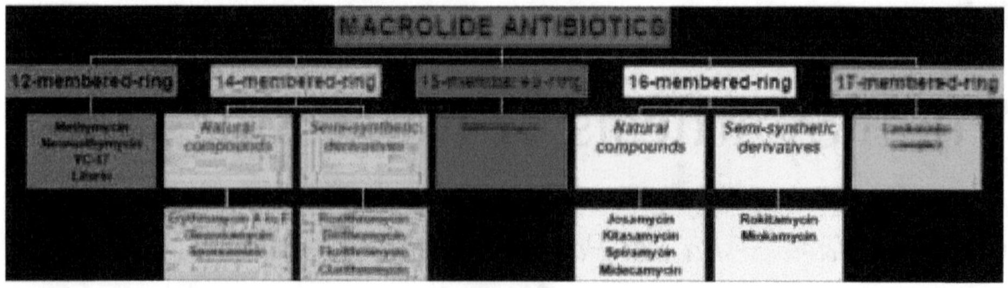

Figure 1.
Classification of macrolide antibiotics.

2. Synthetic strategy for macrolide synthesis

In the polyoxomacrolide ring, generally we will observe the 1,3-diol systems as a core. There are two synthetic approaches for the edifice of 1,3-diols which are illustrated here. They are asymmetric aldol reaction and the other one is asymmetric epoxide and epoxide ring-opening.

2.1 Aldolization

Asymmetric synthesis of β-hydroxy ketones by aldol reactions of ketones with aldehydes is the general and efficient method for the synthesis of 1,3-diol systems and is of great interest in the field of total synthesis. By using the range of chiral ketones, highly diastereoselective *syn* and *anti* aldol products are produced using various boron enolates [1–3]. Some of the reagents shown below (**Figure 2**) direct the relative and absolute stereochemistry of C—C bond formation between various achiral and chiral ketones, thus providing a ubiquitous synthetic tool for macrolide synthesis (**Figure 3**).

A highly efficient and extensively used method for diastereoselective aldol reactions is the Evans aldol reaction using boron enolate derived from a chiral imide [4, 5]. Upon treatment of imide **1** with *n*-Bu_2BOTf and *i*-Pr_2NEt in CH_2Cl_2 followed by addition of aldehyde, aldol reaction proceeds smoothly in stereoselective manner through the chelation transition state to attain 1,2-*syn*-aldol adduct **2** in high yield and with excellent diastereoselectivity. After the reaction, the chiral auxiliary is cleaved by hydrolysis to acid, then reduction to aldehyde or alcohol, conversion to

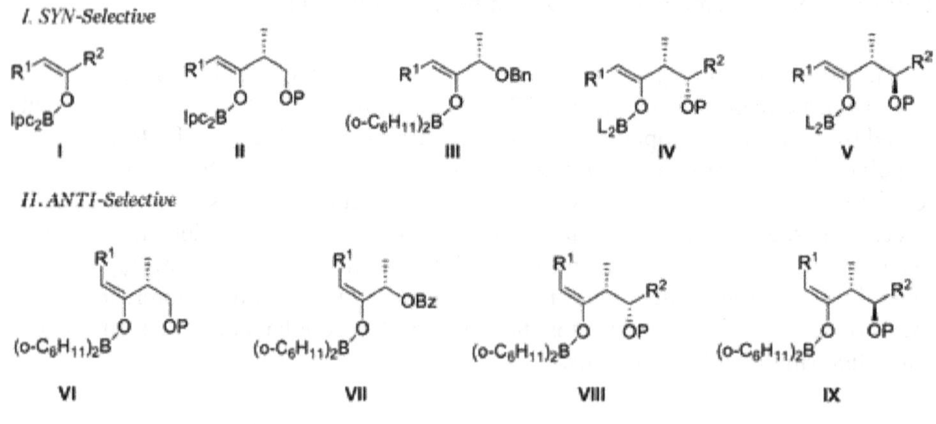

Figure 2.
Reagents for asymmetric Aldol reactions.

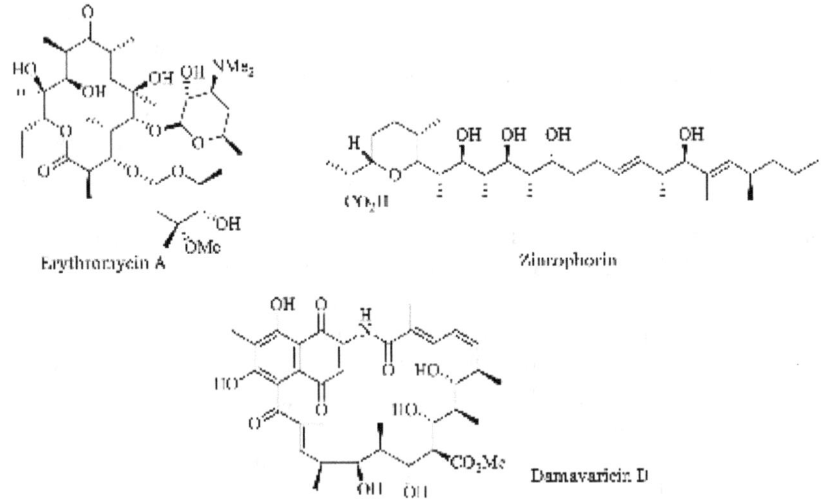

Figure 3.
Macrolide antibiotics: evidence for the chemists interested in the stereochemistry of the Aldol reaction.

Weinreb amide, etc. In contrast, addition of a Lewis acid to the boron enolate provides either *anti*-diol **3** or non-Evans 1,2-*syn*-aldol **4** with excellent diastereoselectivity [6] (**Figure 4**).

2.2 Asymmetric epoxidation and dihydroxylation

The asymmetric epoxidation of allylic alcohols introduced by Katsuki and Sharpless in 1980 has tremendous applications in the synthesis of various

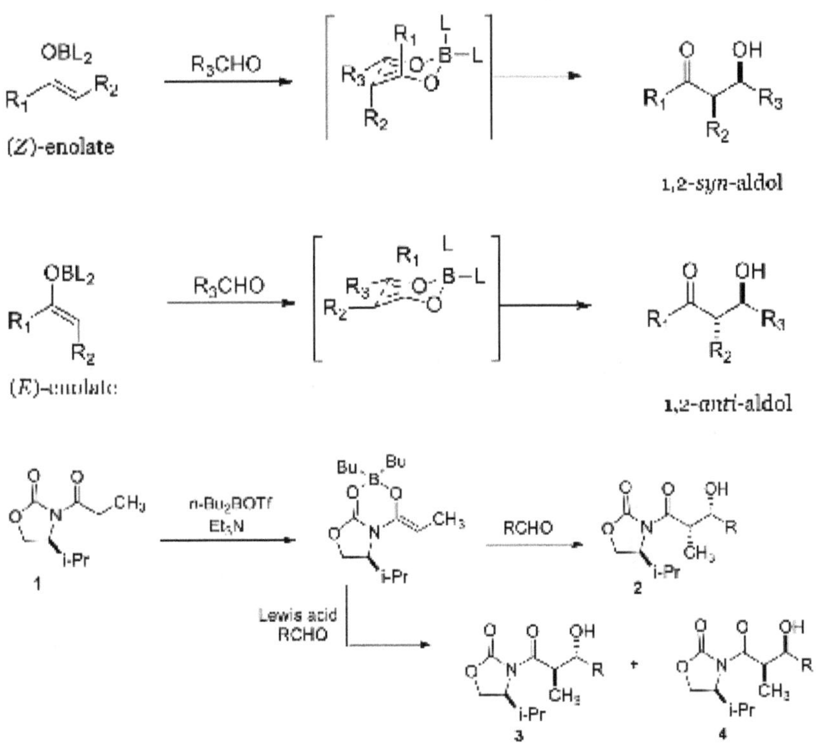

Figure 4.
Evan's Aldol strategy.

compounds [7]. The Sharpless asymmetric epoxidation (AE) is the efficacious reagent in the synthetic organic chemistry particularly in the synthesis of variety of natural products. Epoxidation is carried out from allylic alcohols **5** with *tert*-butyl hydroperoxide in the presence of $Ti(OiPr)_4$. The resulting epoxide stereochemistry is determined by the enantiomer of the chiral tartrate ester (usually diethyl tartrate or diisopropyl tartrate) employed in reaction. When (−)-diester is used, β-epoxide **6** is obtained, while (+)-diester produces α-epoxide 7 (**Figure 5**).

The Sharpless dihydroxylation [8] is another tool used in the enantioselective preparation of 1,2-diols (**9a/9b**) from olefins (**8**). This reaction is performed with osmium catalyst and a stoichiometric oxidant (e.g., $K_3Fe(CN)_6$ or NMO). Enantioselectivity is produced by the addition of enantiomerically-enriched chiral ligands [$(DHQD)_2$PHAL also called AD-mix-β, $(DHQ)_2$PHAL also called AD-mix-α or their derivatives] (**Figure 6**). These reagents are also commercially available as stable and not so expensive.

Stereoselective ring-opening of 2,3-epoxy alcohols **10** is extremely valuable for the synthesis of different functionalized compounds [9]. A wide range of nucleophiles such as secondary amine, alcohol, thiol, azide and carboxylic acid predominantly at C-3 position to give 1,2-diol **11** (**Figure 7**).

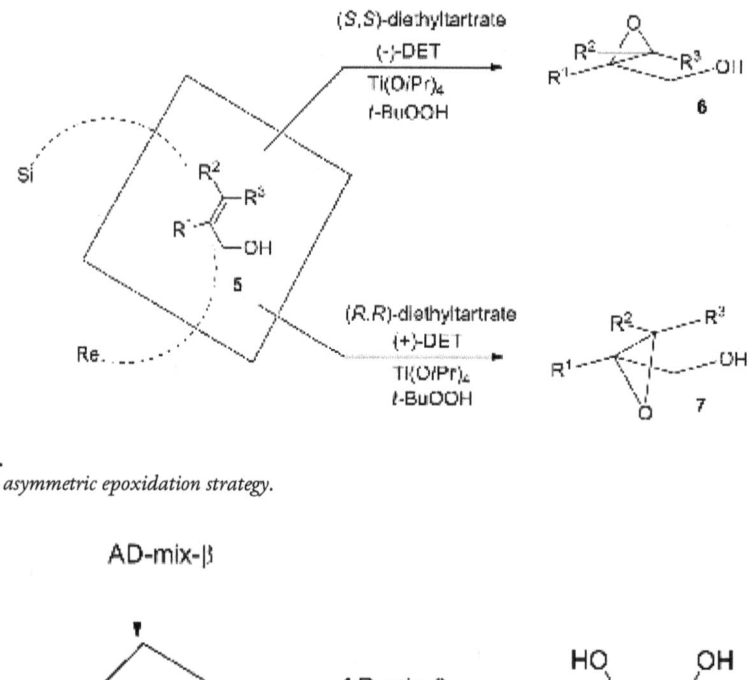

Figure 5.
Sharpless asymmetric epoxidation strategy.

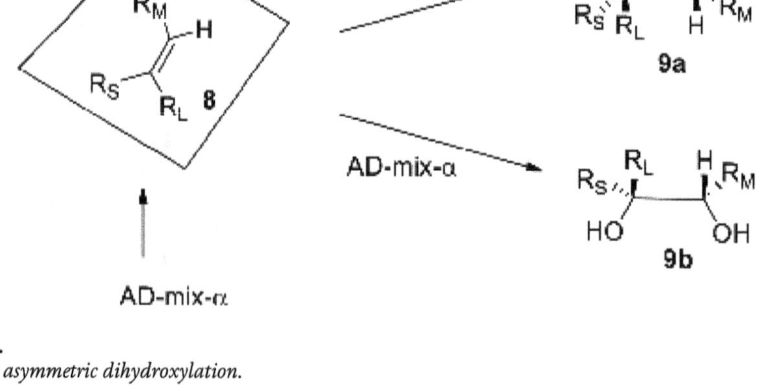

Figure 6.
Sharpless asymmetric dihydroxylation.

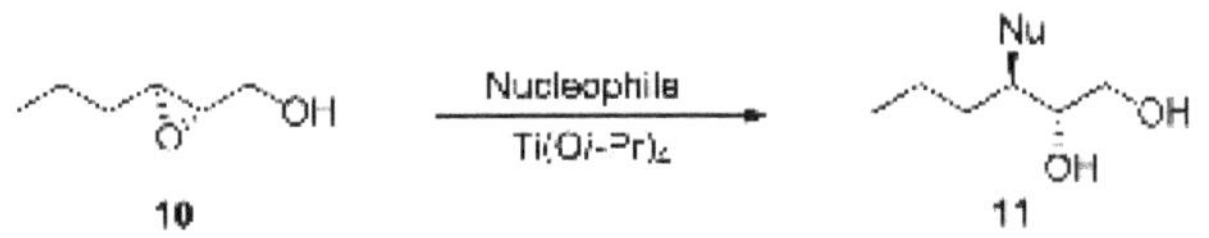

Nucleophiles: Et_2NH, i-PrOH, PhSH, $PhCO_2H$, $TMSN_3$.

Figure 7.
Stereoselective ring-opening of epoxy alcohols.

The logic of macrocyclization in natural product synthesis can be investigated by different strategies; some of them are Prins reaction [10], lactonization [11, 12], ring closing metathesis [13], Wittig reaction [14], Horner Wadsworth Emmons (HWE) reaction [15], Julia-Kocienski reaction [16], metal-mediated cross coupling reaction [17], etc. However, it is true that there is no universal macrocyclization method is reliable in the total synthesis of natural products.

2.3 Macrolactonization

Macrolactonization is the one of the effective and popular methods in the synthesis of macrolactones. The method is based on the lactonization of the corresponding seco-acid. Thus various methods are reported in the literature for the macrolactone synthesis, some of the most commonly used methods are Corey-Nicolaou [18], Shiina [19], Yamaguchi [20], Mitsunobu [21], Keck-Boden [22], and Mukaiyama [23] macrolactonizations (**Figure 8**).

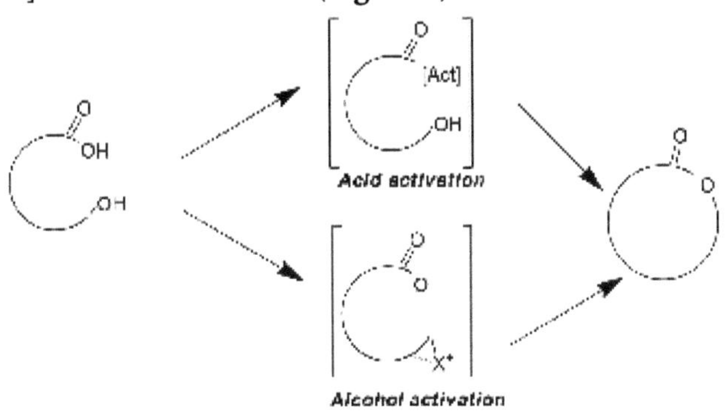

Corey-Nicolaou method:

Reagent: 2,2-dipyridyl disulfide, PPh_3

Shiina method:

Reagent: 2-methyl-6-nitrobenzoic anhydride
DMAP

Yamaguchi method:

Reagent: 2,4,6-trichlorobenzoyl chloride,
Et_3N

Mitsunobu method:

Reagent: PPh_3 and
Diethyl azodicarboxylate (DEAD) or
Diisopropyl azodicarboxylate (DIAD)

Keck & Boden method:

Reagent: DCC, DMAP and DMAP.HCl

Mukaiyama method:

Reagent: 2-halo-N-alkyl pyridinium salts
base

Figure 8.
Some of the popular methods used for macrolactonization.

Hansen et al. [24] reported the synthesis of (−)-aplyolide A from **12** in which they adopted the Corey-Nicolaou macrolactonisation as the key step with 78% yield (**Figure 9**).

Narasaka et al. [25] used the Mukaiyama method for the effective construction of macrocycle ring from corresponding seco-acid **13** in the synthesis of Prostaglandin F-lactone (**Figure 10**).

Enev et al. [26] in his studies towards the total synthesis of laulimalide, crucial Yamaguchi macrolactonization was employed on the ynoic seco-acid **14** and then reducing the triple bond obtained the desired macrolactone **15** (**Figure 11**).

In the synthetic studies towards the synthesis of colletodial, Keck et al. [27] effectively used DCC-DMAP protocol for the macrolactonization of **16** to precursor of colletodial **17** (**Figure 12**).

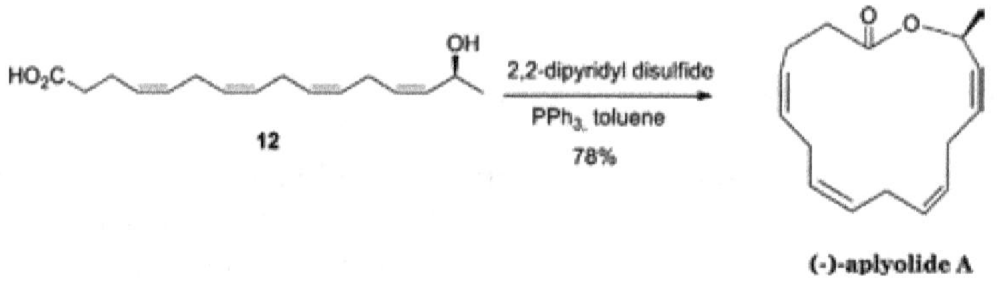

Figure 9.
Application of Corey-Nicolaou macrolactonisation.

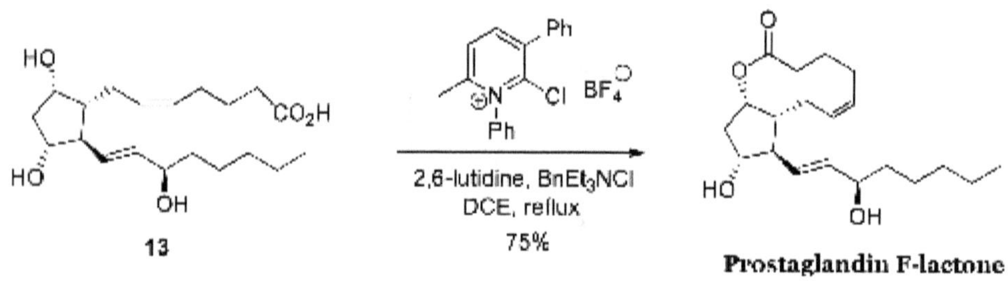

Figure 10.
Mukaiyama method in the synthesis of prostaglandin F-lactone.

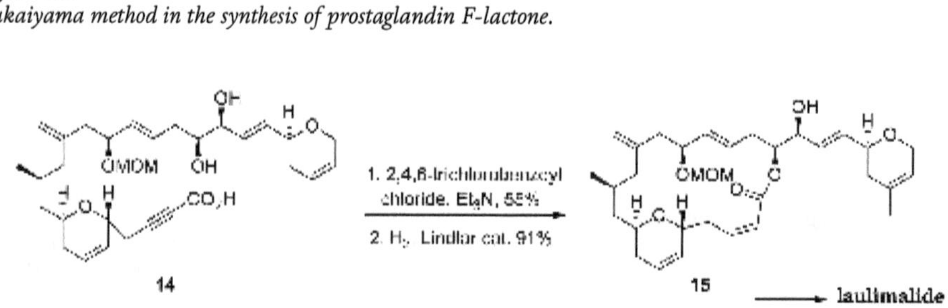

Figure 11.
Yamaguchi protocol in the synthesis of laulimalide.

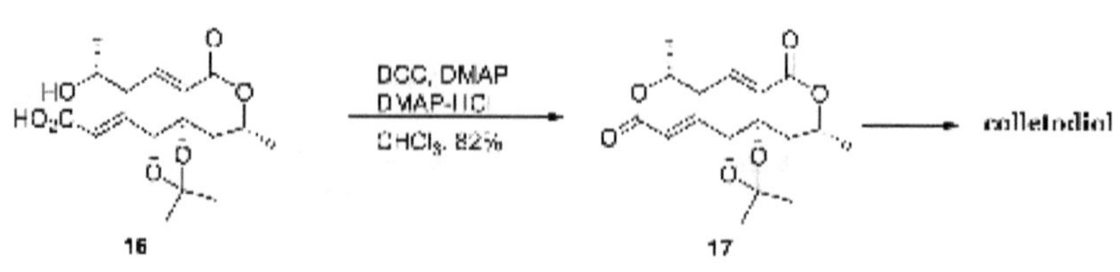

Figure 12.
Keck et al. lactonisation for the synthesis of colletodial.

Mitsunobu macrolactonization protocol based on the activation of the seco-acid alcohol **18** to **19** using diisopropyl azodicarboxylate (DIAD) and triphenylphosphine is used in the total synthesis of natural product (+)-amphidinolide K by Williams and Meyer [28] (**Figure 13**).

In the total synthesis of iejimalide by Schweitzer et al. [29], Shiina macrolactonization (2-methyl-6-nitro benzoic anhydride/DMAP) is used as the key step for the construction of macrolactone **21** in moderate yield. Even the yield is somewhat low, other methods failed to construct the lactone while Shiina protocol worked successfully from **20** (**Figure 14**).

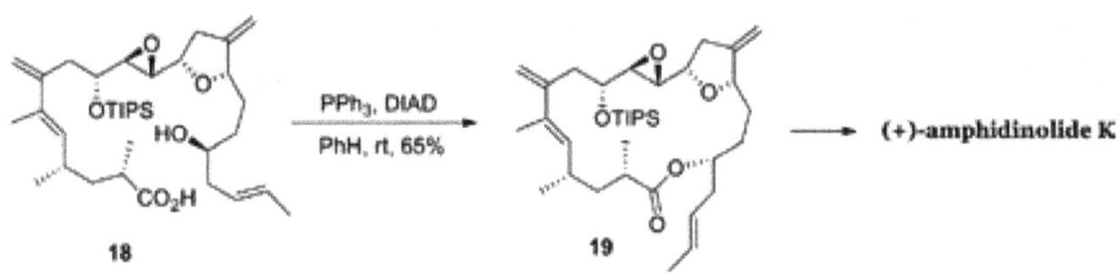

Figure 13.
Mitsunobu esterification in the total synthesis of (+)-amphidinolide K.

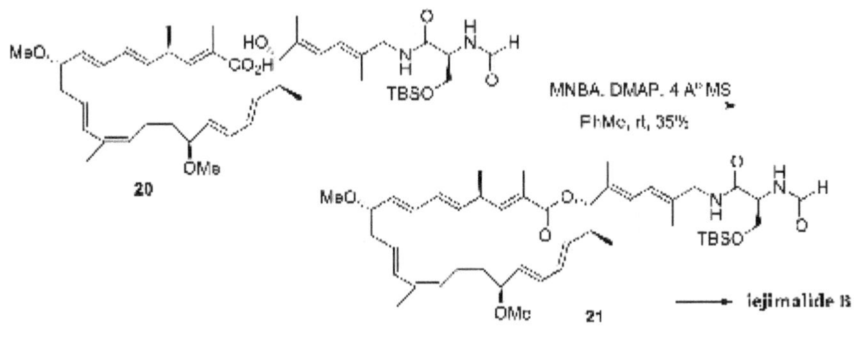

Figure 14.
Shiina macrolactonization towards the synthesis of iejimalide B.

2.4 Ring-closing olefin metathesis

In recent years, ring closing metathesis (RCM) has become one of the most paramount tools in synthetic organic chemistry especially in the field of total synthesis of macrolide natural products [13, 30, 31]. Furthermore, RCM is becoming the most popular way to construct large rings and has the advantage of being compatible with a wide range of functional groups such as ketones, ethers, esters, amides, amines, epoxides, silyl ethers, alcohols, thioesters, etc. In view of this, among the several reagents developed by Grubbs, Shrock, and Chauvin, the catalysts **A–D** represents two generations of ruthenium complexes, while **E** is the molybdenum Shrock catalyst (**Figure 15**). **A** is popularly known as Grubbs first generation catalysts, **B** and **C** are Grubbs second generation catalysts and **D** is

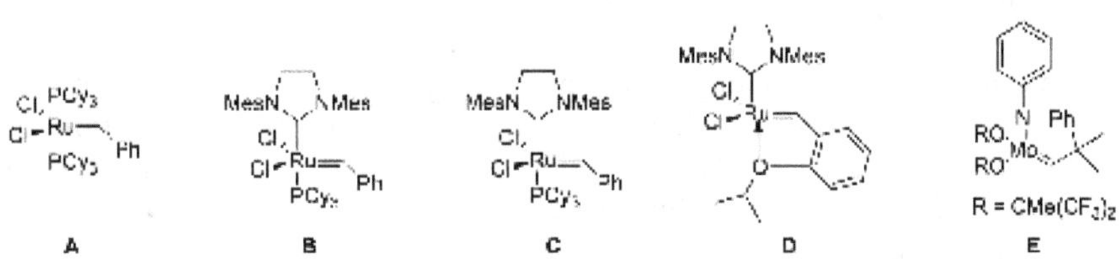

Figure 15.
Various catalysts for ring closing metathesis.

Hoveyda-Grubbs catalyst. The choice of the catalysts can be used in the synthetic organic transformations based on the reactivity of the substrate, and other reaction condition parameters. Substitution in the aromatic ring of **D** has given rise to a new family of third generation catalyst.

Here, some of the applications of ring closing metathesis in the total synthesis of macrolides salicylihalamide A [32], *trans*-resorcylide [33], (+)-lasiodiplodin [34], oximidine III [35], and Sch 38516 [36] by various metathesis catalysts have been illustrated (**Figure 16**).

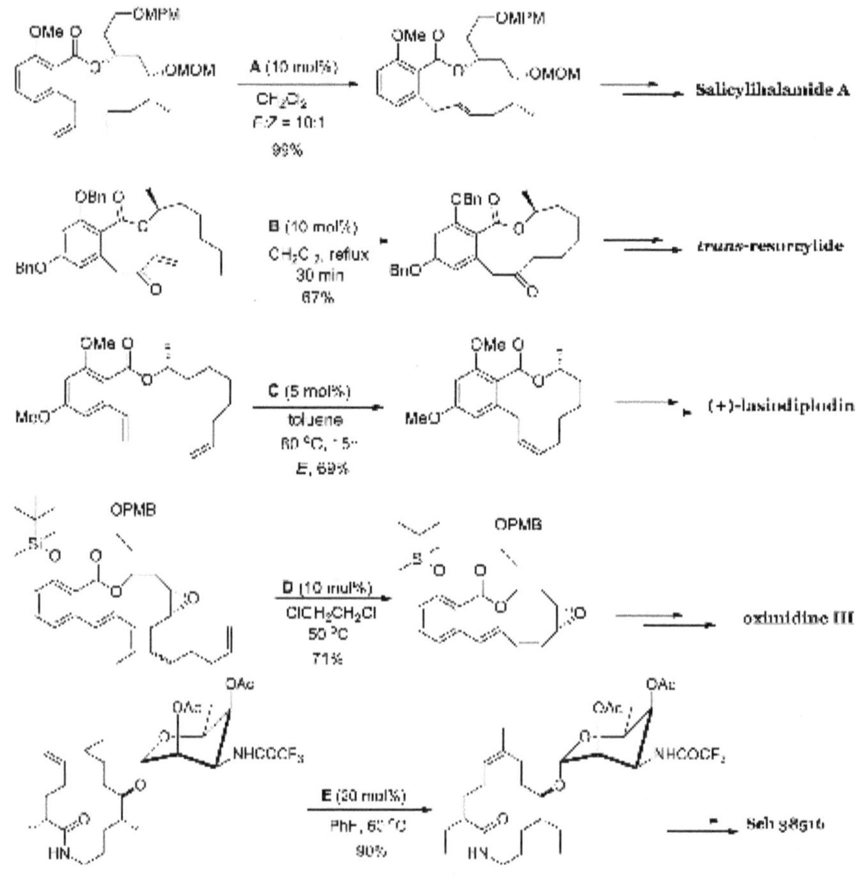

Figure 16.
Some of the applications of ring closing metathesis in total synthesis of macrolides.

2.5 Palladium catalyzed coupling reactions

Palladium-catalyzed coupling reactions have gained more attention in recent years in the field of organic chemistry. In this course, Suzuki reaction using organoboron compounds [37], Heck reaction using alkenes [38], Stille reaction with organostannate [39, 40], Sonogashira reaction with terminal alkyne [41] and Tsuji-Trost reaction with π-allylpalladium intermediate [42, 43], etc. are the most frequently employed reactions in the total synthesis of macrolide natural products. Some of them are depicted here.

Tortosa and co-workers [44] in the synthesis of (+)-superstolide A, Suzuki macrocyclisation approach is used for the construction of 24-membered macrocyclic octane **23** (**Figure 17**).

The first application of the Heck cyclisation to a macrocyclic substrate was reported by Ziegler and co-workers [45] in 1981 during the synthesis of aglycone of the macrocyclic antibiotic carbomycin B. They achieved the cyclisation to the model substrate **25** in 55% yield, by slow addition to a solution of $PdCl_2(MeCN)_2$, Et_3N and formic acid in MeCN at ambient temperature (**Figure 18**).

Stille macrocyclisation as illustrated below used as the key step in the total synthesis of the biselyngbyolide A by Tanabe et al. [46] using $Pd_2(dba)_3$ and lithium chloride in DMF (**Figure 19**).

The utility of Sonogashira macrocyclisation in the first total synthesis of penarolide sulfate A1, an α-glucosidase inhibitor is demonstrated by Mohapatra and co-workers [47]. The macrocyclisation was successfully achieved from compound **28** with catalytic $Pd(PPh_3)_4$ and CuI in Et_2NH at room temperature (**Figure 20**).

Towards the total synthesis of antibiotic natural product A26771B, Trost and co-workers [48] effectively constructed the macrolactone **31** (**Figure 21**) by the use of bidentate phosphine ligand (1,4-bis(diphenylphosphino)butane (DPPB)).

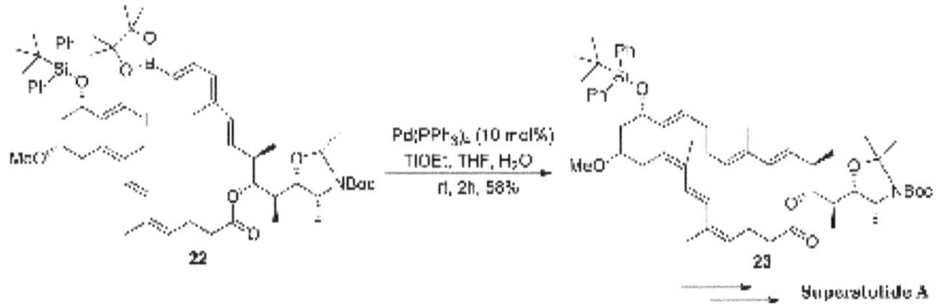

Figure 17.
Suzuki macrocyclisation in the synthesis of superstolide A.

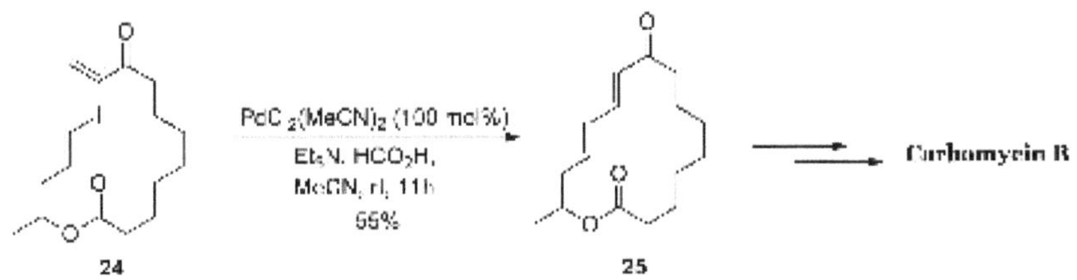

Figure 18.
Heck cyclisation in the synthesis of carbomycin B.

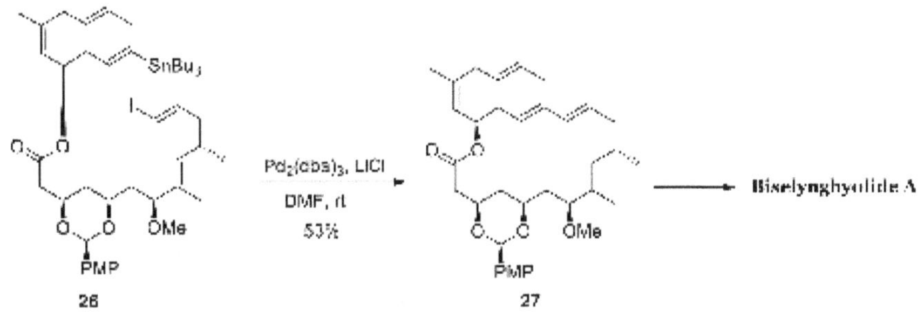

Figure 19.
Stille macrocyclisation in the total synthesis of biselyngbyolide A.

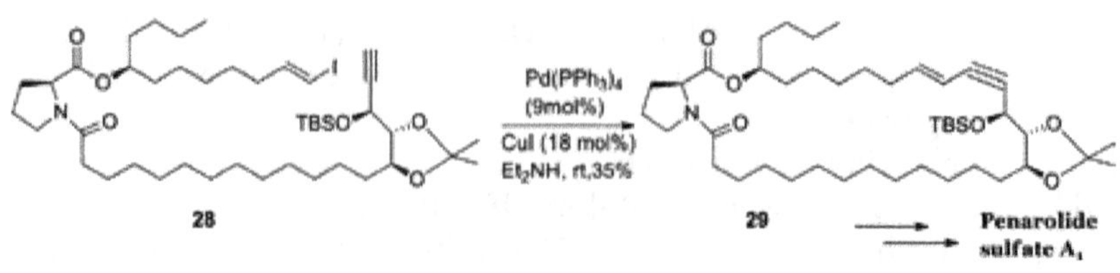

Figure 20.
Sonogashira macrocyclisation in the synthesis of penarolide sulfate A1.

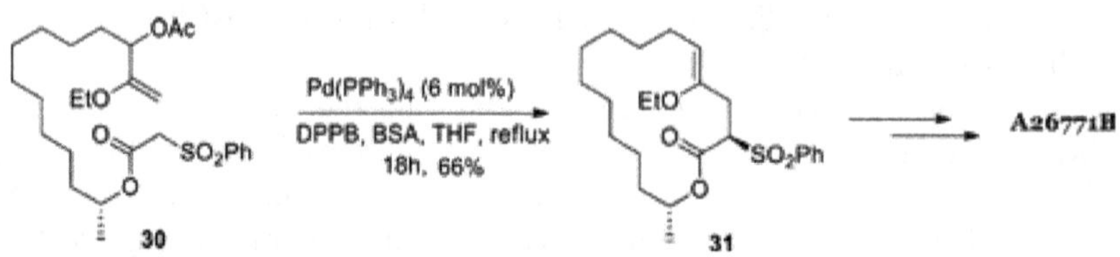

Figure 21.
Tsuji-Trost lactonization for the synthesis of A26771B.

In the end game of total synthesis of macrolides, glycosidation to the aglycon also have more significance. Thus, a wide variety of methods are reported for glycosidation in the literature [49, 50].

3. Total synthesis of selected macrolides

In this section, the total synthesis of selected macrolides is discussed: (+)-neopeltolide (**32**), aspergillide D (**33**) and briefly about miyakolide (**34**) and acutiphycin (**35**).

3.1 (+)-Neopeltolide

(+)-Neopeltolide is a 14-membered macrolide isolated from north coast of Jamaica by Wright and coworkers from a deep water sponge [51]. It was tested for *in vitro* antiproliferative activity against several cancer cell lines comprising A549 human lung adenocarcinoma, NCI/ADR-RES ovarian sarcoma and P388 murine leukemia and shows IC_{50} values 1.2, 5.1 and 0.56 nM. Besides this, neopeltolide also exhibits anti-fungal activity against *Candida albicans* [52]. The complexity of the structure with six chiral centres, tetrahydropyran ring, and an oxazole-bearing unsaturated side chain and its efficacious biological activity led to several total syntheses, few of them are discussed below.

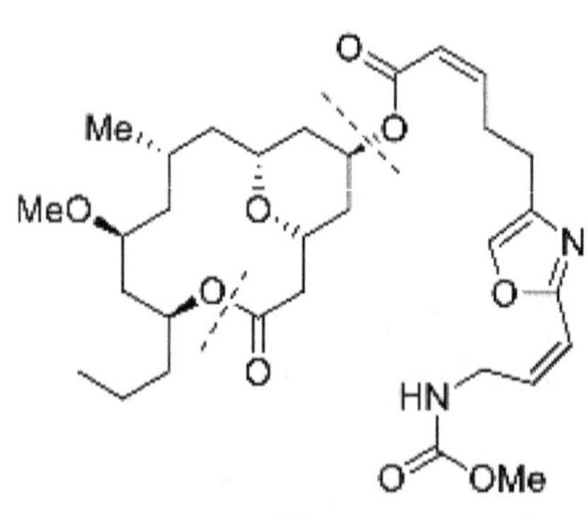

32

In 2013, Ghosh et al. [53] in their total synthesis adopted the retrosynthetic pathway as follows. Disconnection of O—C bond of the oxazole side chain would give acid which can undergo Mitsunobu esterification. Yamaguchi macrolactonization of acid would in turn give the desired macrolactone. The tetrahydropyran ring in acid could be constructed via a hetero Diels-Alder reaction between aldehyde and silyloxy diene ether using Jacobsen's chromium catalyst.

The synthesis of the macrolactone ring of (+)-neopeltolide began with commercially available 3-methyl glutaric anhydride as shown in the scheme. 3-methyl glutaric anhydride, **36** was desymmetrized using PS-30 'Amano' lipase to obtain acid. The resulting acid was treated with borane-dimethyl sulfide complex to afford alcohol, **37**. Alcohol **37** was oxidized to corresponding aldehyde by Swern oxidation and then protected to its acetal, **38**. Ester of **38** was then reduced to alcohol and on Swern oxidation obtained aldehyde and the resulting aldehyde was subjected to Brown's allylation protocol using (+)-Ipc_2BOMe and allyl magnesium bromide to attain alcohol, **39**. Alcohol **39** was methylated with MeI, and on Lemieux-Johnson oxidation gave aldehyde and on Brown's allylation protocol afforded alcohol, **40**. Acetal protection was deprotected and the aldehyde was converted to α,β-unsaturated ketone, **41** using standard Horner-Wadsworth-Emmons olefination conditions. Secondary alcohol in **41** was then protected with TESOTf to obtain the silyloxy diene, **42** in excellent yield (**Figure 22**).

After the completion of requisite silyloxy diene, hetero-Diels Alder reaction of tosyl oxyacetaldehyde, **43** with **42** using chiral chromium catalyst (**44**) gave tetrahydropyranone, **45** in 83% yield (**Figure 23**). After protection of ketone group in **45** as ketal and displaced the tosylate to nitrile **46** using NaCN in DMF. Nitrile **46** was hydrolysed to acid and on deprotection of ketone to afford ketone **47**. Intramolecular Yamaguchi macrolactonization attained the key macrolactone **48** in 40% yield. Olefin **48** was subjected to hydrogenation with 10% Pd/C to give saturated compound and on reduction with $NaBH_4$/EtOH to attain alcohol **49**. Next, the synthesis of unsaturated oxazole side chain **50** is started with known alkyne with LDA and Bu_3SnCl to obtain the alkynyl stannate, which on hydrozirconation gave the carbamate in 38% yield. Crucial Stille cross coupling of carbamate with iodooxazole using $Pd(MeCN)_2Cl_2$ in DMF gave oxazole which can be easily

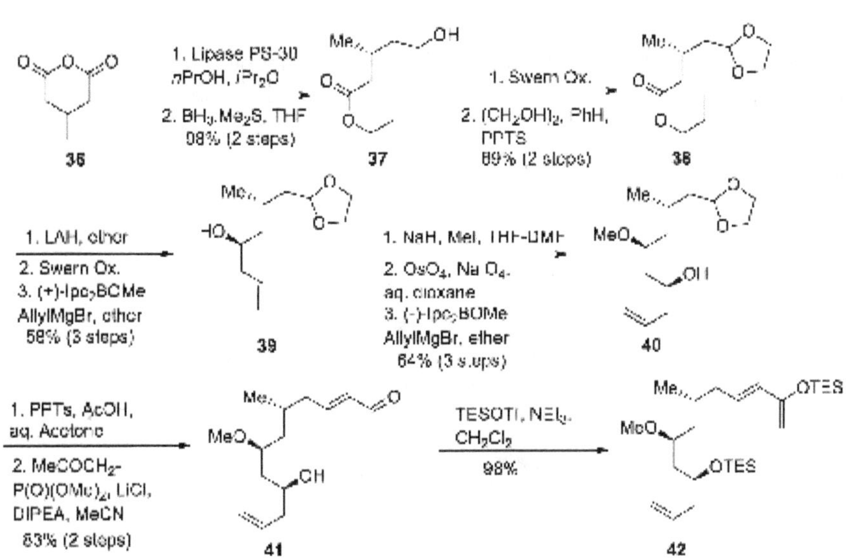

Figure 22.
Synthesis of silyloxy diene 42 fragment.

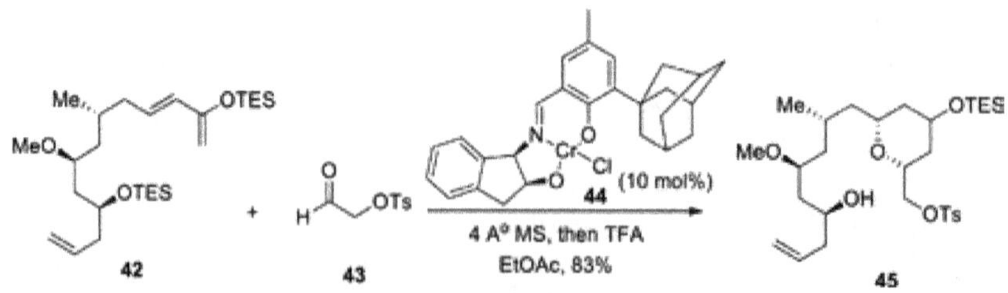

Figure 23.
Hetero-Diels-Alder reaction for the synthesis of tetrahydropyrarone moiety.

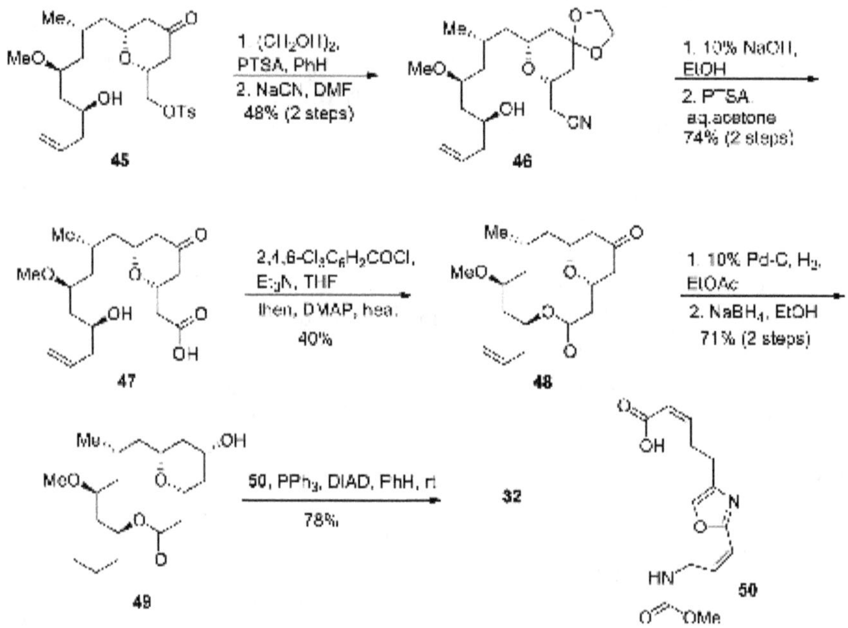

Figure 24.
Completion of synthesis of (+)-neopeltolide.

converted to desired side chain **50**. The target neopeltolide compound was furnished by standard Mitsunobu esterification of **49** with acid **50** (**Figure 24**).

3.2 Paterson strategy

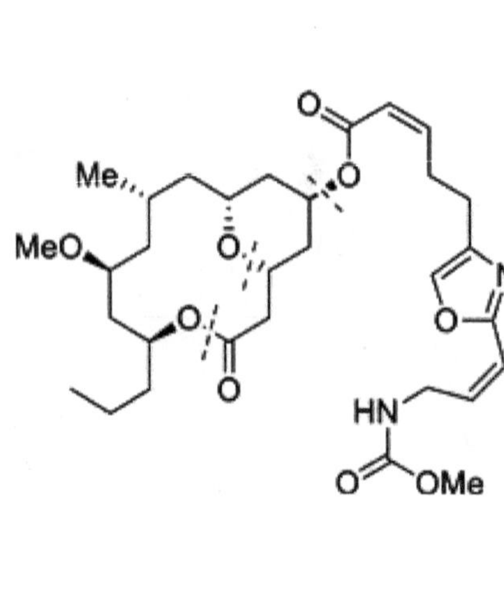

32

Paterson and coworkers [54] reported the synthesis of neopeltolide as follows. Aldehyde was synthesized starting from known β-keto ester, **51**, which on treatment with (*S*)-BINAP-Ru (II) catalyst under Noyori asymmetric hydrogenation to afford (13*S*)-alcohol. The alcohol on TBS protection and DIBAL reduction of the ester produced the enantiopure aldehyde **52.** Aldehyde **52** was subjected to Brown's methallylation using 2-methyl propene and (−)-Ipc_2BOMe furnished the desired C11 alcohol with 94:6 dr, and the alcohol was methylated into the methyl ether **53** by NaH, MeI. Methyl ether **53** was subjected to ozonolysis to obtain methyl ketone and on Horner Wadsworth Emmons reaction with trimethyl phosphonoacetate to attain ester **54** in E/Z isomers (75,25). Esters **54** were reduced to its alcohol by DIBAL-H and on subsequent oxidation with Dess-Martin periodinane produced aldehyde **55.** Next, organo catalytic hydride reduction of enal **55** using MacMillan strategy with imidzolidinone catalyst **56.** TFA (20 mol%) and Hantzch ester furnished as 1,4-reduction product **57** with 76:24 of epimers at C9 stereocentre (**Figure 25**). Further, Jacobsen asymmetric hetero Diels-Alder reaction between **57** and known 2-silyloxy diene **58** produced *cis*-tetrahydropyranone **60** in 60% yield using chiral tridentate chromium (III) catalyst **59.** On PMB deprotection and further oxidation of alcohol to corresponding acid followed by TBS deprotection furnished seco-acid **61.** Macrolactonization of **61** under standard Yamaguchi conditions afforded macrolactone **62** in 80% yield. Reduction of macrolactone **62** to the alcohol with $NaBH_4$ in MeOH followed by Mitsunobu esterification with the oxazole side chain **50** achieved (+)-neopeltolide (**32**) in 52% yield (**Figure 26**).

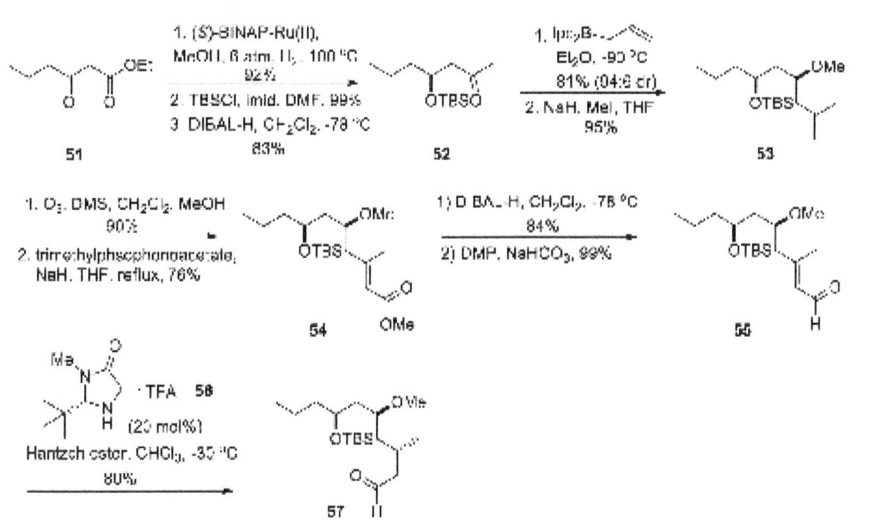

Figure 25.
Synthesis of aldehyde 57.

3.3 Ulanovskaya strategy

The synthesis of neopeltolide by Ulanovskaya et al. [55] is depicted as follows, Prins desymmetrization of diene **63** followed by benzyl protection and Wacker oxidation of alkene afforded ketone **64.** Formation of boron enolate from ketone and on addition of aldehyde **65** gave the anticipated aldol product with >98:2 diastereoselectivity, which was treated with $Ph_3P{=}CH_2$ (Wittig methylenation) followed by cleavage of dioxolone by acidic work-up afforded ketone **66** in 75%

yield. Ketone **66** was selectively reduced to *syn*-alcohol using Et_2BOMe and $NaBH_4$ followed by ester hydrolysis gave acid which was subjected to Yamaguchi macrolactonization to furnish desired macrolactone **67**. Alkene in **67** was hydrogenated using Pd/C to afford desired alcohol **68** as a major product. Alcohol **68** was subjected to Mitsunobu conditions, followed by hydrolysis with K_2CO_3 in MeOH to get the inversion product. Subsequent methylation with MeO_3BF_4 & hydrogenolysis of benzyl ether achieved desired macrolide **49**. The final coupling of fragment **49** with oxazole side chain **50** with standard Mitsunobu conditions furnished target (+)-neopeltolide **32** (**Figure 27**).

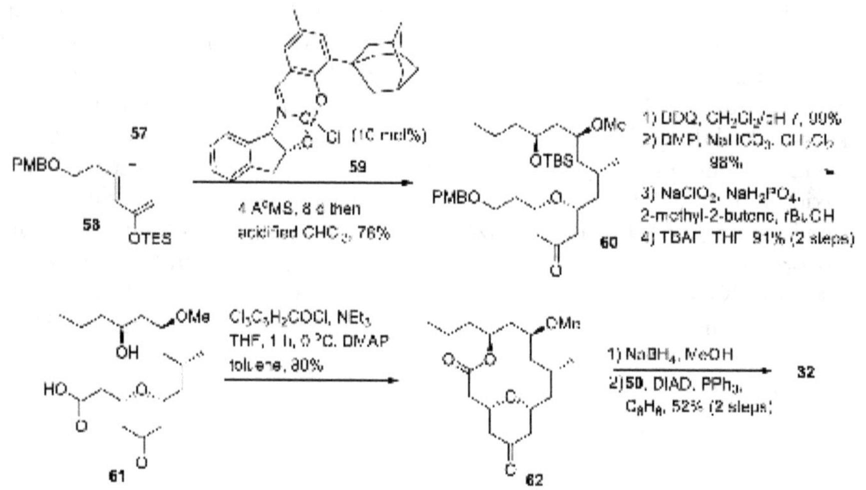

Figure 26.
Total synthesis of (+)-neopeltolide.

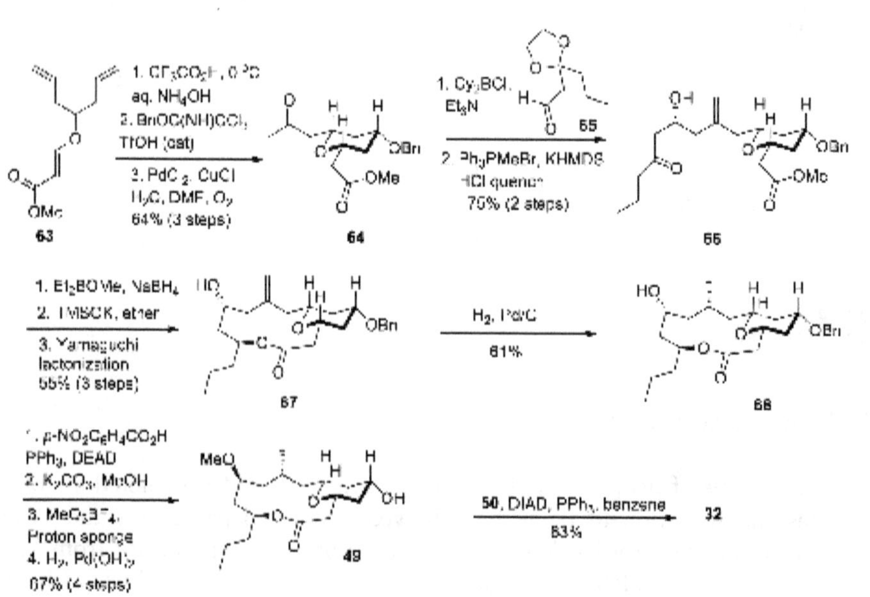

Figure 27.
Alternate synthesis of (+)-neopeltolide.

3.4 Aspergillide-D

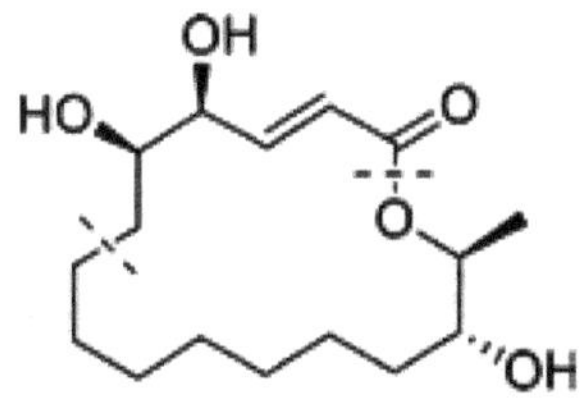

33

Bao and coworkers in 2013 isolated 16-membered macrolide, aspergillide D, from the extract of *Aspergillus* sp. SCSGAF 0076 [56]. Aspergillide D macrolactone contains four chiral centres, α,β-unsaturation, three hydroxyl groups and the first total synthesis was reported by Jena et al. in 2017 as follows [57].

The retrosynthetic analysis of aspergillide D was depicted as shown above, macrolactone could be synthesized from seco acid via intramolecular Shiina esterification. For the total synthesis of Aspergillide D, the acid fragment was synthesized from commercially available D-ribose which was transformed to lactol **69** by using three step sequence, that is, catalytic amount of H_2SO_4 & acetone to form acetonide which on reduction with $NaBH_4$ and on oxidative cleavage of the diol with $NaIO_4$. The lactol was subjected to Wittig type olefination using PPh_3=CH_2 and the obtained primary alcohol **70** was oxidized to carboxylic acid **71** by using TEMPO/BAIB conditions (**Figure 28**). The synthesis of alcohol fragment was started with mono-PMB protection **73** of commercially available 1,8-octane diol **72** and the other alcohol was converted to racemic allyl alcohol **74** by Swern oxidation and subsequent treatment of aldehyde with vinyl magnesium bromide in the presence of CuI. The allylic alcohol **74** was subjected to standard Sharpless kinetic resolution conditions by using (−)-DIPT & Ti(O*i*Pr)$_4$ to obtain enantiomeric epoxy alcohol **75**. Upon MOM protection **76** to the secondary alcohol **75** and PMB deprotection produced **77**, which on oxidation with Dess-Martin periodinane to afford aldehyde. Aldehyde was converted to olefin **78** by treating PPh_3=CH_2 in THF. **78** was cleaved to alcohol **79** by reduction with LAH in THF (**Figure 29**).

Acid **71** and alcohol **79** fragments were coupled together under Yamaguchi esterification conditions afforded diene ester **80** in 65% yield. Intramolecular RCM was employed on diene ester by using Grubbs' second generation catalyst in refluxing CH_2Cl_2 to produce the requisite macrolactone **81**. Double bond in **81** was hydrogenated by using PtO_2 in MeOH to attain saturated lactone **82**. Lactone **82** was reduced with DIBAL-H to afford lactol which on further treatment with Ph_3P=$CHCO_2Et$ in C_6H_6 afforded α,β-unsaturated ester **83**. The ester was converted to carboxylic acid **84** by LiOH in THF/H_2O which on adopting key Shiina's

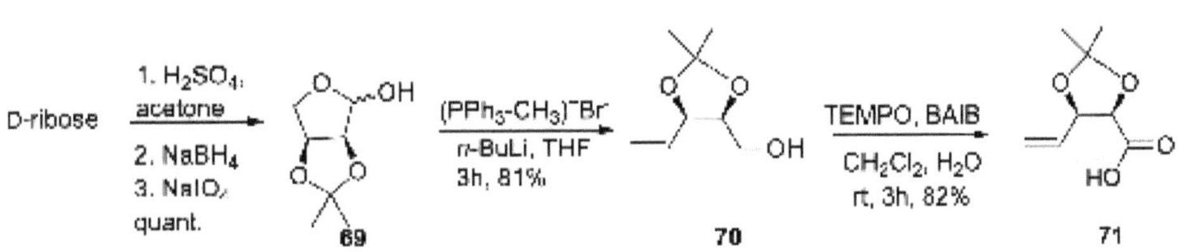

Figure 28.
Synthesis of acid fragment 71.

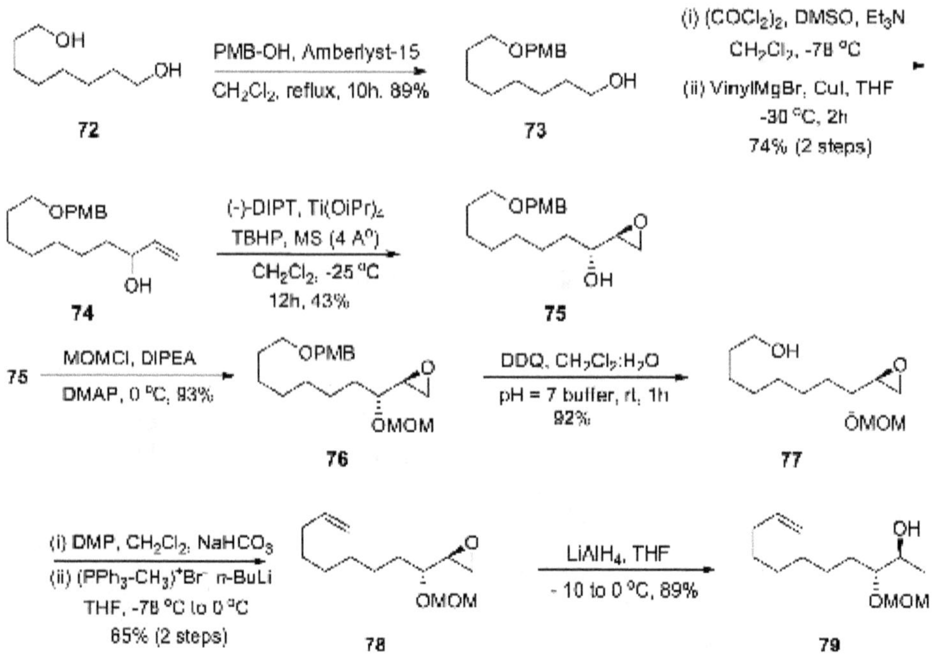

Figure 29.
Synthesis of alcohol fragment 79.

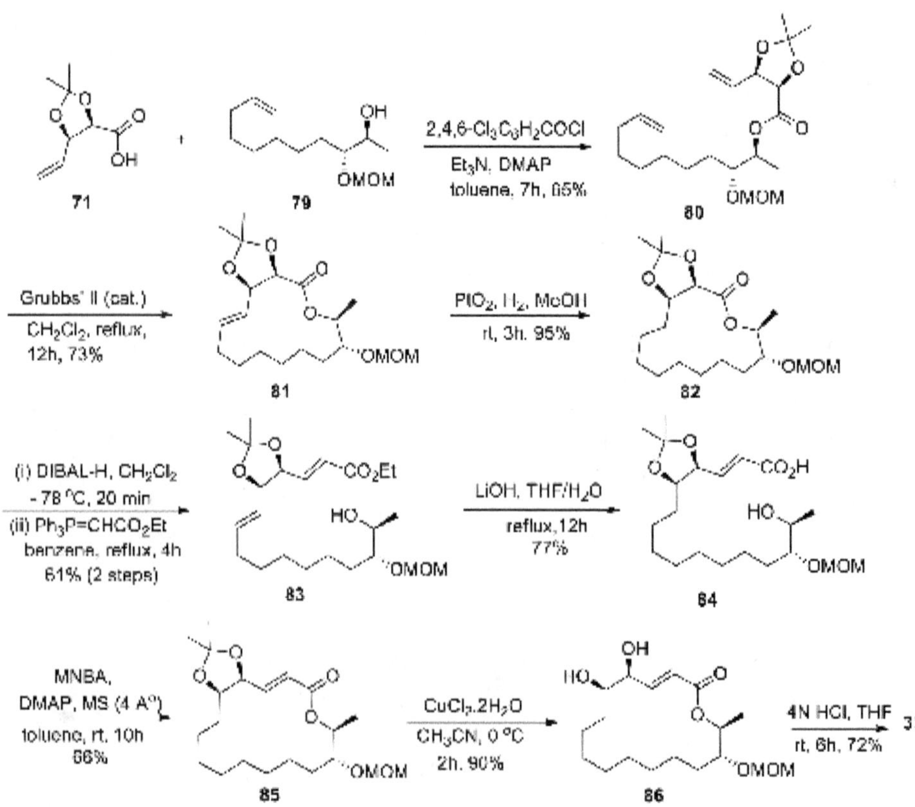

Figure 30.
Completion of synthesis of aspergillide D.

macrolactonization protocol to provide the desired mactrolactone **85** in 51% yield. On deprotection of acetonide with $CuCl_2.2H_2O$ gave diol **86** and removal of MOM group, the synthesis of aspergillide D **33** was achieved (**Figure 30**).

3.5 Miyakolide

Evan's strategy of bond connections & key reactions in the synthesis of **34** is illustrated [58].

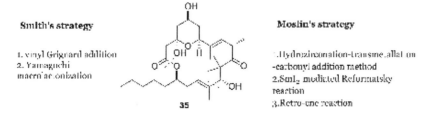

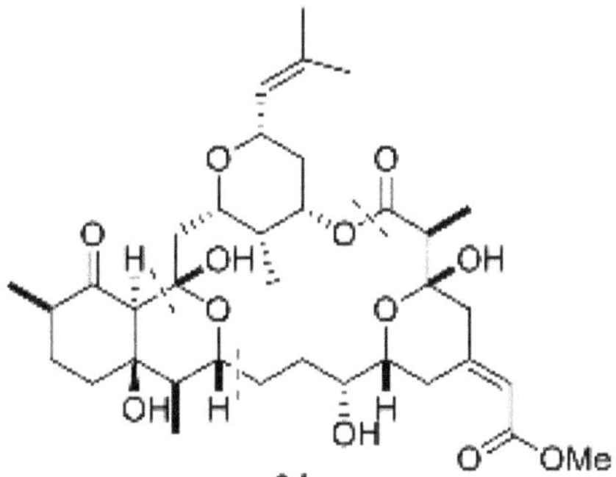

34

Evan's strategy

1. aldol reaction
2. [3+2] dipolar cycloaddition
3. Yamaguchi macrolactonization

3.6 Acutiphycin

Smith's strategy [59] & Moslin's strategy of acutiphycin [60] is shown below (**Figure 31**).

Smith's strategy

1. vinyl Grignard addition
2. Yamaguchi macrolactonization

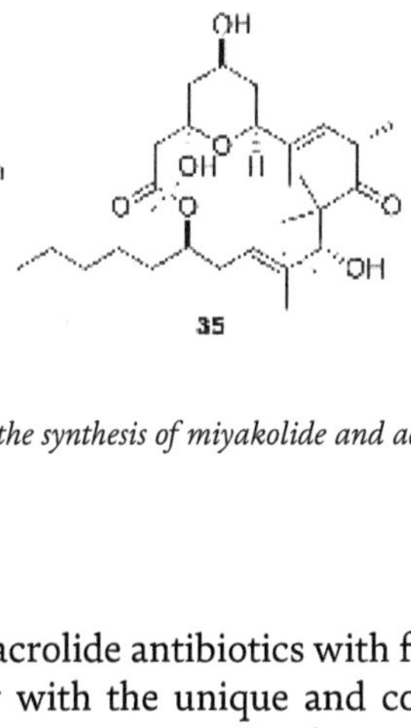

35

Moslin's strategy

1.Hydrozirconation-transmetallation-carbonyl addition method
2.SmI_2 mediated Reformatsky reaction
3.Retro-ene reaction

Figure 31.
Key reactions and strategies in the synthesis of miyakolide and acutiphycin.

4. Conclusions

A number of new macrolide antibiotics with fascinating biological activities have been isolated everyday with the unique and complex structures have been determined with extensive spectroscopic studies. Toward the total synthesis of such macrolide antibiotics, very efficient synthetic strategies and various new methodologies are also developed. Recent advances in macrolide synthesis based on newly developed strategies and methodologies are noteworthy. Further synthetic studies on macrolide antibiotics will make an immense contribution to progress in both organic and medicinal chemistry.

Acknowledgements

The author wishes to thank Vice Chancellor, Dean R & D, VFSTRU for constant support and encouragement. The author wishes to express his gratitude to Prof. V.

Anuradha and other staff members of chemistry department, Vignan for the fruitful discussion.

Author details

Chebolu Naga Sesha Sai Pavan Kumar
Division of Chemistry, Department of Sciences and Humanities, Vignan's Foundation for Science, Technology and Research, Guntur, Andhra Pradesh, India

References

[1] Cowden CJ, Paterson I. Asymmetric Aldol reactions using boron enolates. Organic Reactions. 1997;**51**:1-200. DOI: 10.1002/0471264180.or051.01

[2] Kim BM, Williams SF, Masamune S. The Aldol reaction: Group III enolates. In: Trost BM, Fleming I, editors. Comprehensive Organic Synthesis. Oxford, UK: Pergamon Press; 1991. pp. 301-320

[3] Paterson I, Cowden CJ, Wallace DJ. Stereoselective Aldol reactions in the synthesis of polyketide natural products. In: Otera J, editor. Modern Carbonyl Chemistry. Weinheim, Germany: Wiley-VCH; 2000. pp. 249-297

[4] Evans DA, Nelson JV, Vogel E, Taber TR. Stereoselective Aldol condensations via boron enolates. Journal of the American Chemical Society. 1981; **103**:3099-3111. DOI: 10.1021/ja00 401a031

[5] Gage JR, Evans DA. Diastereoselective Aldol condensation using a chiral oxazolidinone auxiliary: (2S*3S*)-3-hydroxy-3-phenyl-2-methylpropanoic acid. Organic Syntheses. 1990;**68**:83-91. DOI: 10.15227/orgsyn.068.0083

[6] Walker MA, Heathcock CH. Extending the scope of the Evans asymmetric aldol reaction: Preparation of anti and "Non-Evans" Syn aldols. Journal of Organic Chemistry. 1991;**56**: 5747-5750. DOI: 10.1021/jo00020a006

[7] Katsuki T, Sharpless KB. The first practical method for asymmetric epoxidation. Journal of the American Chemical Society. 1980;**102**:5974-5976. DOI: 10.1021/ja00538a077

[8] Hentges SG, Sharpless KB. Asymmetric induction in the reaction of osmium tetroxide with olefins. Journal of the American Chemical Society. 1980; **102**:4263-4265. DOI: 10.1021/ja00214a053

[9] Caron M, Sharpless KB. Titanium isopropoxide-mediated nucleophilic openings of 2,3-epoxy alcohols. A mild procedure for regioselective ring-opening. The Journal of Organic Chemistry. 1985;**50**:1557-1560. DOI: 10.1021/jo00209a047

[10] Crane EA, Scheidt KA. Prins type macrocyclizations as an efficient ring-closing strategy in natural product synthesis. Angewandte Chemie, International Edition. 2010;**49**: 8316-8326. DOI: 10.1002/anie.201002809

[11] Parenty A, Moreau X, Campagne J-M. Macrolactonizations in the total synthesis of natural products. Chemical Reviews. 2006;**106**:911-939. DOI: 10.1021/cr0301402

[12] Parenty A, Moreau X, Niel G, Campagne J-M. Update 1 of: Macrolactonizations in the total synthesis of natural products. Chemical Reviews. 2013;**113**:PR1-PR40. DOI: 10.1021/cr300129n

[13] Gradillas A, Perez-Castells J. Macrocyclization by ring-closing metathesis in the total synthesis of natural products: Reaction conditions and limitations. Angewandte Chemie International Edition. 2006;**45**: 6086-6101. DOI: 10.1002/anie. 200600641

[14] Nicolaou KC, Harter MW, Gunzner JL, Nadin A. The Wittig and related reactions in natural product synthesis. Liebigs Annalen. 1997;**1997**(7): 1283-1301. DOI: 10.1002/jlac.19971997 0704

[15] Bisceglia JA, Orelli LR. Recent applications of the Horner-Wadsworth-

Emmons reaction to the synthesis of natural products. Current Organic Chemistry. 2012;**16**:2206-2230. DOI: 10.2174/138527212803520227

[16] Chatterjee B, Bera S, Mondal D. Julia-Kocienski olefination: A key reaction for the synthesis of macrolides. Tetrahedron: Asymmetry. 2014;**25**:1-55. DOI: 10.1016/j.tetasy.2013.09.027

[17] Ronson TO, Taylor RJK, Fairlamb IJS. Palladium-catalysed macrocyclisations in the total synthesis of natural products. Tetrahedron. 2015; **71**:989-1009. DOI: 10.1016/j.tet.2014.11.009

[18] Corey EJ, Nicolaou KC. Efficient and mild lactonization method for the synthesis of macrolides. Journal of the American Chemical Society. 1974;**96**: 5614-5616. DOI: 10.1021/ja00824a073

[19] Shiina I, Kubota M, Ibuka R. A novel and efficient macrolactonization of ω-hydroxycarboxylic acids using 2-methyl-6-nitrobenzoic anhydride (MNBA). Tetrahedron Letters. 2002;**43**: 7535-7539. DOI: 10.1016/S0040-4039 (02)01819-1

[20] Inanaga J, Hirata K, Saeki H, Katsuki T, Yamaguchi M. A rapid esterification by means of mixed anhydride and its application to large-ring lactonization. Bulletin of the Chemical Society of Japan. 1979;**52**: 1989-1993. DOI: 10.1246/bcsj.52.1989

[21] Kurihara T, Nakajima Y, Mitsunobu O. Synthesis of lactones and cycloalkanes. Cyclization of ω-hydroxy acids and ethyl α-cyano-ω-hydroxycarboxylates. Tetrahedron Letters. 1976;**17**:2455-2458. DOI: 10.1016/0040-4039(76)90018-6

[22] Boden EP, Keck GE. Proton-transfer steps in Steglich esterification: A very practical new method for macrolactonization. The Journal of Organic Chemistry. 1985;**50**:2394-2395. DOI: 10.1021/jo00213a044

[23] Mukaiyama T, Usui M, Saigo K. The facile synthesis of lactones. Chemistry Letters. 1976;**5**:49-50. DOI: 10.1246/cl.1976.49

[24] Hansen TV, Stenstrom Y. First total synthesis of (−)-aplyolide A. Tetrahedron: Asymmetry. 2001;**12**: 1407-1409. DOI: 10.1016/S0957-4166 (01)00250-6

[25] Narasaka K, Maruyama K, Mukaiyama T. A useful method for the synthesis of macrocyclic lactone. Chemistry Letters. 1978;**7**:885-888. DOI: 10.1246/cl.1978.885

[26] Enev VS, Kaehlig H, Mulzer J. Macrocyclization via allyl transfer: Total synthesis of laulimalide. Journal of the American Chemical Society. 2001;**123**: 10764-10765. DOI: 10.1021/ja016752q

[27] Keck GE, Boden EP, Wiley MR. Total synthesis of (+)-colletodial: New methodology for the synthesis of macrolactones. The Journal of Organic Chemistry. 1989;**54**:896-906. DOI: 10.1021/jo00265a033

[28] Williams DR, Meyer KG. Total synthesis of (+)-amphidinolide K. Journal of the American Chemical Society. 2001;**123**:765-766. DOI: 10.1021/ja005644l

[29] Schweitzer D, Kane JJ, Strand D, McHenry P, Tenniswood M, Helquist P. Total synthesis of iejimalide B. An application of the Shiina macrolactonization. Organic Letters. 2007;**9**:4619-4622. DOI: 10.1021/ol702129w

[30] Majumdar KC, Rahaman H, Roy B. Synthesis of macrocyclic compounds by ring closing metathesis. Current Organic Chemistry. 2007;**11**:1339-1365. DOI: 10.2174/138527207782023166

[31] Grubbs RH, Miller SJ, Fu GC. Ring-closing metathesis and related processes in organic synthesis. Accounts of Chemical Research. 1995;**28**:446-452. DOI: 10.1021/ar00059a002

[32] Wu Y, Esser L, De Brabander JK. Revision of the absolute configuration of salicylihalamide A through asymmetric total synthesis. Angewandte Chemie, International Edition. 2000;**39**:4308-4310. DOI: 10.1002/1521-3773(20001201)39:23%3C4308::aid-anie4308%3E3.0.CO;2-4

[33] Garbaccio RM, Stachel SJ, Baeschlin DK, Danishefsky SJ. Concise asymmetric synthesis of radicicol and monocillin I. Journal of the American Chemical Society. 2001;**123**:10903-10908. DOI: 10.1021/ja011364+

[34] Furstner A, Seidel G, Kindler N. Macrocycles by ring-closing metathesis, XI: Syntheses of (R)-(+)-lasiodiplodin, zeranol and truncated salicylihalamides. Tetrahedron. 1999;**55**:8215-8230. DOI: 10.1016/S0040-4020(99)00302-6

[35] Wang X, Bowman EJ, Bowman BJ, Porco JA. Total synthesis of the salicylate enamide macrolide oximidine III: Application of relay ring-closing metathesis. Angewandte Chemie International Edition. 2004;**43**:3601-3605. DOI: 10.1002/anie.200460042

[36] Xu Z, Johannes CW, Salman SS, Hoveyda AH. Enantioselective total synthesis of antifungal agent Sch 38516. Journal of the American Chemical Society. 1996;**118**:10926-10927. DOI: 10.1021/ja9626603

[37] Miyaura N, Suzuki A. Stereoselective synthesis of arylated (E)-alkenes by the reaction of alk-1-enylboranes with aryl halides in the presence of palladium catalyst. Journal of the Chemical Society, Chemical Communications. 1979;**1979**(19):866-867. DOI: 10.1039/C39790000866

[38] Heck RF, Nolley JP. Palladium-catalyzed vinylic hydrogen substitution reactions with aryl, benzyl, and styryl halides. The Journal of Organic Chemistry. 1972;**37**:2320-2322. DOI: 10.1021/jo00979a024

[39] Heravi MM, Mohammadkhani L. Recent applications of Stille reaction in total synthesis of natural products: An update. Journal of Organometallic Chemistry. 2018;**869**:106-200. DOI: 10.1016/j.jorganchem.2018.05.018

[40] Milstein D, Stille JK. Palladium-catalyzed coupling of tetraorganotin compounds with aryl and benzyl halides. Synthetic utility and mechanism. Journal of the American Chemical Society. 1979;**101**:4992-4998. DOI: 10.1021/ja00511a032

[41] Sonogashira K, Tohda Y, Hagihara N. A convenient synthesis of acetylenes: Catalytic substitutions of acetylenic hydrogen with bromoalkenes, iodoarenes and bromopyridines. Tetrahedron Letters. 1975;**16**:4467-4470. DOI: 10.1016/S0040-4039(00)91094-3

[42] Trost BM, Fullerton TJ. New synthetic reactions. Allylic alkylation. Journal of the American Chemical Society. 1973;**95**:292-294. DOI: 10.1021/ja00782a080

[43] Trost BM. Cyclizations via palladium-catalyzed allylic alkylations [new synthetic methods (79)]. Angewandte Chemie International Edition in English. 1989;**28**:1173-1192. DOI: 10.1002/anie.198911731

[44] Tortosa M, Yakelis NA, Roush WR. Total synthesis of (+)-superstolide A. Journal of the American Chemical Society. 2008;**130**:2722-2723. DOI: 10.1021/ja710238h

[45] Ziegler FE, Chakraborty UR, Weisenfeld RB. A palladium-catalyzed carbon-carbon bond formation of

conjugated dienones: A macrocyclic dienone lactone model for the carbomycins. Tetrahedron. 1981;**37**: 4035-4040. DOI: 10.1016/S0040-4020(01)93278-8

[46] Tanabe Y, Sato E, Nakajima N, Ohkubo A, Ohno O, Suenaga K. Total synthesis of biselyngbyolide A. Organic Letters. 2014;**16**:2858-2861. DOI: 10.1021/ol500996n

[47] Mohapatra DK, Bhattasali D, Gurjar MK, Khan MI, Shashidhara KS. First asymmetric total synthesis of penarolide sulfate A1. European Journal of Organic Chemistry. 2008;**2008**(36):6213-6224. DOI: 10.1002/ejoc.200800680

[48] Trost BM, Brickner SJ. Palladium-assisted macrocyclization approach to cytochalasins: A synthesis of antibiotic A26771B. Journal of the American Chemical Society. 1983;**105**:568-575. DOI: 10.1021/ja00341a043

[49] Toshima K, Tatsuta K. Recent progress in O-glycosylation methods and its application to natural products synthesis. Chemical Reviews. 1993;**93**: 1503-1531. DOI: 10.1021/cr00020a006

[50] Danishefsky SJ, Bilodeau MT. Glycals in organic synthesis: The evolution of comprehensive strategies for the assembly of oligosaccharides and glycoconjugates of biological consequence. Angewandte Chemie International Edition in English. 1996; **35**:1380-1419. DOI: 10.1002/anie.199613801

[51] Wright AE, Botelho JC, Guzman E, Harmody D, Linley P, McCarthy PJ, et al. Neopeltolide, a macrolide from a lithistid sponge of the family Neopeltidae. Journal of Natural Products. 2007;**70**:412-416. DOI: 10.1021/np060597h

[52] Altmann KH, Carreira EM. Unraveling a molecular target of macrolides. Nature Chemical Biology. 2008;**4**:388-389. DOI: 10.1038/nchembio0708-388

[53] Ghosh AK, Shurrush KA, Dawson ZL. Enantioselective total synthesis of macrolide (+)-neopeltolide. Organic & Biomolecular Chemistry. 2013;**11**: 7768-7777. DOI: 10.1039/C3OB41541D

[54] Paterson I, Miller NA. Total synthesis of the marine macrolide (+)-neopeltolide. Chemical Communications. 2008;**2008**(39): 4708-4710. DOI: 10.1039/B812914B

[55] Ulanovskaya OA, Janjic J, Suzuki M, Sabharwal SS, Schumacker PT, Kron SJ, et al. Synthesis enables identification of the cellular target of leucascandrolide A and neopeltolide. Nature Chemical Biology. 2008;**4**:418-424. DOI: 10.1038/nchembio.94

[56] Bao J, Xu XY, Zhang XY, Qi SH. A new macrolide from a marine-derived fungus *Aspergillus* sp. Natural Product Communications. 2013;**8**:1127-1128. DOI: 10.1177%2F1934578X1300800825

[57] Jena BK, Reddy GS, Mohapatra DK. First asymmetric total synthesis of aspergillide D. Organic & Biomolecular Chemistry. 2017;**15**:1863-1871. DOI: 10.1039/c6ob02435a

[58] Evans DA, Ripin DHB, Halstead DP, Campos KR. Synthesis and absolute assignment of (+)-miyakolide. Journal of the American Chemical Society. 1999; **121**:6816-6826. DOI: 10.1021/ja990789h

[59] Smith AB III, Chen SS-Y, Nelson FC, Reichert JM, Salvatore BA. Total syntheses of (+)-acutiphycin and (+)-trans-20,21-didehydroacutiphycin. Journal of the American Chemical Society. 1997;**119**:10935-10946. DOI: 10.1021/ja972497r

[60] Moslin RM, Jamison TF. Highly convergent total synthesis of (+)-acutiphycin. Journal of the American Chemical Society. 2006;**128**: 15106-15107. DOI: 10.1021/ja0670660

6

Torrefaction of Sunflower Seed: Effect on Extracted Oil Quality

Jamel Mejri, Youkabed Zarrouk and Majdi Hammami

Abstract

The aim of this work is to study the effect of heat treatment on the lipidic profile of sunflower seed oil. It determined and compared the contents of bioactive components in seed oils extracted with n-hexane (Soxhlet method) from raw and roasted sunflower. The influence of torrefaction on fatty acid composition, triglyceride composition, and peroxide value (PV) has been studied. Thermal oxidation assays were carried out, and samples were evaluated by measuring induction time. Oleic acid was the main unsaturated fatty acid. Concerning triglyceride composition, OOL + LnOO, OOO + PoPP, POP and OOO + PoPP, OOL + LnOO, POP were the main, respectively, for raw and roasted samples. The seed oil samples extracted from the roasted sample exhibited a higher peroxide value (213.68 meq.O_2/kg) than the raw sample (5.79 meq.O_2/kg). The acid values were, respectively, 3.24 and 1.81 mg of KOH/g of oil for roasted and raw samples. On the other hand, induction time for raw sample was higher (16.23 h) than the roasted sample one (2.67 h).

Keywords: torrefaction, sunflower, seed oil, oxidation

1. Introduction

Lipids are major components of a man's diet. Their high quantities may be found in plant seeds distributed in many regions of the world. They can provide oils with a high concentration of monounsaturated fatty acids that prevent cardiovascular diseases by several mechanisms [1]. Several oleaginous seeds exist in the world. Some seeds are eaten as they are, such as sunflower seeds; others are used in the extraction of oil [2]. Sunflower (*Helianthus annuus* L.) is cultivated for its seeds' high oil content. Oil represents up to 80% of its economic value [3]. Abd EL-Satar et al. [4] concluded from their works that wider plant spacing and increasing nitrogen fertilization levels in addition to cultivars with high yield potential increase the plant's ability to take the needs of nutrients and solar radiation; this leads to an increase in photosynthesis, which reflected the increasing economic yield. Solvent extraction is one of the traditional techniques of extracting vegetable oil from oil seeds. Oil seeds are put in contact with a suitable solvent, in its pure form, for extracting the oil from the solid matrix to the liquid phase [5]. In many cases, chemical studies that employ a series of chemical compounds and/or sensory descriptors are used to characterize edible oil and fats [6]. In Tunisia roasted sunflower seeds, called "glibettes," are frequently consumed. Roasting enhances the organoleptic characteristics of seeds and gives them a taste and a pleasant smell. A huge number of papers on studies of different oils and fats are published every year. However, the effect of this heat treatment on the composition and nutritional qualities has not been studied.

There is no published work. The main objective of this study was to determine the TG, total FA composition, peroxide value (PV), acid value, and oxidative stability of the sunflower seed oil before and after torrefying. This study can be used to understand the causes of certain diseases related to the consumption of oxidized fat.

2. Experimental

2.1 Sunflower seed samples

Sunflower seeds (*Helianthus annuus* L.) are grown in Beja region (latitude 36°43′32″; longitude 9°10′54″; elevation 248 m), located in the northwest of Tunisia. After harvesting the seeds are stored in a dry place at room temperature, protected from light. Then the seeds were roasted at an artisan (called Hammas). The temperature and processing time are, respectively, 180°C and 10 min. Sunflower seeds were placed in a bowl and covered with salted water. Thus, they will absorb some of the water and will not dry too much during cooking. Seeds were drained and salted water was emptied. The oven was preheated to about 180° C. The seeds were arranged in a thin layer on the plate for better cooking. Seeds were baked and broiled for about 10 min. Occasionally, seeds were stirred in order to grill them evenly. Seeds may develop a slight crack in the middle during torrefaction. The still hot seeds were cooled and stored in an airtight box.

2.2 Seed oil extraction

The fat content was measured with a Soxhlet extractor apparatus with 250 ml of hexane at 60°C for 6 h, and then the solvent was removed by evaporation. The seed oil obtained was drained under a nitrogen stream (N_2) and was then stored in a freezer at −20°C until analysis.

2.3 Fatty acid composition

Fatty acid composition was determined by the analytical methods described by the European Parliament and the European Council in EEC regulation 2568/91 (1991) [7]. Fatty acids were converted to fatty acid ethyl esters (FAMEs) before being analyzed by shaking off a solution of 0.2 g of oil and 3 ml of hexane with 0.4 ml of 2 N methanolic potassium hydroxide. The FAMEs were then analyzed in a Hewlett-Packard model 4890D gas chromatograph furnished with an HP-INNOWAX-fused silica capillary column (cross-linked PEG), 30 m × 0.25 mm × 0.25 μm, and a flame ionization detector (FID). Inlet and detector temperatures were held at 230 and 250°C, respectively. The initial oven temperature was held at 120°C for 1 min, and then it was raised to 240°C at a rate of 4.0°C/min for 4 min. The FAME-injected volume was 1 μl, and nitrogen (N_2) was used as the carrier gas at 1 ml/min with a split inlet flow system at a 1:100 split ratio. Next, heptadecanoic acid C17:0 was added as an internal standard before methylation in order to measure the amount of fatty acids. Eventually, fatty acid contents were calculated using a 4890A Hewlett-Packard integrator.

2.4 Triacylglycerol composition

Triacylglycerol in different samples were determined according the International Olive Council [8]. The chromatographic separation of TAGs was performed using an Agilent 1100-reverse phase high-performance liquid chromatography (HPLC) system (Agilent Technologies, Waldbronn, Germany) equipped with an Inertsil ODS-C18 (5 µm, 4.5 × 250 mm) column. Elution was performed by using the mixture of acetonitrile/acetone (50:50, v/v) at a flow rate of 1 mL/min at 30°C. The working solutions of triacylglycerols (1%, w/v) were prepared in the elution mixture and injected into the column to determine their specific retention times. Identification of the peaks was carried out using a soybean oil chromatogram as reference. The mean of the data was calculated from three biological repeats obtained from three independent experiments.

2.5 Peroxide value, acid value, and thermal oxidation

Official methods of the American Oil Chemists' Society [9] were used for the determination of the peroxide value (method Cd 8-53) and the acid value (method Cd 8-53). The oxidative stability of the oils was determined using a Rancimat 743 Metrohm apparatus (Metrohm Co., Basel, Switzerland). This instrument was used for automatic determination of the oxidation stability of oils and fats. The level of stabilization was measured by the oxidative-induction time using 3.5 ± 0.01 g samples of oils. The temperature was set at 100°C, the purified airflow passing through at a rate of 10 l/h. During the oxidation process, volatile acids were formed in the deionized water and were measured conductometrically [10]. Samples of oils were placed in the apparatus and analyzed simultaneously. The samples were placed at random. The induction times were recorded automatically by the apparatus' software and taken as the break point of the plotted curves [11].

3. Results and discussions

3.1 Yield oil

The extraction yields are 43 and 52%, respectively, for raw and roasted seeds. Thus, we get a gain in yield of 9%. This gain is due to the roasting. Hydrolytic and proteolytic enzymes disrupt the structure of the cell and improve extraction yields. Oil yield depends on the cell disruption during the extraction process. Oil was located inside the cell. Various factors can influence the efficiency of the extraction process such as size of the solid particles, agitation, ratio of liquid/solid, extraction duration, pH, and temperature. Since the optimal temperature value coincides with the optimum protein degradation value, extraction of oil can be considered as a process aimed at degrading proteins which results in the release of the oil. However, the quality of the oil obtained depends on the operating conditions of extraction [12]. The yield extraction can be improved using other methods such as the Folch method. Hence, oils extracted using polar solvents such as a combination of chloroform and methanol may cause extraction of polar materials (phospholipids). In addition, neutral triacylglycerols can affect the oil yield extraction [1]. The effect of extraction time and temperature can also be significant for oil yield. However, several researchers have studied aqueous extraction of oil from sunflower. Evon et al. [3] have studied the feasibility of an aqueous process to extract sunflower seed oil using a corotating twin-screw extruder. The best oil extraction yield obtained was approximately 55%.

3.2 Fatty acid composition

Table 1 shows fatty acid composition of sunflower seed oil compared to those of literature. Oleic, linoleic, palmitic, and stearic acids were found as major fatty acids of sunflower seed oils. Their contents are 46.64, 38.11, 8.81, and 5.48%, respectively, for the raw sunflower seed. According to the work of [14], this composition depends on the environmental conditions during grain filling. The main environmental factors driving oil fatty acid composition are temperature and solar radiation. For oil quality purposes, oleic and linoleic are the most important fatty acids because they constitute almost 85% of the total fatty acids in sunflower oil. Sunflower fatty acid composition has been modified by breeding and mutagenesis parameters for minimum and maximum oleic acid percentage [15]. The roasted sunflower seed fatty acid contents were found to be 44.91, 36.95, 9.13, and 7.26%, respectively, for oleic, linoleic, palmitic, and stearic acids. Linoleic acid is the fatty acid most susceptible to degradation in sunflower oils [16]. The high amount of linoleic acid present in sunflower seed oil can make it more susceptible to oxidation and consequently cause higher cytotoxicity due to the production of free radicals. Diminution of unsaturated fatty acid was detected, caused by thermal treatment. Two news fatty acids appear: arachidic (0.91%) and behenic acid (0.83%). These fatty acids were detected in sunflower seeds in low amount [12]. They were 0.23 and 1.35%, respectively, for arachidic and behenic acid. Authors confirmed that the amount of arachidic and behenic acid were, respectively, 0.33 and 0.52% [17].

Sunflower seed oil is very nutritional because of its oleic acid content. The oleic acid content is varied: 46.64% in our study, 85.8% in [12], and 24.86% in [13]. It showed that fatty acid composition is highly variable [16, 18]. The palmitic acid, oleic acid, and linoleic acid contents ranged, respectively, from 5.3 to 27.9%, 31.6 to 84%, and 2.4 to 56.8%. Sunflower seed oil was fully liquid at the ambient temperature, as it is very rich in monounsaturated (oleic) and polyunsaturated (linoleic) fatty acids. Sunflower seed oil gives better functional properties such as good spreadability at refrigeration temperatures because of its high content of PUFA [19].

		The present study		**[12]**	**[13]**
Fatty acid content (%)	**Symbol**	**Raw**	**Roasted**		
Myristic acid	C14:0			0.05	—
Palmitic acid	C16:0	8.81	9.13	3.48	0.068
Palmitoleic acid	C16:1	0.45	—	—	6.12
Stearic acid	C18:0	5.48	7.26	3.65	3.41
Oleic acid	C18:1	46.64	44.91	85.8	24.86
Linoleic acid	C18:2	38.11	36.95	4.96	63.18
Linolenic acid	C18:3	0.51	—	—	0.082
Arachidic acid	C20:0	—	0.91	0.23	—
Behenic acid	C22:0	—	0.83	1.46	—
Lignoceric acid	C24:0	—	—	0.30	—
SFA		**14.29**	**18.13**	**9.17**	**3.478**
MUFA		**47.09**	**44.91**	**85.80**	**30.98**
PUFA		**38.60**	**36.65**	**4.96**	**63.262**
PUFA/SFA		**2.70**	**2.03**	**0.54**	**18.18**

SFA, saturated fatty acid; MUFA, monounsaturated fatty acid; PUFA, polyunsaturated fatty acid. Bold entries are to express the sum.

Table 1.
Fatty acid composition of sunflower seed oil.

3.3 Triglyceride composition

The compositions of triglycerides (TGs) expressed as the equivalent carbon number (ECN) found in sunflower seed oil samples are reported in **Table 2**. The main triglycerides found in the sunflower seed oil samples analyzed were OOL + LnOO, OOO + PoPP, POP and OOO + PoPP, OOL + LnOO, POP, respectively, for raw and roasted samples. These accounted for more than 62 and 66% of the total area of peaks in the chromatogram, respectively, for raw and roasted samples.

The level of OOL + LnOO, OOO + PoPP, the main TG in sunflower seed oil samples, was remarkably high, with a concentration of 25.90, 24.50 and 21.30, and 26.90%, respectively, for raw and roasted samples. The OOL + LnOO content of raw sunflower seed oil is greater than that in the roasted sample. However, the OOO + PoPP content is lower in the raw sunflower seed oil one. The next three TG fractions are POP, OOLn + PLL, and SOL with contents of 11.91, 10.80, and 10.34% and 18.17, 7, and 9.62%, respectively, for raw and roasted samples.

3.4 Peroxide value, acid value, and thermal oxidation

Peroxide value is a measure of the concentration of peroxides and hydroperoxides formed in the initial stages of lipid oxidation. Peroxide value is one of the most widely used tests for the measurement of oxidative rancidity in oils and fats [20]. The quality parameters of a crude oil included (i) the acid value, expressed in mg of KOH/g of oil, which is an indication of the free fatty acid content of the oil, and (ii) the peroxide value, expressed in terms of meq.O_2/kg of oil [21]. The results of peroxide value, acid value, and Rancimat test are shown in **Table 3**. Peroxide value increases considerably from 5.79 to 213.68 meq.O_2/kg, respectively, for raw and roasted oil samples. This is due to the high linoleic acid content, which is the fatty acid most susceptible to degradation in sunflower oils. Thermal oxidation assays of

TAG	ECN	Raw	Roasted
LLL	ECN 42	0.30	0.98
PoLL + OLLn + PoOLn	ECN 42	0.28	0.27
PLLn	ECN 42	0.51	0
OLL + PoOL	ECN 44	0.15	0.17
OOLn + PLL	ECN 44	10.80	7.00
PPLn + PPoPo	ECN 44	0.20	0
OOL + LnOO	ECN 46	25.90	21.30
PoOO	ECN 46	5.00	4.12
OOO + PoPP	ECN 48	24.50	26.90
SOL	ECN 48	10.34	9.62
POO	ECN 48	0.64	0.67
POP	ECN 50	11.91	18.17
SOO	ECN 50	4.21	3.00
POS + SLS	ECN 50	4.26	7.77

P, palmitic; Po, palmitoleic; S, stearic; O, oleic; L, linoleic; Ln, linolenic; and A, arachidic acids.

Table 2.
Triacylglycerol composition of sunflower seed oil.

Sample	Peroxide value (meq.O_2/kg)	Acid value (mg of KOH/g of oil)	Induction time (h)
Raw	5.79	1.81	16.23
Roasted	213.68	3.24	2.67

Table 3.
Peroxide value and oxidative stability of sunflower seed oil.

sunflower seed oil were carried out. The new compounds formed were evaluated [16]. Results showed that the levels of all the new compounds analyzed strongly depended on the degree of oil unsaturation and unsaturated oils with low content of linoleic acid, and high content of palmitic acid behaved exceptionally well. The linoleic acid is most susceptible to polymerization. The saturated fatty acids show a great importance in delaying oil polymerization [16].

The acid value (AV) expresses the extent of hydrolytic changes in the sunflower oils. The acid values were 1.81 mg of KOH/g of oil for the raw sample and 3.24 mg of KOH/g of oil for the roasted one. This increase of acid value indicates that TG hydrolysis occurred during the heat treatment. However, it can be consider that the operating conditions did not change oil quality significantly. The acid value remained stable at less than 3.5 mg of KOH/g of oil. The characteristic of crude sunflower oil based on specification from the American Fats and Oils Associations shall be pure with free fatty acid of 3% maximum or acid value below 6 mg of KOH/g of oil [21]. It showed that the feedstock sunflower oils possessed high free fatty acid [22]. Hydrolysis reactions of triglyceride with enzymatic and chemical pathways produce the free fatty acid (FFA). FFA is one of the important quality parameters. The formation of free fatty acid chain due to hydrolysis may lead to sensorial characterization [23]. The stability of sunflower seed oil expressed as the oxidation induction time was about 2.67 and 16.23 h, respectively, for raw and roasted seeds. This value may be justified by the high contents of MUFA and PUFA [24, 25]. Induction time values were quite different according to the oil composition (degradation), in proportion to the heat treatment. A high oxidation stability (33–45 h) of date seed oil measured by Rancimat was justified by the relatively low content of PUFA and the high content of natural antioxidants, such as phenolic compounds. Authors indicated that the species containing linoleic acid were oxidized more rapidly than those containing oleic acid [24, 26]. TAG polymers are the most characteristic compounds formed at high temperature, their rate of formation being dependent on the content of polyunsaturated fatty acids [27].

4. Conclusion

From the results and discussion of the study conducted, it can be concluded that the operating condition of torrefaction had an important influence on the oil extraction yield and the quality of oil extracted. Higher oil extraction yield was reached with increased temperature (torrefaction). The oil extraction yield of 52% was obtained under operating conditions of 180°C and 10 min. However, torrefaction process produced oil of bad quality. Changes of fatty acid composition, triglyceride composition peroxide value, acid value, and oxidative stability were observed. During torrefaction process oxide species were produced under the effect of high temperature. Thus, we can understand some diseases appeared to the customer of roasted sunflower seed (glibettes).

Author details

Jamel Mejri[1*], Youkabed Zarrouk[2] and Majdi Hammami[3]

1 Higher School of Engineers of Medjez el Bab (ESIM), University of Jendouba, Tunisia

2 Field Crops Laboratory, National Agronomic Research Institute of Tunisia (INRAT), Tunisia

3 Aromatic and Medicinal Plants Laboratory, Biotechnology Center of Borj-Cedria Hammam-Lif, Tunisia

*Address all correspondence to: jamel.mejri.faq@gmail.com

References

[1] Kozłowska M, Gruczynska E, Scibisz I, Rudzinska M. Fatty acids and sterols composition and antioxidant activity of oils extracted from plant seeds. Food Chemistry. 2016;**213**:450-456

[2] Bilgic E, Yaman S, Haykiri-Acma H, Kucukbayrak S. Limits of variations on the structure and the fuel characteristics of sunflower seed shell through torrefaction. Fuel Processing Technology. 2016;**144**:197-202

[3] Evon P, Vandenbossche V, Pontalier PY, Rigal L. Direct extraction of oil from sunflower seeds by twin-screw extruder according to an aqueous extraction process: Feasibility study and influence of operating condition. Industrial Crops and Products. 2007;**26**:351-359

[4] Abd EL-Satar MA, Abd-EL-Halime Ahmed A, Ali Hassan TH. Response of seed yield and fatty acid compositions for some sunflower genotypes to plant spacing and nitrogen fertilization. 2017;**4**:241-252

[5] Dutta R, Sarkar U, Mukherjee A. Extraction of oil from crotalaria Juncea seeds in a modified Soxhlet apparatus: Physical and chemical characterization of a prospective bio-fuel. Fuel. 2014;**116**:794-802

[6] Aranda F, Gomez-Alonso S, Rivera Del Alamo RM, Salvador MD, Fregapane G. Triglyceride, total and 2-position fatty acid composition of Cornicabra virgin olive oil: Comparison with other Spanish cultivars. Food Chemistry. 2004;**86**:485-492

[7] EEC. Characteristics of olive and olive pomace oils and their analytical methods. Regulation EEC/2568/1991. Official Journal of the European Communities. 1991;**L248**:1-82

[8] Aued-Pimentel S, Takemoto E, Kumagai EE, Cano CB. Calculation of the difference between the actual and theoretical ECN 42 triacylglyceride content to detect adulteration in olive oil samples commercialized in Brazil. 2008;**31**:31-34

[9] AOCS, editor. Official Methods and Recommended Practices of the American Oil Chemist's Society. 5th ed. Champaign, USA: AOCS Press; 1997

[10] Halbault L, Barbé C, Aroztegui M, De La Torre C. Oxidative stability of semisolid excipient mixtures with corn oil and its implication in the degradation of vitamin A. International Journal of Pharmaceutics. 1997;**147**:31-41

[11] Symoniuk E, Ratusz K, Krygier K. Kinetics parameters of refined and cold-pressed rapeseed oils after oxidation by Rancimat. Italian Journal of Food Science. 2017;**29**:276-287

[12] Amalia Kartika I, Pontalier PY, Rigal L. Extraction of sunflower oil by twin screw extruder: Screw configuration and operating condition effects. Bioresource Technology. 2006;**97**:2302-2310

[13] Suria R, Azwani ML, Siti AH. Changes on the solid fat content of palm oil/sunflower oil blends via interesterification. Malaysian Journal of Analytical Sciences. 2013;**17**:164-170

[14] Echarte MM, Puntel LA, Aguirrezabal LAN. Assessment of the critical period for the effect of intercepted solar radiation on sunflower oil fatty acid composition. Field Crops Research. 2013;**149**:213-222

[15] Angeloni P, Mercedes Echarte M, Pereyra Irujo G, Izquierdo N, Aguirrezábal L. Fatty acid composition of high oleic sunflower hybrids in a

changing environment. Field Crops Research. 2017;**202**:146-157

[16] Marmesat S, Morales A, Velasco J, Dobarganes MC. Influence of fatty acid composition on chemical changes in blends of sunflower oils during thermoxidation and frying. Food Chemistry. 2012;**135**:2333-2339

[17] de Mello Silva Oliveiraa N, Resendea MR, Moralesa DA, de ragão Umbuzeiroa G, Boriollo MFG. In vitro mutagenicity assay (Ames test) and phytochemical characterization of seeds oil of *Helianthus annuus* Linné (sunflower). Toxicology Reports. 2016;**3**:733-739

[18] Perez EE, Carelli AA, Crapiste GH. Temperature-dependent diffusion coefficient of oil from different sunflower seeds during extraction with hexane. Journal of Food Engineering. 2011;**105**:180-185

[19] Noor Lida HMD, Sundram K, Siew WL, Aminah A, Mamot S. TAG composition and solid fat content of palm oil, sunflower oil, and palm kernel olein blends before and after chemical interesterification. JAOCS. 2002;**79**:1137-1144

[20] Zhang Y, Yang L, Zu Y, Chen X, Wang F, Liu F. Oxidative stability of sunflower oil supplemented with carnosic acid compared with synthetic antioxidants during accelerated storage. Food Chemistry. 2010;**118**:656-662

[21] Amalia Kartika I, Pontalieb PY, Rigal L. Twin-screw extruder for oil processing of sunflower seeds: Thermo-mechanical pressing and solvent extraction in a single step. Industrial Crops and Products. 2010;**32**:297-304

[22] Saydut A, Erdogan S, Kafadar AB, Kaya C, Aydin F, Hamamci C. Process optimization for production of biodiesel from hazelnut oil, sunflower oil and their hybrid feedstock. Fuel. 2016;**183**:512-517

[23] Öğütcü M, Yılmaz E. Influence of different antioxidants and pack materials on oxidative stability of cold pressed poppy seed oil. La rivista italiana delle sostanze grasse. 2017;**XCIV**:45-52

[24] Besbes S, Blecker C, Deroanne C, Drira NE, Attia H. Date seeds: Chemical composition and characteristic profiles of the lipid fraction. Food Chemistry. 2004;**84**:577-584

[25] Rezig L, Chouaibi M, Msaada K, Hamdi S. Chemical composition and profile characterization of pumpkin (*Cucurbita maxima*) seed oil. Industrial Crops and Products. 2012;**37**:82-87

[26] Guinda Á, Dobarganes MC, Ruiz-Mendez MV, Mancha M. Chemical and physical properties of a sunflower oil with high levels of oleic and palmitic acids. European Journal of Lipid Science and Technology. 2003;**105**:130-137

[27] Dobarganes MC. Formation and analysis of high molecular-weight compounds in frying fats and oil. OCL. 1998;**5**:41-47

7

Synthesis of s- and p-Element Organosiloxanes by Mechanochemical Activation

Vitaliy Viktorovich Libanov, Alevtina Anatol'evna Kapustina and Nikolay Pavlovich Shapkin

Abstract

The interaction of some organosilicon compounds with oxides, hydroxides, and organic derivatives of s- (beryllium, magnesium, calcium) and p-elements (boron, aluminum, gallium, and tin) under conditions of mechanochemical activation was studied. Based on polyphenylsilsesquioxane and boric acid, the conditions for the synthesis of polyelementorganosiloxanes were selected, which included the activation time, carrier rotation speed, and the ratio of the nozzle mass to the payload mass. The influence of the nature of the heteroatom and the organic substituent of the heteroatom on the process has been studied. The effect of organic substituents at the silicon atom and the functionality of the organosilicon derivative on the mechanochemical interaction with acetylacetonates of boron difluoride and tin dichloride were also studied. The mechanisms of solid-phase reactions are proposed.

Keywords: mechanochemical activation, mechanisms of solid-phase reactions, organosilicon compounds, acetylacetonates of s- and p-elements

1. Introduction

Currently, the urgent task of chemistry is the search for new environmentally friendly methods for the synthesis of chemical compounds and materials based on them. One of these methods is mechanochemical, which excludes the use of solvents not only at the synthesis stage but in some cases when isolating the target product. In addition to the fact that mechanochemical activation excludes the use of solvents at the synthesis stage, the generated mechanical energy leads to the breaking of bonds and the creation of various radicals that cannot be formed in solution. Therefore, as a result of mechanochemical reactions, new compounds are formed that cannot be synthesized under the conditions of the use of solvents. This is due to the fact that most solvents interfere with the reagents or irreversibly bind to the product, changing their structure and reactivity.

In addition to the environmental component, this method has a number of economic advantages due to the short reaction time (in the case of using planetary activators, 3 minutes) and energy costs. Mechanochemical synthesis and mechanochemical activation are now widely used in various industries, for example, the production of catalysts [1], processing of materials, pharmaceuticals, utilization of carcinogenic organic derivatives, pyrite concentrates, etc. Mechanochemical activation can be used to decompose polyamides [2], to synthesize nanotubes [3], and to destroy toxic chlorine compounds, for example, DDT ([2,2-bis (4-chlorophenyl)-1,1,1-trichloroethane]) [4].

Recently, mechanochemical activation is often used in medicine and pharmacology, since the use of this method leads to an increase in the solubility of drugs and, as a consequence, to an increase in bioavailability.

It is promising to use the method of mechanochemical activation for the synthesis of organoelement, in particular organosilicon compounds.

Organosilicon compounds have a number of practically important properties, such as heat resistance, adhesion to metals, good electrical insulation characteristics, mechanical strength, and resistance to cold and water. Thus, organosilicon compounds containing indium ions have high catalytic activity [5] and containing magnesium atoms in their structure can be used to modify the surface of layered silicate to obtain an antifriction coating [6, 7]; siloxanes that simultaneously contain magnesium and titanium atoms in their structure are used as catalysts [8]. Siloxanes containing calcium ions in their structure are widely introduced into medicine [9, 10], while siloxane esters containing lithium salts are used in lithiumion batteries [11].

Organosilicon polymers are used in electronics and semiconductor instrumentation [12]. Siloxanes modified with boron and aluminum compounds increase the fire resistance and mechanical properties of materials [13], improve the hydrophobic qualities of materials [14, 15], and are used as heat-conducting composites [16]. Thymine-modified siloxanes can be widely used in optoelectronics and impart biological properties of materials [17]. The introduction of borsiloxanes into rubbers greatly increases their elasticity [18], and the high corrosion resistance of materials is acquired when tin atoms are introduced into siloxanes [19].

Possible applications of polymetallorganosiloxanes are also described in review [20].

2. The choice of synthesis conditions based on polyphenylsilsesquioxane and boric acid

The conditions for the synthesis of polyelementorganosiloxanes containing s- and p-atoms were selected on the basis of the reaction of polyphenylsilsesquioxane (PPSSO) and boric acid. This choice is due to two important points. First, due to the presence of hydroxyl groups and coordination of water in the initial organosilicon compound, a heterofunctional condensation reaction can occur, and secondly, boric acid itself can split the siloxane chain.

To select the conditions of mechanochemical synthesis, the following activation parameters were varied: synthesis time, number of mill turns (rotation speed), and ratio of nozzle mass to payload mass.

Mechanochemical activation was carried out in the planetary monomill "Pulverisette 6." A stainless steel beaker and grinding balls made of the same material were used as a reactor. The activation time ranged from 30 seconds to 7 minutes, the rotational speed of the mill drove from 100 to 600 rpm.

The estimated reaction scheme for the interaction of PPSSO and boric acid is as follows:

$$x(C_6H_5SiO_{1.5})n + nH_3BO_3 \rightarrow [(C_6H_5SiO_{1.5})x(BO_{1.5})]_n + {}_{1.5n}H_2O, \quad (1)$$

where the original x = 1, 2 or 3.

After milling, reaction mixture was extracted with toluene. As a result of the syntheses, two fractions were isolated. The toluene-insoluble fractions were gray-white powdery substances, which, according to elemental and X-ray phase analysis and IR spectroscopy, were unreacted boric acid. Soluble fractions of all syntheses according to gel permeation chromatography were high molecular weight compounds (≥ 5000).

Mass fractions of soluble fractions in polymers slightly increased with an increase in activation time. At the same time, a slight increase in the content of boron in the polymer chain occurred. The Si/B ratio in the soluble fractions differed from the predetermined one and was approaching (or equal to) to 2:1. By fractional precipitation from solutions of polymers, only one fraction was isolated, which indicates the homogeneity of the obtained compounds.

Since the obtained Si/B ratios were approximately equal to 2, a number of syntheses were carried out in which PPSSO and boric acid were introduced under similar conditions in the initial ratios of 2:1.

The Si/B ratios obtained in this case are close to the given value. This fact makes it possible to assume that the ratios obtained under mechanochemical activation conditions are determined by the nature of the heteroatom introduced into the siloxane chain to a greater extent than the initial Si/B ratio. The ratio Si/B = 2:1 may indicate the formation of a cyclolinear structure with stable six-membered boronsi-loxane fragments.

An increase in the initial Si/B ratio to 3:1 led to the formation of soluble fractions with a larger mass fraction, which is associated with an increase in the length of the siloxane fragment in the polymer chain.

Based on the data of elemental analysis, gel permeation chromatography, and IR spectroscopy, we can conclude that polymers consist of two types of structural units:

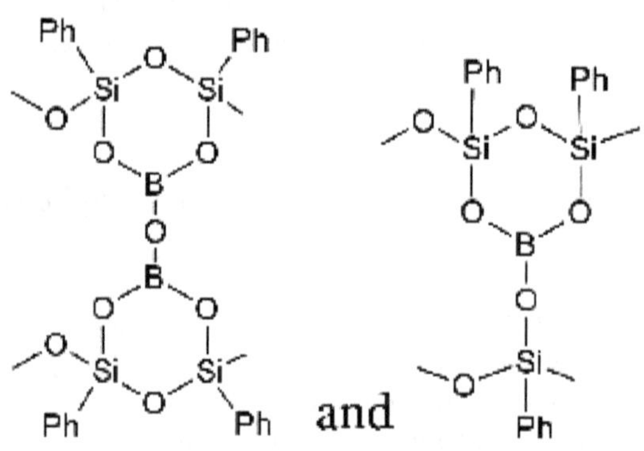

Since the optimal ratio for polyboronphenylsiloxanes obtained by mechano-chemical activation was 2:1, to study the influence of activation parameters for all subsequent syntheses, the initial Si/B ratios were taken to be 2:1.

To study the effect of activation time on the entry of boron into the polymer chain, we carried out syntheses in which the activation time was 30 seconds and 2, 4, 6, and 7 minutes.

As a parameter showing the entry of boron into the polymer chain, we used the concept of "degree of conversion," which was calculated by the formula:

$$\alpha = \frac{n_1}{n_2},$$

where n_1 is the amount of boron included in the polymer chain and n_0 is the initial amount of boron.

For greater clarity, the dependence of the degree of conversion on the activation time [$\alpha = f(t)$] is given in **Figure 1**.

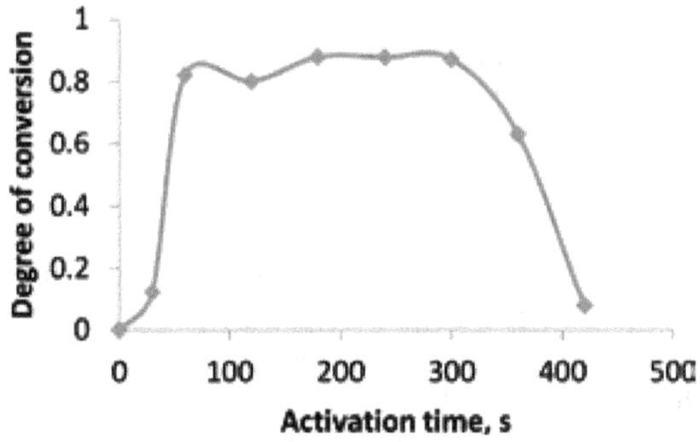

Figure 1.
The dependence of the degree of conversion on activation time.

Analyzing the obtained dependence, it can be seen that the degree of boron conversion remains practically unchanged in the time interval from 3 to 5 minutes. Therefore, in order to save energy, it is advisable to carry out syntheses precisely at 3 minutes of activation. After 5 minutes of activation, both the degree of conversion and the boron content in the polymer chain decrease, since an increase in the activation time leads to a break in the -O-B- bond in the Si-O-B fragment and the boron is removed from the polymer chain.

The entry of boron into the siloxane chain is also affected by such an important parameter as the rotational speed of the mill. With an increase in the rotation frequency, the mass fraction of the soluble fraction increases, and the degree of entry of boron into the polymer chain increases. Due to the fact that the mill mode at a frequency of 100–300 revolutions per minute is abrasive, under the influence of grinding media, the friction force between the balls and the reaction mixture increases, and, as a result, the local temperature increases. This can also explain the fact that, as a result of activation, the insoluble fraction is a mixture of metaboric acid, unreacted PPSSO, and silicon oxide formed as a result of the separation of the phenyl radical. The data obtained do not contradict to the previously described in the literature [21, 22]. **Figure 2** shows the dependence of the percentage of boron and the degree of conversion on the rotational speed of the mill.

With an increase in the ratio of the mass of the nozzle to the mass of the payload, an increase in the impact force from the grinding bodies occurs.

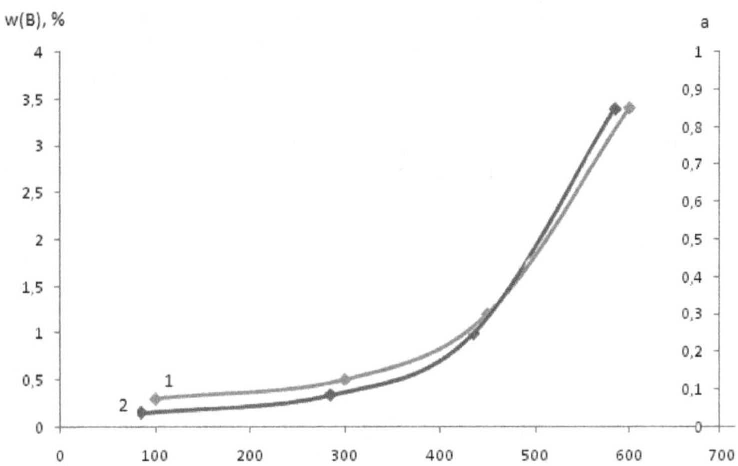

Figure 2.
Dependence of boron content w(B) (1) and degree of conversion a (2) on carrier speed.

An increase in the ratio of the mass of the nozzle to the mass of the payload to 3.04 leads to a decrease in the entry of boron into the polymer chain and to a decrease in the mass fractions of soluble fractions with an increase in the activation time. Apparently, an increase in impact force does not favor the entry of boron into the polymer chain.

Thus, the decisive factors on which the degree of conversion and the percentage of boron included in the polymer chain depend on are the rotational speed of the mill, the activation time, and the ratio of the nozzle mass to the payload mass. Moreover, the optimal conditions for the selected model reaction are conditions under which the synthesis time is 3 minutes, the ratio of the mass of the nozzle to the mass of the payload is 1.8, and the frequency is 10 Hz (600 revolutions per minute), providing shock operation.

3. Syntheses based on PPSSO and oxides of s- and p-elements

The synthesis of polyelementophenylsiloxanes based on PPSSO and oxides of s- and p-elements was carried out under the optimized conditions defined in Section 2 of this work. Beryllium, magnesium, and calcium oxides were used as starting compounds containing an s-element heteroatom.

The proposed reaction scheme for the interaction of PPSSO and alkaline earth metal oxides is as follows:

$$x(C_6H_5SiO_{1.5})_n + nMO \rightarrow [(C_6H_5SiO_{1.5})_x(MO)]_n, \quad (2)$$

where the original x = 1 or 2.

The beryllium atom did not enter the polymer chain at all, which can be explained by the extremely high energy of the crystal lattice and Gibbs energy. Beryllium oxide has a very high thermal conductivity, which at 100°C is 209.3 W/(mK). This can also explain the fact that the resulting high local temperature at the impact boundary of grinding media does not go to the excitation of radical ions but to the heating of the crystal.

The low percentage of magnesium entering the polymer chain can be explained from the standpoint described by Butyagin [23]. A magnesium oxide crystal consists of chemically inert Mg2+ and O2- ions with electron shells like noble gases. The energy of electrostatic interaction for small doubly charged ions is quite high, so the crystal of magnesium oxide is sufficiently strong and inert. The crystal lattice energy of this oxide is 3810 kJ/mol. In calcium oxide, this energy is not much less (3520 kJ/mol), which also explains its inertness in the reactions initiated by mechanical activation.

Thus, the synthesis of polymetallophenylsiloxanes based on PPSSO and oxides of beryllium, magnesium, and calcium by this method under the selected conditions is not possible.

The opposite picture is observed for oxides of p-elements.

The proposed reaction scheme for the interaction of PPSSO with oxides of elements of group XIII is as follows:

$$2x(C_6H_5SiO_{1.5})_n + nM_2O_3 \rightarrow 2[(C_6H_5SiO_{1.5})x(MO_{1.5})]_nM = B, Al, Ga \quad (3)$$

The relative mass fraction of soluble fractions with increasing time is first increased and then slightly decreased. The ratio of n Si/B in soluble fractions differed from the predetermined one and approached 2:1 values, which confirms our earlier assumption that stable six-membered rings are formed.

A comparison of the results obtained during the mechanochemical activation of PPSSO with boron oxide and similar syntheses into which boric acid was introduced showed that using boric acid, products of a composition closer to the specified one are obtained. Apparently, the passage of the process was facilitated by the presence of protons in the system. This assumption is consistent with the conclusions of the authors of the work, which showed that "acidic" protons in mechanochemical reactions diffuse, as the most mobile, to the surface oxygen atoms (Lewis main centers). The consequence of this is the formation of water molecules and new metal-oxygen-metal bonds.

The interaction of PPSSO with alumina under conditions of mechanochemical activation in a vibrating ball mill was first described by us [24].

In the planetary ball mill, three syntheses based on PPSSO and alumina were performed, which differ in the initial ratios Si/Al = 1:1, 2:1, and 3:1.

It was found that regardless of the initial ratio, soluble polyaluminophenylsiloxanes with a Si/Al ratio of approximately 4 are obtained, which corresponds to the optimal coordination number of the aluminum atom.

A comparison of the results obtained in the planetary-type activator with the results in a vibration mill shows that the use of a more energy-intensive activator, such as the planetary monomill, leads to an increase in the mass fraction of the soluble fraction and to obtain a higher ratio of silicon to aluminum in the obtained soluble products with the same initial ratio of reagents.

The introduction of a more basic gallium oxide into the reaction led to the formation of products with trace amounts of metal, which indicates the effect of the nature of the introduced oxide on the ability to cleave the siloxane bond. Thus, the cleavage of the siloxane bond under the action of an oxide under the conditions of mechanochemical activation occurs easier, with a higher acidity of the corresponding oxide.

Based on the synthesized PPSSO with oxides of alkaline earth metals, boron, aluminum, and gallium, we can conclude that the heteroatom incorporation into the siloxane chain decreases with an increase in the basic properties of the oxide and an increase in the ionicity and strength of the crystal lattice:

$$B > Al > Ga.$$

This conclusion is consistent with our conclusion [25], on the interaction of PPSSO with tin and germanium oxides under conditions of mechanochemical activation.

4. Study of the interaction of organosilicon compounds with different functionalities and boron difluoride acetylacetonate

In order to study the effect of the nature of the initial derivatives of the heteroatom on the process, boron difluoride acetylacetonate was used as the starting material. The initial Si/B ratios were 1:1, 2:1, and 3:1.

The choice of boron difluoride acetylacetonate is due to the great reactivity of this compound: in addition to reactive fluorine atoms, the interaction can be carried out at the hydrogen atom located in the gamma position of the acetylacetonate ring [26–28].

The proposed reaction scheme for the interaction of PPSSO and boron difluoride acetylacetonate is as follows:

$$x(PhSiO_{1.5})_n + nF_2BAcAc + n/_2H_2O \rightarrow [(PhSiO_{1.5})_{x-1}(PhSiFO)BOAcAc]_n + nHF \quad (4)$$

where the original x = 1, 2 or 3.

Linear polyboronphenylsiloxanes containing fluorine atom at silicon and acetylacetonate group at boron have been synthesized by mechanochemical activation. Mechanochemical activation of polyphenylsilsesquioxane and boron difluoride acetylacetonate taken into molar ratio of 1:1 was shown to lead to the formation of polyboronphenylsiloxane with the given Si/B ratio.

The increase of the starting ratio Si/B enhances side processes and formation of polydisperse products with the Si/B ratio different from the desired one.

For the initial ratio Si/B = 1:1 by various physicochemical methods of analysis, including MALDI TOF, it was found that the synthesis is carried out according to the following scheme:

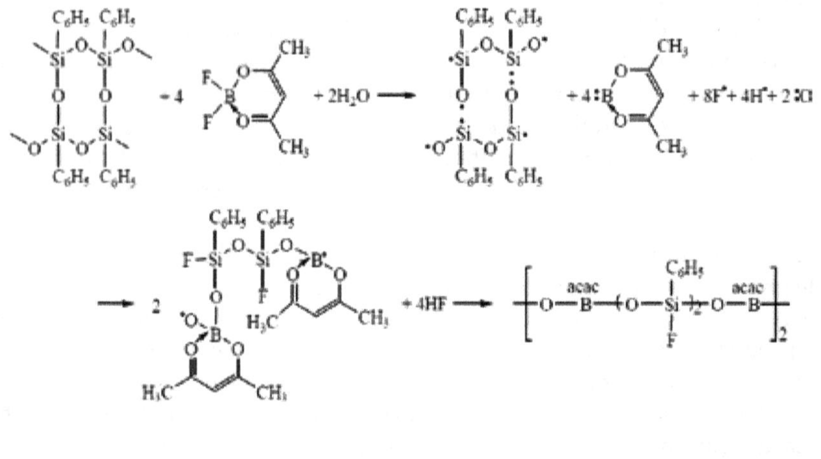

(5)

This synthesis was described in detail in previous work [29].

To study the effect of the functionality of the organosilicon derivative, diphenylsilanediol, triphenylsilanol, and octa(phenylsilsesquioxane) (Ph8T8 cubane) were used.

According to gel permeation chromatography, the soluble fraction of synthesis based on diphenylsilanediol is a low molecular weight substance with a relative molecular weight of 660. It was found that at the first stage of the synthesis, the starting materials condensed with the release of hydrogen fluoride, and then the reaction was carried out by hydroxide groups:

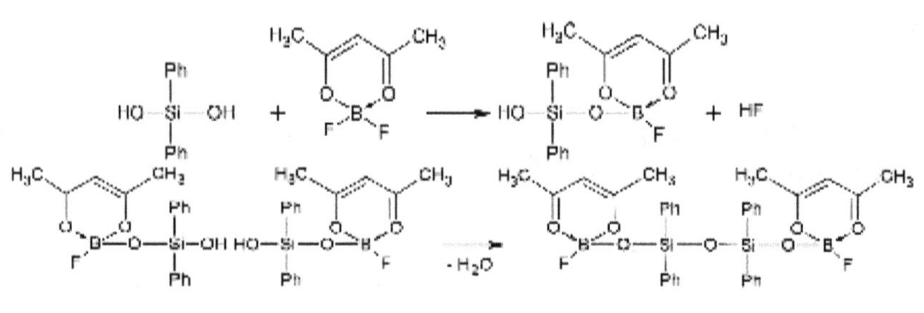

(6)

With a decrease in the functionality of the organosilicon derivative to unity (the use of triphenylsilanol), the formation of various degradation products of the starting compounds was observed. Only 1,3-difluoro-1,1,3,3-tetraphenyldisiloxane was isolated from a mixture of the resulting substances in an individual form, and the remaining substances were determined by chromato-mass-spectrometric analysis. Thus, tetraphenylsilane, triphenylfluorosilane, triphenylformylsilane, triphenylsilyl formic acid ester, and other products were discovered.

This synthesis was described in detail in previous work [30].

In continuation of the study of the influence of the functionality of the organosilicon compound on the course of mechanochemical activation, the

Ph8T8 cubane was introduced into the synthesis (it does not have functional reaction groups).

The interaction of the cubane with boron difluoride acetylacetonate did not occur due to the absence of water or hydroxyl groups, which were present in PPSSO, diphenylsilanediol, and triphenylsilanol and, according to the previously proposed scheme, taking part in the synthesis process.

The addition of the calculated amount of water to the "cubane-boron difluoride acetylacetonate" system resulted in the formation of a polymer product, the same as described for synthesis with PPSSO.

5. Interaction of PPSSO and acetylacetonates of alkaline earth metal and some p-elements

During the mechanochemical activation of PPSSO and beryllium, magnesium, and calcium oxides, the atoms of these elements did not enter the siloxane chain, which was associated with a high degree of ionicity of the corresponding oxides and, as a consequence, the high strength of the crystal lattice. Therefore, milder compounds, organic salts, and 2,4-pentanedionates were used.

In the case of the use of beryllium and magnesium acetylacetonates, the occurrence of a heteroatom did not occur. Apparently, during the activation process, these acetylacetonates were destroyed with the formation of stable oxides. However, when using a calcium salt, the heteroatom enters the siloxane chain (low metal percentage still remains). Thus, as in the interaction of PPSSO with alkaline earth metal oxides, an increase in the atom size facilitated the process of heterolytic polycondensation under these conditions, increasing the polarization of the compound and breaking of the bond in it.

When aluminum tris-acetylacetonate is introduced into the reaction, products with a Si/Al ratio of 3:1 are formed (compared with the synthesis based on aluminum oxide). In the case of using gallium tris-acetylacetonate as the starting compound, a polymer product with a given Si/Ga ratio is obtained; however, its yield is not high (less than 26%).

A comparison of the results of syntheses based on aluminum and gallium acetylacetonates shows that an increase in the basicity of the heteroatom leads to a decrease in the mass fraction of polyelementorganosiloxanes; however, the Si/M ratio is closer to the specified one.

As a result of the mechanochemical activation of PPSSO and bis(acetylacetonate) tin dichloride (initial ratio Si/Sn = 1:1), three fractions are distinguished. The toluene-insoluble fraction is tin and silicon oxides, as well as the unreacted starting components. Fraction 2, obtained by precipitation of hexane from a toluene-soluble fraction, is a polymer product. The resulting ratio of Si/Sn differs from the set and is 1.96:1:

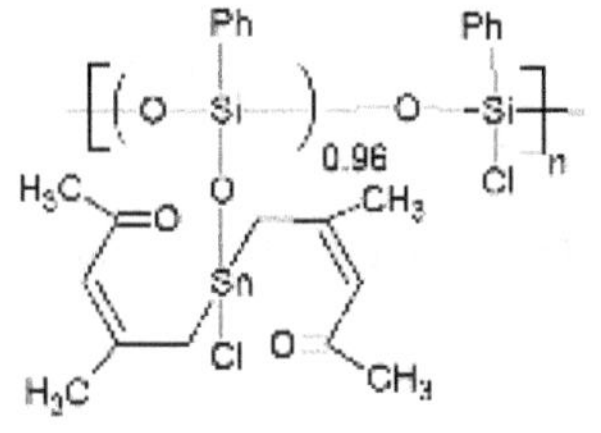

The third fraction is a monomeric compound corresponding to the following structural formula:

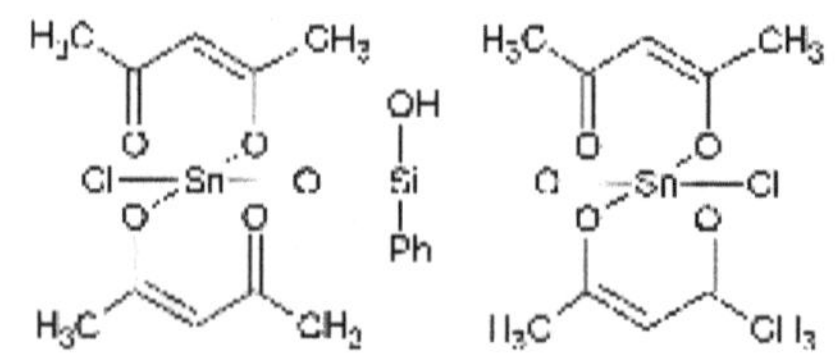

6. Conclusions

1. By the mechanochemical activation, low and high molecular weight elementoorganosiloxanes containing atoms of boron, aluminum, gallium, and tin can be synthesized. As starting materials, boron oxide, boric acid, boron difluoride acetylacetonate, aluminum and gallium acetylacetonates, and bis(acetylacetonate) tin dichloride can be used.

2. The composition and structure of the products obtained depend on the conditions of mechanochemical activation, as well as the initial ratio of the reacting substances.

3. An increase in the activation time and the ratio of the mass of the nozzle to the mass of the payload leads to the breaking of the siloxane bond and the removal of the heteroatom from the polymer chain by increasing the impact force from the grinding bodies.

4. During the mechanochemical activation of polyphenylsilsesquioxane and element oxides, the incorporation of the p-element heteroatom into the siloxane chain decreases in the series B > Al > Ga with an increase in the basic properties of the oxide, an increase in the ionicity and strength of the crystal lattice, and the s-element heteroatom in the series Ca > Mg > Be with a decrease in its radius and an increase in the strength of the crystal lattice of heteroatom oxide.

5. During the mechanochemical activation of polyphenylsilsesquioxane and organic derivatives of s- and p-elements, a heteropolycondensation reaction occurs, the result of which depends on the polarity of the M-O-C bond: when alkaline earth metal acetylacetonates are used as the starting compounds, the polycondensation ability increases with the size of the atom included in the corresponding metal acetylacetonate and, accordingly, bond polarity: $Ca(AcAc)_2 > Mg(AcAc)_2 > Be(AcAc)_2$.

6. The course of the mechanochemical process, as well as the composition and structure of the reaction products, depends on the functionality of the starting compounds:

   a. A decrease in the functionality of the organoboron derivative results in compounds with a lower Si/B ratio.

   b. A decrease in the functionality of the organosilicon derivative leads to its lower reactivity and the formation of low molecular weight reaction products.

7. The presence of hydroxyl groups and crystallization water affects the course of mechanochemical activation.

8. When using bifunctional halogen-containing organic derivatives of p-elements, the products of mechanochemical activation have a linear structure. Products of both polymer and monomeric nature are formed.

Author details

Vitaliy Viktorovich Libanov*, Alevtina Anatol'evna Kapustina and Nikolay Pavlovich Shapkin
Department of General, Inorganic and Elementorganic Chemistry, School of Natural Sciences, Far Eastern Federal University, Vladivostok, Russia

*Address all correspondence to: libanov.vv@dvfu.ru

References

[1] Isupova LA, Tsybulya SV, Kryukova GN, Alikina GM, Boldyreva NN, Vlasov AA, et al. Mechanochemical synthesis and catalytic properties of the calcium ferrite Ca2Fe2O5. Kinetics and Catalysis. 2002;**43**:122-129. DOI: 10.1023/A:1014217716883

[2] Tanaka Y, Zhang Q , Saito F. Mechanochemical decomposition of an aromatic polyamide film. Industrial and Engineering Chemistry Research. 2003;**42**:5018-5023. DOI: 10.1021/ie0300833

[3] Chen Y, Gerald J, Williams J, Willis P. Mechanochemical synthesis of boron nitride nanotubes. Journal of Metastable and Nanocrystalline Materials. 1999;2-6:173-178. DOI: 10.4028/www.scientific. net/MSF.312-314.173

[4] Hall AK, Harrowfield JM, Hart RJ, McCormick PG. Mechanochemical reaction of DDT with calcium oxide. Environmental Science & Technology. 1996;**30**:3401-3407. DOI: 10.1021/es950680j

[5] Adam Ahmed F. Indium incorporated silica from rice husk and its catalytic activity. Microporous and Mesoporous Materials. 2007;**103**:284-295. DOI: 10.1016/j.micromeso.2007.01.055

[6] Shapkin NP, Leont'ev LB, Leont'ev AL, Shkuratov AL. The control of forming of compositional wear resistance, metal-ceramic coatings on the surfaces of friction of details. Tehnicheskie nauki. 2012;**11**:630-635

[7] Shapkin NP, Leont'ev LB, Leont'ev AL, Makarov VN. Method of making antifriction composition. Sovremennie materialy i tehnologii. 2014;**4**:147-148

[8] Liu J-C. Mx-Oy-Siz bonding models for silica-supported Ziegler-Natta catalysts. Applied Organometallic Chemistry. 1999;**13**:295-302

[9] Nakamura J, Kasuga T. Mx-Oy-Siz bonding models for silica-supported Ziegler-Natta catalysts. Journal of Materials Chemistry B. 2014;**2**:1250-1254

[10] Yamada S, Ota Y, Nakamura J, Sakka Y, Kasuga T. Enhancement of crystalline plane orientation in silsesquioxane-containing vaterite particles towards tuning of calcium ion release. Journal of the Ceramic Society of Japan. 2014;**122**:1010-1015

[11] Ha J, Im S, Lee S, Jang H, Ryu T, Lee C, et al. Liquid type of fluorosulfonyl lithium salts containing siloxane for Li-ion electrolyte. Journal of Industrial and Engineering Chemistry. 2016;**37**:319-324

[12] Ton'shin AM, Kamaritskii BA, Spektor VN. The technology, structure, and properties of organosilicon dielectrics-polyorganosilasesquioxanes. Russian Chemical Reviews. 1983;**52**:775-803

[13] Khelevina OG. Modification of siloxane coatings with boron and aluminum compounds. Russian Journal of Applied Chemistry. 2012;**85**:277-284. DOI: 10.1134/S1070427212020218

[14] Kujawa J, Kujawski W, Koter S, Rozicka A, Cerneaux S, Persin M, et al. Efficiency of grafting of Al2O3, TiO2 and ZrO2 powders by perfluoroalkylsilanes. Colloids and Surfaces A: Physicochemical and Engineering Aspects. 2013;**420**:64-73

[15] Santiago A, Gonzalez J, Iruin J, Fernandez-Berridi MJ, Munoz ME, Irusta L. Urethane/siloxane copolymers with hydrophobic properties. Macromolecular Symposia (Conference Paper). 2012;**321-322**:150-154

[16] Im H, Kim J. Enhancement of the thermal conductivity of aluminum oxide-epoxy terminated poly(dimethyl siloxane) with a metal oxide containing polysiloxane. Journal of Materials Science. 2011;**46**:6571-6580

[17] Enea R, Apostol I, Damian V, Hurduc N, Iordache I. Photo-sensible (thymine containing) azo-polysiloxanes: Synthesis and light induced effects. Journal of Physics: Conference Series. 2008;**100**(Part 1):1-4

[18] Li X, Zhang D, Xiang K, Huang G. Synthesis of polyborosiloxane and its reversible physical crosslinks. RSC Advances. 2014;**4**:32894-32901. DOI: 10.1039/C4RA01877J

[19] Kunst SR, Ludwig GA, Santana JA, Sarmento VHV, Bertoli PP, Menezes TL, Ferreira JZ, Malfatti CF. Elaboration and characterization of hybrid films siloxane-PMMA prepared by sol-gel process on tin plates: Influence of pH sol Ciencia e Tecnologia dos Materiais. 2014;**26**: 33-38

[20] Zdanov AA, Sergienko NV, Trankina ES. Siloxane polymers with skeleton metallosiloxane fragments and their chemical transformations. Rossiyskiy himicheskiy zurnal. 2001;**25**:44-48

[21] Dubinskaya AM. Transformations of organic compounds under the action of mechanical stress. Russian Chemical Reviews. 1999;**68**:637-652. DOI: 10.1070/RC1999v068n08ABEH000435

[22] Dubinskaya AM, Streletskii AN. Formation of low-molecular compounds during mechanical degradation of macromolecules. Visokomolekulyarnie soedineniya. 1983;**24A**:1924-1930

[23] Butyagin PY. Structural disorder and mechanochemical reactions in solids. Russian Chemical Reviews. 1984;**53**:1025-1038

[24] Kapustina AA, Shapkin NP, Libanov VV, Guliaeva LI. The use of oxides and acids to obtain polyelementorganosiloxanes under conditions of mechanochemical activation [thesis]. Moscow, RAS. VIII International Symposium "Fundamental and Applied Problems of Science"; 2013;**4**: 42-51

[25] Kapustina AA, Shapkin NP, Ivanova EB, Lyakhina AA. Mechanochemical synthesis of poly(germanium and tin organosiloxanes). Russian Journal of General Chemistry. 2005;**75**:571-574

[26] Svistunova IV, Shapkin NP, My Z. Boron difluoride acetylacetonate sulfenyl(selenyl) halides. Russian Journal of General Chemistry. 2010;**80**:2430-2437. DOI: 10.1134/S1070363210120054

[27] Bukvetskii BV, Svistunova IV, Gelfand NA. Study of the crystal structure of binuclear boron difluoride acetylacetonate. Journal of Structural Chemistry. 2014;**55**(2):290-294. DOI: 10.1134/S0022476614020140

[28] Fedorenko EV, Tretyakova GO, Mirochnik AG, Beloliptsev AY, Svistunova IV, Sazhnikov VA, et al. Nitrogen-containing analog of bibenzoylmethanate of boron difluoride: Luminescence, structure, quantum chemical modeling, and delay fluorescence. Journal of Fluorescence. 2016;**26**:1839-1847. DOI: 10.1007/s10895-016-1876-2

[29] Kapustina AA, Shapkin NP, Libanov VV. Preparation of polyboronphenylsiloxanes by mechanochemical activation. Russian Journal of General Chemistry. 2014;**84**:1320-1324. DOI: 10.1134/S1070363214070123

[30] Libanov VV, Kapustina AA, Shapkin NP, Rumina AA. Mechanochemical interaction of boron difluoride acetylacetonate with organosilicon derivatives of different functionality. SILICON. 2019;**11**:1489-1495. DOI: 10.1007/s12633-018-9969-y

8

N,N-Dialkyl Amides as Versatile Synthons for Synthesis of Heterocycles and Acyclic Systems

Andivelu Ilangovan, Sakthivel Pandaram and Tamilselvan Duraisamy

Abstract

N,N-Dialkyl amides such as *N,N*-dimethylformamide (DMF), *N,N*-dimethylacetamide (DMA), are common polar solvents, finds application as a multipurpose reagent in synthetic organic chemistry. They are cheap, readily available and versatile synthons that can be used in a variety of ways to generate different functional groups. In recent years, many publications showcasing, excellent and useful applications of *N,N*-dialkyl amides in amination (R-NMe_2), formylation (R-CHO), as a single carbon source (R-C), methylene group (R-CH_2), cyanation (R-CN), amidoalkylation (-R), aminocarbonylation (R-$CONMe_2$), carbonylation (R-CO) and heterocycle synthesis appeared. This chapter highlights important developments in the employment of *N,N*-dialkyl amides in the synthesis of heterocycles and functionalization of acyclic systems. Although some review articles covered the application of DMF and/or DMA in organic functional group transformations, there is no specialized review on their application in the synthesis of cyclic and acyclic systems.

Keywords: amination, amidation, amidoalkylation, aminocarbonylation, cyanation, dialkyl amides, formylation, heterocycles

1. Introduction

The great advantage of DMF, DMA and other *N,N*-dialkylamides are their versatility as reaction medium, polar and aprotic nature, high boiling point, cheap and ready availability. DMF can react as electrophile or a nucleophile and also act as a source of several key intermediates and take a role in reactions as a dehydrating agent, as a reducing agents [1] or as a catalyst [2–5], stabilizer [6–10]. For the synthesis of metallic compounds DMF can be an effective ligand. *N,N*-dialkylamides could be considered as a combination of several functional groups such as alkyl, amide, carbonyl, dialkyl amine, formyl, N-formyl and highly polar C-N, C═O, and C-H bonds. Due to flexible reactivity of *N,N*-dialkylamides, during the past few years, chemists have succeeded in developing reactions, where DMF and DMA could be used to deliver different functional groups such as amino (R-NMe_2), formyl (R-CHO), methylene (R-CH_2), cyano (R-CN), amidoalkyl ($CH_2N(CH_3)$-C(═O) CH_3-R) aminocarbonyl(R-$CONMe_2$), carbonyl(R-CO), methyl (-Me), a single atoms such as C, O, H etc. (**Figure 1**). Similarly, DMF and DMA could be used in the

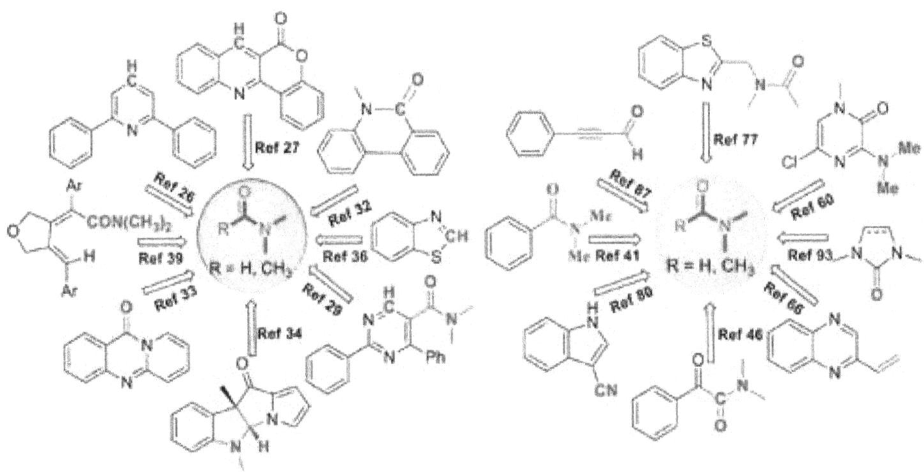

Figure 1.
DMF and DMA as a synthon for the various reactions.

preparation of heterocyclic compound through formylation of active methylene groups, conversion of methyl groups to enamines, and formylation of amino groups to amidines. Further, it can also be utilized as an intermediate in the modification of heterocyclic compounds [11].

A non-exhaustive seminal review by Muzart [1], highlighted different roles of DMF inorganic synthesis covered literature up to 2009, another comprehensive review by Ding and Jiao appeared in 2012 [12] which covered aspects of DMF as a multipurpose precursor in various reactions. Further, specialized review by Batra et al. [13], and other reviews dealing with recent applications of DMF and DMA as a reagent [14] and triple role of DMF as a catalyst, reagent and stabilizer also appeared [15].

In this book chapter we summarized developments on applications of DMF and DMA in reactions such as amination (R-NMe_2) [16], formylation (R-CHO) [17, 18], as a single carbon source (R-C), methylene group (R-CH_2) [19], carbonylation (R-CO), as well as newer reactions such as amidoalkylation (-$CH_2N(CH_3)$-C(=O) CH_3-R) [20], metal catalyzed aminocarbonylation (R-$CONMe_2$) [21], cyanation (R-CN) [22, 23], and formation heterocycles, took place during the past few decades and up to October 2019. Heterocycles are important compounds finding excellent applications as useful materials and medicinally important compounds. Thus unlike other reviews appeared on this subject [1, 12–15], we provided special emphasis on synthesis of heterocyclic compounds and reactions involving DMF and DMA. Thus, first part of this book chapter will cover synthesis of construction of cyclic system, especially heterocycles, the next part will cover the formation of open chain compounds. Although DMF can serve as a reagent in organic reactions such as Friedel-Crafts [24] and Vilsmeier-Haack [25] reactions the actual reagent is derivative of DMF, hence we did not cover such subjects. We hope this book chapter will stimulate further research interest on the application of DMF and DMA in organic synthesis.

2. DMF and DMA as synthon in synthesis of heterocycles

2.1 Construction of pyridine ring

Guan and co-workers reported synthesis of symmetrical pyridines from ketoxime carboxylates using DMF as a one carbon source in the presence of ruthenium catalyst and $NaHSO_3$ as an additive (**Figure 2**). A series of ketoxime acetates **2** reacted smoothly with DMF to give corresponding pyridine derivatives **3**. Replacement $NaHSO_3$ with other oxidants led to decrease in the yield. The reaction condition was optimized by use of various additives and catalysts. The desired product was obtained in good yield, in the presence of $NaHSO_3$, Ru(cod)Cl_2 and at 120°C. Both electron withdrawing and electron donating group attached to the aryl rings gave the corresponding symmetrical pyridines. But the yield decreased due to steric effect by the orthosubstituents.

A possible mechanism for the reaction was proposed. Oxidation of DMF by Ru (II) gives an iminium species **A** and Ru(0). Followed by which oxidative addition of ketoxime acetate to Ru(0) generates an imino-Ru(II) complex **B**, undergoes tautomerization to afford enamino-Ru(II) complex **C**. Then, nucleophilic addition of **C** to species **A** produces an imine intermediate **D**. Condensation of imine intermediate **D** with a second ketoxime acetate gives intermediate **E**. Nucleophilic substitution of **E** by $NaHSO_3$ followed by intramolecular cyclization of the intermediate **F** forms a dihydropyridine intermediate **G**. Finally, Ru-catalyzed oxidative aromatization of **G** by oxygen provided the product **H** [26].

Su et al., reported cyclisation of 4-(phenylamino)-2*H*-chromen-2-ones to give novel functionalized 6*H*-chromeno[4,3-b]quinolin-6-ones (**Figure 3**) in the presence of $Cu(OAc)_2.H_2O$/TBPB catalytic system (**Figure 3**). In this reaction, DMF served as the source of methine group.

The reaction proceeded smoothly with electron-donating and electron-withdrawing substituents on the aniline ring and the expected products were obtained in good yields. A plausible mechanism was proposed by the author in. Initially, DMF is converted into iminium ion **A** with the help of Cu/TBPB via radical pathway. Next, reaction of 4-(phenylamino)-2H-chromen-2-ones with active iminium ion **B** gives intermediate **C**. Further, removal of MeNHCHO group afforded **D** which is attacked by $NaHSO_3$ followed by an intramolecular cyclization to afford desired product **5** [27].

In 2015, Deng and co-workers reported the Ru catalyzed multi-component reaction of acetophenones **6**, ammonium acetate (N source) and DMF (one carbon source) to get 2,4-diarylsubstituted-pyridines **7** under O_2 atmosphere (**Figure 4**).

In this reaction DMF, in the presence of Ru/O_2 catalyst, acted as a single carbon source. For better understanding of reaction mechanism, several control experiments were carried out [28] (**Figure 4**). Acetophenone was converted into a methyl

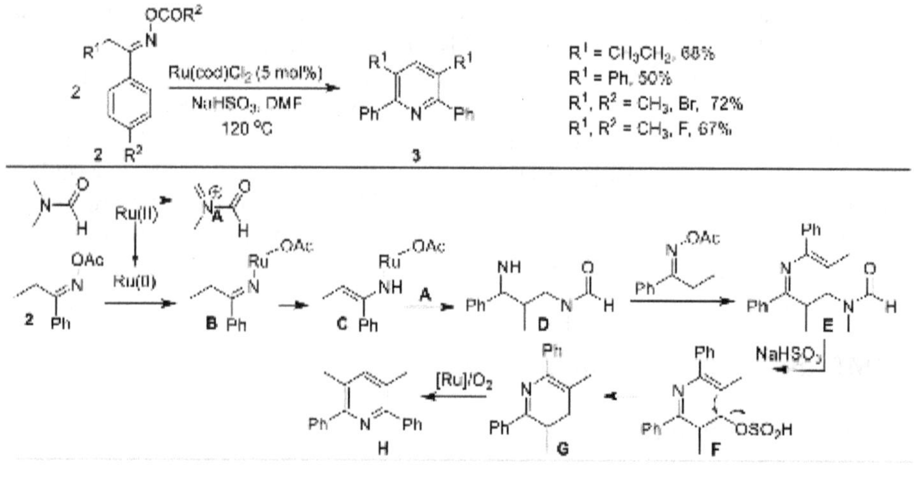

Figure 2.
Pyridine ring formation by DMF using Ru-catalyzed cyclization of aryl ethyl ketoxime acetates.

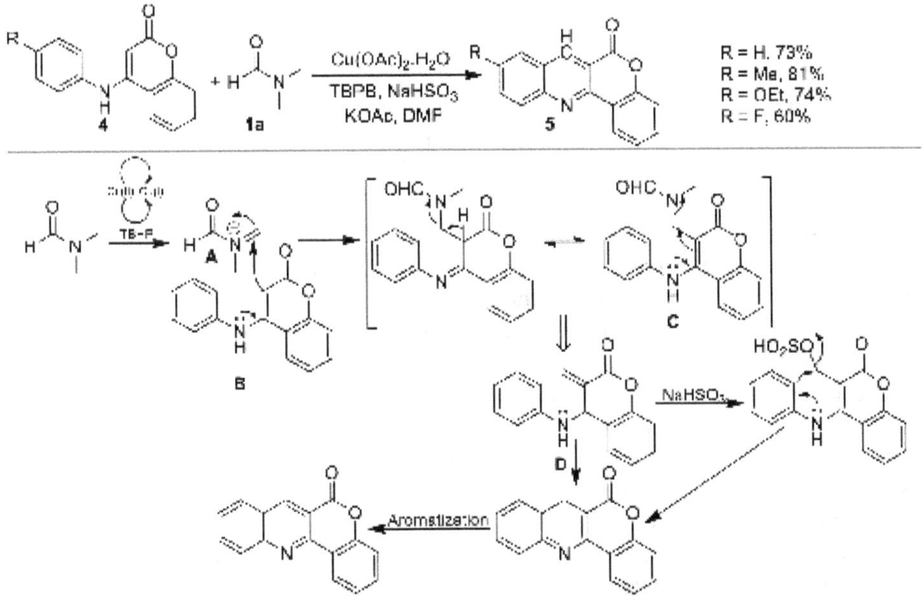

Figure 3.
DMF as a methine source in pyridine ring formation via cyclization of 4-(phenylamino)-2H-chromen-2-ones.

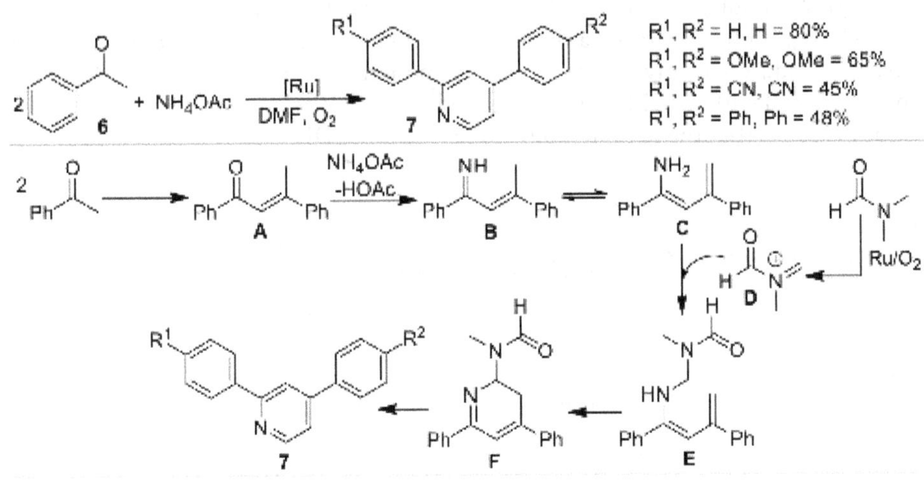

Figure 4.
Ru-catalyzed cyclization of acetophenones with NH4OAc.

ketene intermediate **A** by homo-condensation, which immediately converts into imine intermediate **B**, with the aid of NH_4OAc. Further, tautomerization of imine intermediates lead to the formation of intermediate **C**, which reacted smoothly with iminium species **D** to give intermediate **E** then this can be oxidized by Ru/O_2 to afford intermediate **F**, which further undergoes 6π electron cyclization followed by methylamide elimination to give the desired pyridine.

2.2 Construction of pyrimidine ring

Jiang and co-workers developed the first example of employing *N,N*-dimethylformamide (DMF) as a dual synthon, a one-carbon atom and amide source. A multi-component reaction between amidines **8**, styrene **9**, and *N,N*-dimethylformamide

(DMF) took place in the presence of palladium-catalyst (**Figure 5**) to form pyrimidine carboxamide **10**.

The desired product was obtained in good yield under the optimal reaction condition $Pd(TFA)_2$ (5 mol%), Xantphos (5 mol%) and 70% TBHP (3.0 equiv) in 1.0 mL DMF at 120°C. Benzamidine salts containing electron-releasing or electron-withdrawing group on the benzene ring gave their desired product in moderate to good yield. Addition of radical scavenger, such as TEMPO (2,2,6,6-tetramethylpiperidine-1-oxyl), BHT (2,6-di-tert-butyl-4-methylphenol), and DPE (1,1-diphenylethylene) led to no desired product formation, which indicates the radical pathway is involved in this transformation [29].

Xiong et al., reported a general and highly selective method for annulation of amidines **15** (**Figure 6**).

This is an efficient copper catalyzed synthesis of quinazolines **12** through C-N bond formation reaction between N-H bonds of amidines and $C(sp^3)$-H bond adjacent to sulfur or nitrogen atoms. In addition to DMF and DMA, DMSO, NMP and TMEDA could be used as solvent and as one carbon synthon [30]. This method avoids pre-functionalization of substrates.

In 2017, Fan et al., reported an efficient method for the synthesis of pyrimidines **13** from amidines **8** and ketones **12** through [3 + 2 + 1] type intermolecular cycloaddition reaction, under metal free condition (**Figure 7**). The reaction condition was optimized with different parameters and the suitable condition for multicomponent synthesis of pyrimidines was found to be, treatment of amidines (0.25 mmol), ketone (0.30 mmol), 70% TBHP (3.0 equiv), Cs_2CO_3 (2.0 equiv) in DMF (1.0 mL) at 120°C [31]. Both substituted amidines and substituted ketones worked well under standard condition to give pyrimidines in moderate to good yield. The reaction progressed well with d_7-DMF and the desired isotopic labeled product was obtained. This is evidence that the carbon atom comes from the DMF.

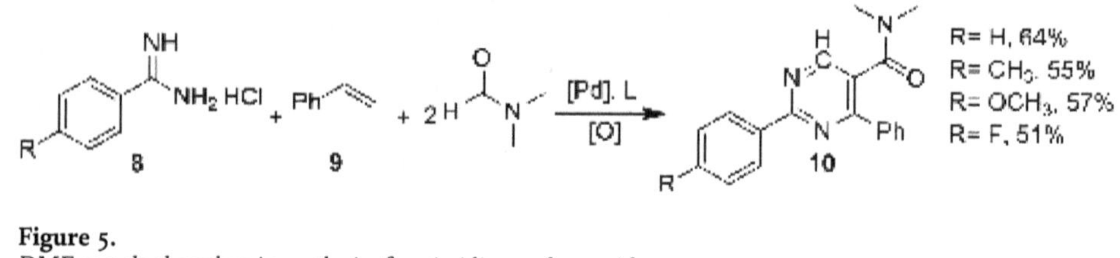

Figure 5.
DMF as a dual synthon in synthesis of pyrimidine carboxamide.

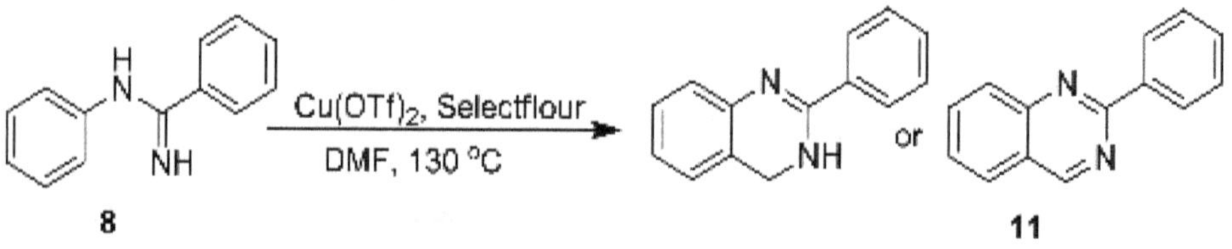

Figure 6.
DMF as a one carbon source in Cu-catalyzed annulations of amidines.

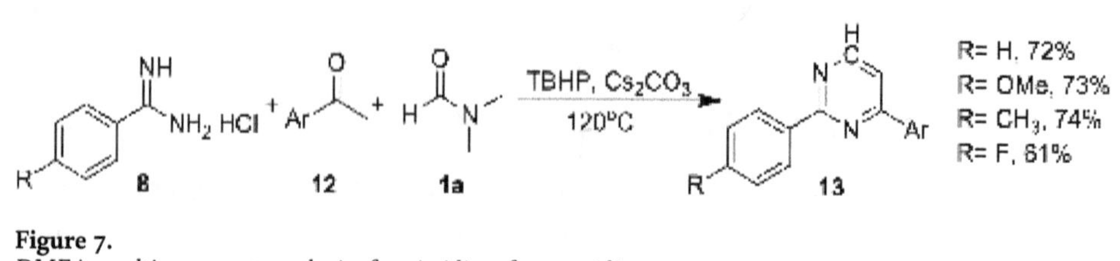

Figure 7.
DMF in multicomponent synthesis of pyrimidines from amidines.

2.3 Construction of quinazolinone ring

In 2016, Das et al., reported Pd/Ag catalyzed direct carbonylation of sp^2C-H bonds of **14** and **16** by employing DMF as one carbon source under oxygen for the synthesis of biologically important motifs pyrido-fused quinazolinone **15** and phenanthridinone **17**, respectively (**Figure 8**).

The reaction was examined using different metal catalyst systems such as Pd-Ag, Cu-Ag, Co-Ag, Ni-Ag and finally Pd-Ag catalytic system was found to be suitable for this transformation [32]. When labeled DMF (CO^{18}) was used as the solvent it has been found that product found not to contain O^{18}. From these results, it can be concluded that incorporated carbonyl group is coming from the methyl group of DMF. Reaction under argon instead of oxygen lead to the poor yield, which indicates "O" atom is coming from oxygen environment.

In 2015, Wu et al., reported C-H bond activation of arenes **14** followed by cyclization wherein DMF was used as the CO synthon, in the presence of $Pd(OAc)_2$-$K_2S_2O_8$ catalytic system under carbon monoxide atmosphere (**Figure 9**). The reaction works at autoclave free condition for the formation of *H*-pyrido[2,1*b*] quinazolin-11-ones **15**.

The reaction was optimized using different oxidant and catalysts under different temperature condition and the desired product was obtained in good yield in the

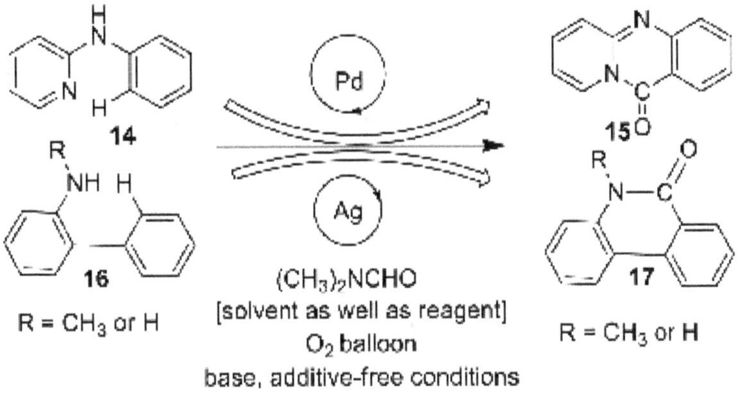

Figure 8.
Pd/Ag catalyzed pyrido carbonylation of N*-phenylpyridin-2-amine.*

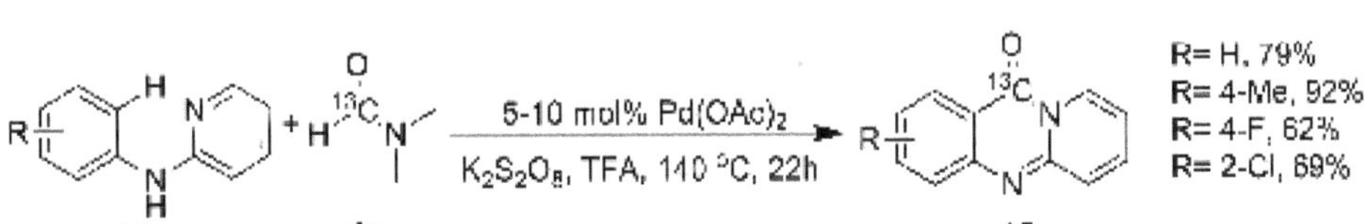

Figure 9.
DMF as CO source in Pd-catalyzed carbonylation.

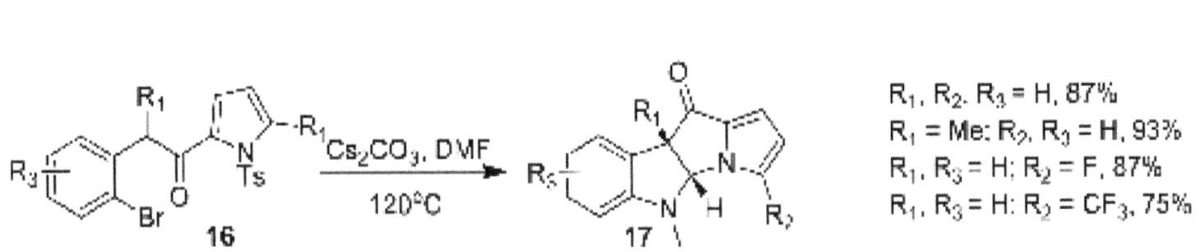

Figure 10.
*Synthesis of dihydropyrrolizino[3,2-*b*]indol-10-one.*

presence of $Pd(OAc)_2$-$K_2S_2O_8$ and DMF/TFA solvent system at 140°C under O_2 atmosphere. When the reaction was conducted with ^{13}CO-labeled DMF (**1a**), the formation of ^{13}C product was detected using gas chromatography (GC). This indicates CO gas has been generated from the carbonyl of DMF with acid as the promoter. This protocol is simple, has broad substrate scope and the products are obtained in excellent yields [33].

2.4 Construction of dihydropyrroline indolone ring

In 2017, Chang and coworkers reported metal, ligand free, base promoted cascade reaction of DMF with *N*-tosyl-2-(2-bromophenylacetyl)pyrroles (**17**) for the synthesis of dihydropyrrolizino[3,2-*b*]indol-10-ones **16** (**Figure 10**) [34].

2.5 Construction of acyl indole ring

Deng et al., reported a metal free approach for the synthesis of 3-acylindoles **18** through a cascade reaction between 2-alkenylanilines **19** with *N,N*-dimethylformamide (DMF) as a one-carbon source (**Figure 11**). This methodology worked with O_2 as a terminal oxidant as well as oxygen donor. The 2-alkenylanilines containing different substitution such as, tosyl groups and other sulfonamides gave the desired 3-acylindoles in low to good yields. Unluckily, the substrate with a primary amine group failed to provide the desired product.

To prove the synthetic utility of this transformation gram scale experiment was conducted under optimized condition, wherein the yield of the corresponding product decreased slightly. Control experiments revealed that DMF acts as carbon source and O_2 is the source of the oxygen. When deuterium labeled DMF was used as solvent, the labeled product was observed. Meantime, to probe the source oxygen atom in the final product a reaction has implemented with ^{18}O-DMF and only non-labeled product was obtained. Thus, author justified that O_2 is the source of the oxygen atom in the final product [35].

2.6 Construction of benzothiazole ring

Liu et al., developed a methodology for the synthesis of *N*-containing heterocycles including benzothiazoles, benzomidazoles, quinazolinone and benzoxazole using combination of $B(C_6F_5)_3$, atmospheric CO_2 and Et_2SiH_2 (**Figure 12**).

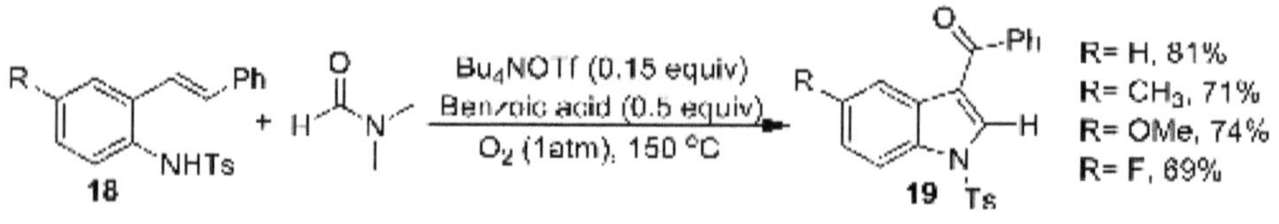

Figure 11.
Formation of 3-acylindoles from 2-alkenylanilines.

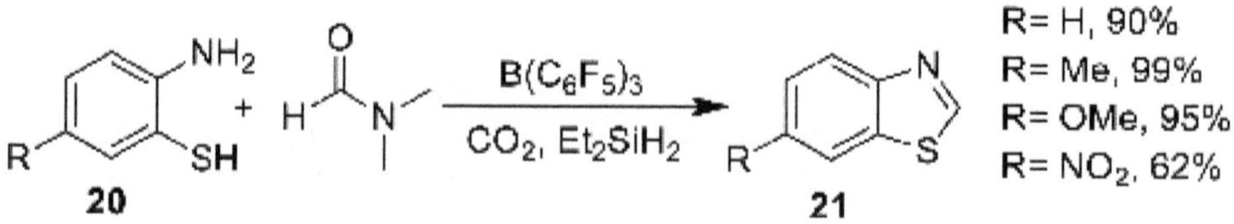

Figure 12.
The cyclization of 2-aminothiophenol with DMF.

This catalytic system was found to be highly effective for the cyclization of 2-aminobenzenethiol **20** or *o*-phenylenediamine **23** with *N,N*-dimethylformamide **1a**, utilizing CO_2 in this process. The reaction condition was optimized with different parameters and the corresponding product was obtained in the presence of 2-aminothiophenol (0.5 mmol), $B(C_6F_5)_3$ (5 mol%), Et_2SiH_2 (2 mmol), DMF (1 mL), CO_2 at 120°C.

To understand the role of CO_2 in this reaction, isotopic labeling reaction were carried out using $^{13}CO_2$, the non-labeled benzothiazole was observed in excellent yield [36]. When this cyclization reaction was carried out using d_7-DMF instead of DMF, deuterated benzothiazole was obtained. This experiment revealed that DMF served as the formylating reagent CO_2 as the promoter.

2.7 Construction of benzimidazole ring

Yadav et al. developed a cost effective synthetic protocol with 100% conversion of o-nitroaniline to benzimidazole using DMF as in-situ source of dimethylamine and CO. Herein, DMF undergoes water gas shift reaction in the presence of $CuFe_2O_4$ as catalyst to produce hydrogen (**Figure 13**). It mainly involves two steps the reduction of o-nitroaniline **22** to o-phenylenediamine **24** followed by cyclization. The ratio of DMF:water affects the conversion of o-nitroaniline to benzimdazole **24** hence the optimized ratio is 2:1 for the best conversion and selectivity. Homogeneous catalyst ($CuCl_2$) didn't show any conversion, CuO showed diminished activity and $CuFe_2O_4$ exhibited better activity. Optimum temperature for the reaction condition was 180°C [37].

A possible mechanism was proposed by author. Thermal degradation of DMF in the presence of water provides CO, which undergoes water gas shift reaction in the presence of catalyst to release hydrogen gas. This H_2 reduces nitro group to form amine group. The formation of o-phenylenediamine was confirmed with the help of GC-MS and HPLC analysis and compared with standard samples. Further, formylation of one of the amine groups took place in the presence CO, then intramolecular cyclisation takes place to give benzimidazole.

2.8 Construction of coumarin ring

Ohshita et al. developed method for the synthesis of coumarins **29** from *ortho*-quinone methide **26** formed *via* [2 + 2] cycloaddition of aryne **25** with DMF. Compound **26** reacted effectively with ester enolates **27** or ketenimine **28** *via* [4 + 2] cycloaddition to provide different coumarins **29** (**Figure 14**) [38].

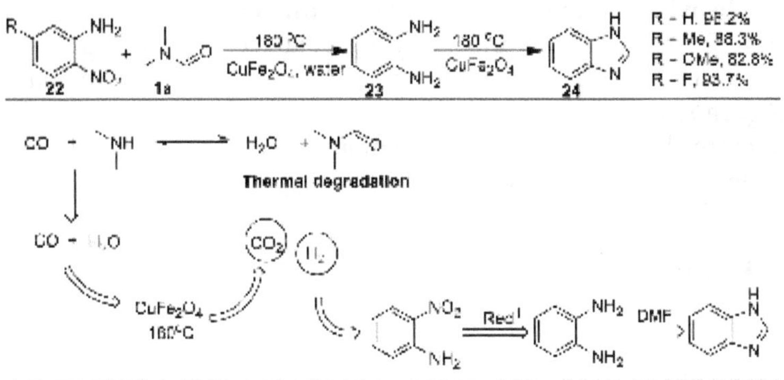

Figure 13.
One-pot synthesis of benzimidazole.

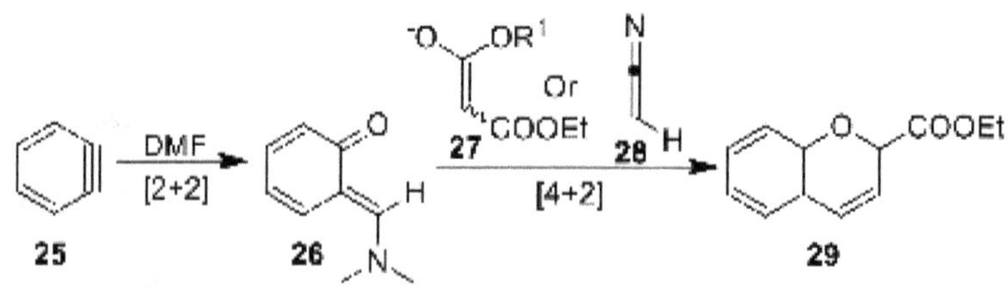

Figure 14.
Synthesis of different coumarin derivatives.

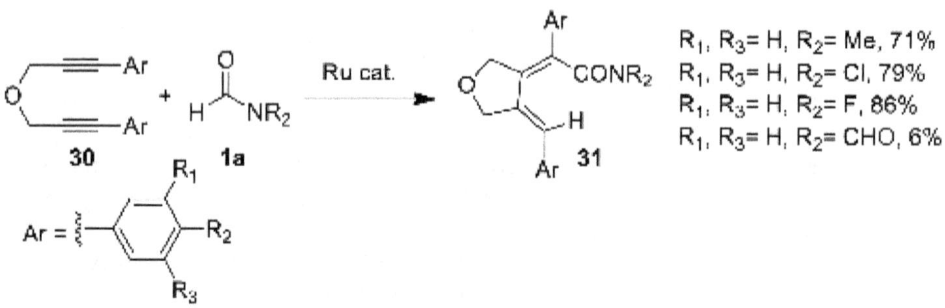

Figure 15.
Hydrocarbamoylative cyclization of 1,6-diynes with DMF.

2.9 Construction of cyclic ether

Yamamoto and coworkers synthesized exocyclicdiene-type α,β,γ,δ-unsaturated amides **31** from hydrocarbamoylative cyclization of 1,6-diynes **30** with formamides under Ru-catalyst with complete stereoselectivity (**Figure 15**) [39].

3. Amidation

Having covered literature on construction of cyclic system, especially heterocycles using DMF or DMA as a next part we cover literature on the formation of open chain compounds.

An excellent method to access benzamides **33** *via* aminocarbonylation of aryl and alkenyl iodides **32**, with DMF as amide source, in the presence of Pd/$POCl_3$ catalytic system, was demonstrated by Hiyama et al. (**Figure 16**) [40].

Similarly, Indolese et al. reported aminocarbonylation of aryl halides **32** with Pd catalyst, triphenylphosphine ligand in CO atmosphere under pressure. DMAP is used as base for this reaction and the yield obtained is very high [41]. It is an important synthetic method since it can also be applied to pyridine and thiophene halides (**Figure 16**).

Furthermore, Lee and co-workers demonstrated the same reaction between aryl bromides/iodides **32** and DMF with the help of inexpensive Nickel acetate

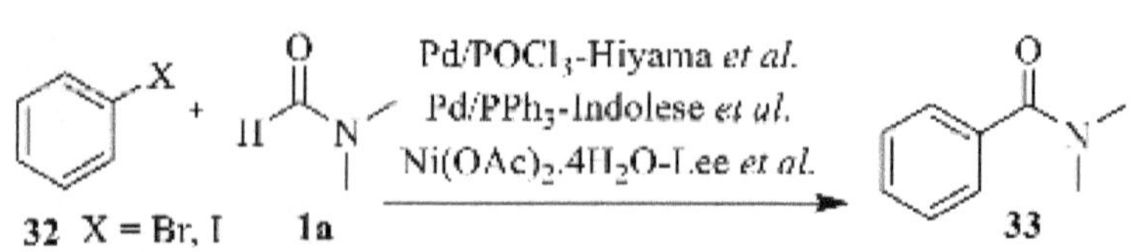

Figure 16.
Metal catalyzed aminocarbonylation of aryl halides using DMF.

tetrahydrate as catalyst and using phosphite ligand and sodium methoxide as base in dioxane solvent (**Figure 16**) [42].

Wang et al., reported a metal-free radical amidation of thiazoles and oxazoles **34** with a series of formamides and *tert*-butyl perbenzoate (TBPB) as radical initiator. By this method, synthesis of high yields of amidated azoles **35** were easily achieved (**Figure 17**) [43].

Wang et al., demonstrated direct amidation of alcohols **36** with formamides in the presence of an I_2/TBHP with sodium hydroxide as a base and DMF as amide source (**Figure 18**) [44]. The same author reported amidation of benzyl amine **38** under the acidic condition [45].

Feng and coworkers proposed green protocol for the synthesis of α-ketoamides **41** through TBAI catalyzed sp^3 C-H oxidative radical/radical cross-coupling. This method is applicable for broad range of substrates [46]. The only by product is water and no CO or CO_2 emission is observed (**Figure 19**).

Similarly, the synthesis of α-ketoamides **41** was achieved with readily available aryl methyl ketones **42** using inexpensive *N,N*-dialkylformamides in the presence of nBu_4NI and aq.TBHP as catalyst and oxidant for radical oxidative coupling process (**Figure 19**). This strategy is a green and metal-free approach developed by Mai et al. [47].

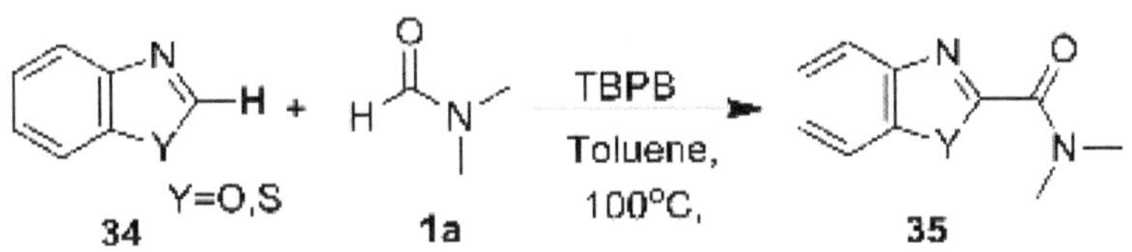

Figure 17.
DMF as a source for aminocarbonylation of azoles.

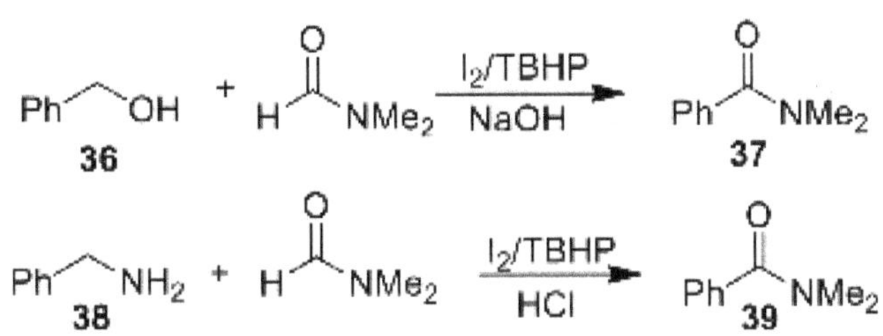

Figure 18.
DMF as a source for aminocarbonylation of alcohol and amines.

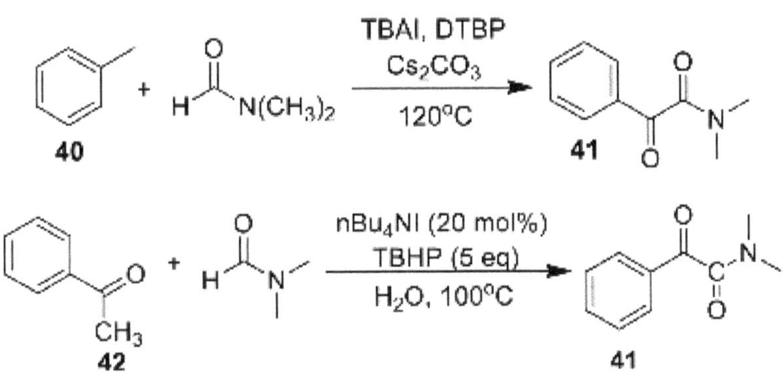

Figure 19.
DMF as aminocarbonylation source in synthesis of α-ketoamides.

In 2016, Xiao and his team developed a simple and efficient technique for the synthesis of amides **33** by cross coupling of carboxylic acids **43** with *N*-substituted formamides in the presence of Ru catalyst and the desired amide was obtained after the release of CO_2 (**Figure 20**). The carbonyl group in the amide product came from benzoic acid and not from N-substituted formamides. This synthetic method is stable, inexpensive, low toxicity and eco-friendly. This method works well with different carboxylic acid derivatives and *N*-substituted formamides [48].

Similarly, Tortoioli and co-workers demonstrated one-pot synthesis of dialkyl amides under metal free condition through the reaction between benzoic acid and DMF in presence of propyl phosphonic anhydride (T_3P) with acid additives [49]. This mild method has been applied to the synthesis of dihydrofolate reductase inhibitor, triazinate (**Figure 21**).

Bhat et al. reported direct carbamoylation of heterocycles **44** *via* direct dehydrogenative aminocarbonylation under transition metal-free condition **45** (**Figure 22**). Persulfate which is played the role of an efficient oxidant, good radical initiator, mild and eco-friendly low cost reagent and formamides NMF and DMF acted as reagent to form primary to tertiary carboxamides [50].

Bhisma et al. gave an efficient copper catalyzed synthesis of phenol carbamates **47** from dialkylformamides as aminocarbonyl surrogate and phenols possessing directing groups such as benzothiazoles, quinoline and formyl at ortho-position (**Figure 23**). It's a cheap and eco-friendly reaction with tolerance of wide range of functional groups and phosgene free route to carbamates [51].

Phan and coworkers under oxidative condition synthesized organic carbamates **49** through C-H activation using metal organic framework $Cu_2(BPDC)_2(BPY)$ (BPDC = 4,4′-biphenyldicarboxylative, BPY = 4,4′-bipyridine) as heterogeneous catalyst for cross dehydrogenative coupling of DMF with 2-substituted

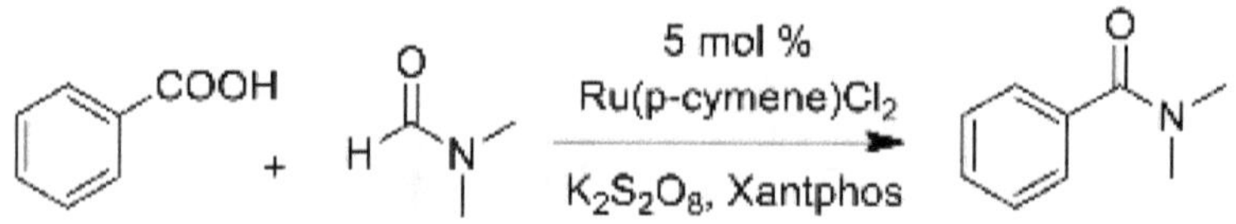

Figure 20.
DMF in Ru-catalyzed amidation of carboxylic acids.

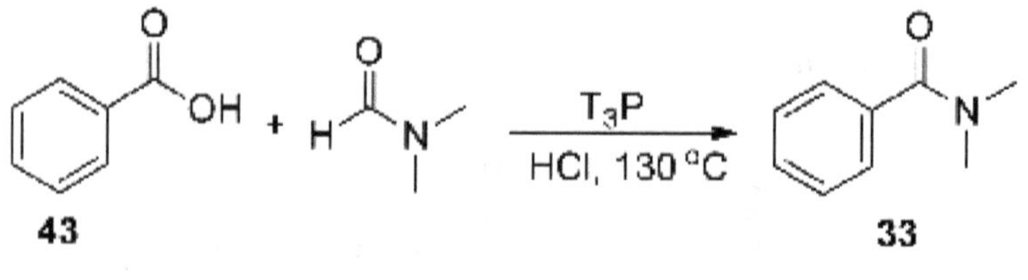

Figure 21.
Amidation of benzoic acid with DMF.

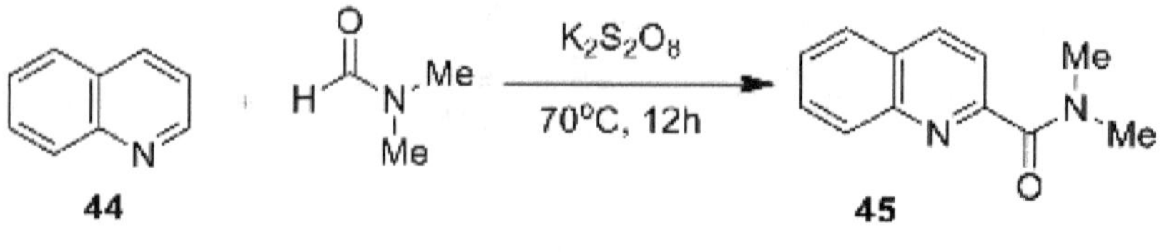

Figure 22.
DMF as source for aminocarbonylation of quinoline.

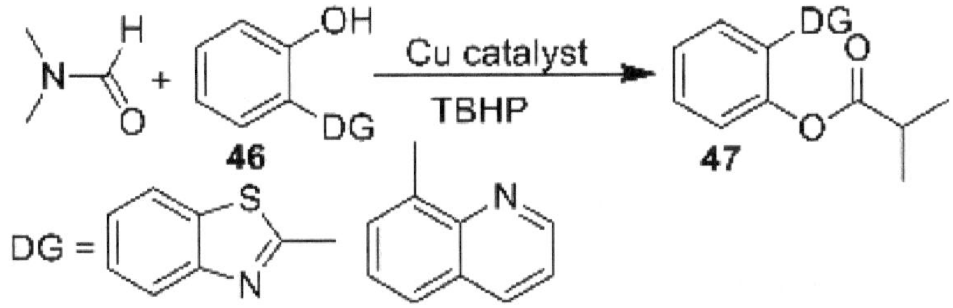

Figure 23.
Carbamate synthesis from phenols and formamides.

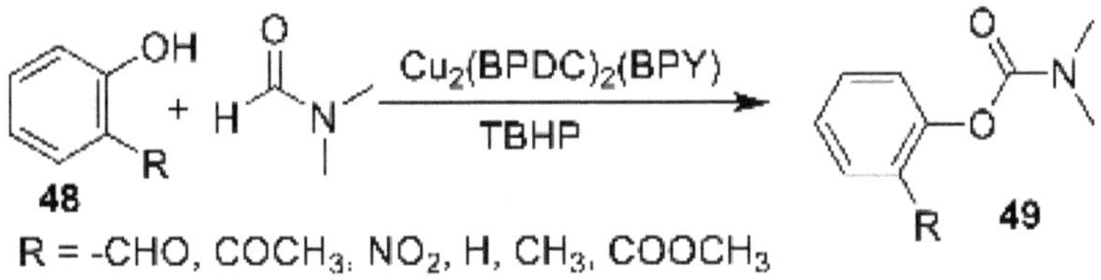

Figure 24.
CDC reaction of phenol with DMF.

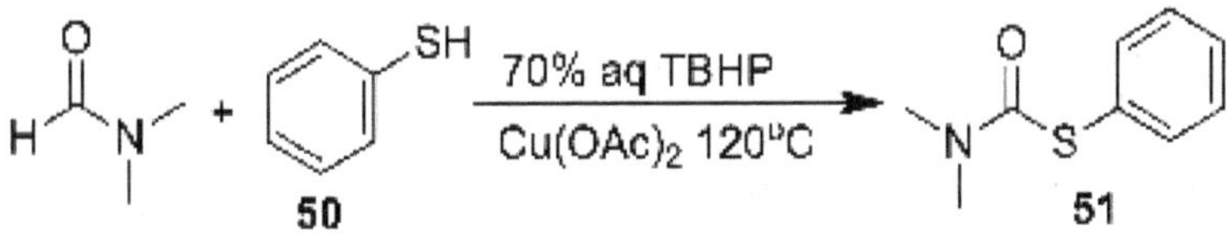

Figure 25.
Synthesis of S-phenyldialkylthiocarbamate.

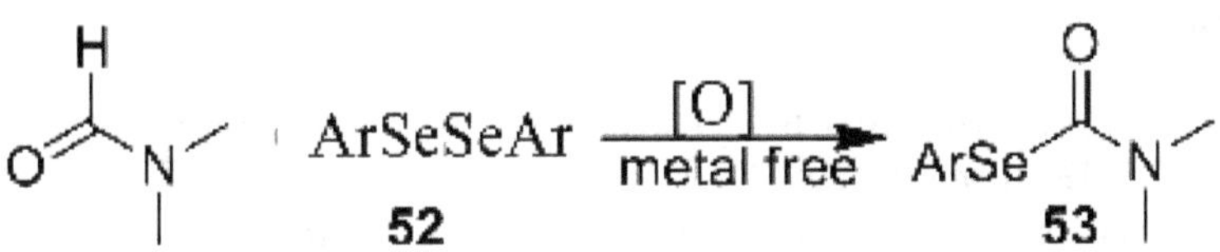

Figure 26.
Oxidative C-Se coupling of formamides and diselenides.

phenols **48** (**Figure 24**). This catalyst has higher catalytic activity and it is easily recoverable and reusable [52].

Yuan et al., synthesized S-phenyldialkylthiocarbamate **51** compounds under solvent free conditions through TBHP promoted radical pathway, in which direct oxidation of acylC-H bond of formamides took place in the presence of $Cu(OAc)_2$ to form the reaction intermediate for oxidative coupling reaction of formamides with thiols **50** (**Figure 25**) [53]. This protocol is efficient and green.

Kamal and coworkers proposed an efficient and greener methodology for the synthesis of selenocarbamates **53** by oxidative coupling reaction between formamides and diselenides **52** under metal free conditions (**Figure 26**). By using simple reaction condition, a metal-free approach to direct C-Se bond formation occurred at carbonyl carbon by using TBHP and molecular sieves. It uses non-functionalized substrate which is an advantage of this reaction [54].

Reddy and coworkers synthesized chiral symmetrical urea derivatives **54** through copper catalyzed C-H/N-H coupling of formamides (both mono and di) with different amines **53** (primary, secondary and substituted aromatic amines) using TBHP as an oxidant and it involves a radical pathway (**Figure 27**).

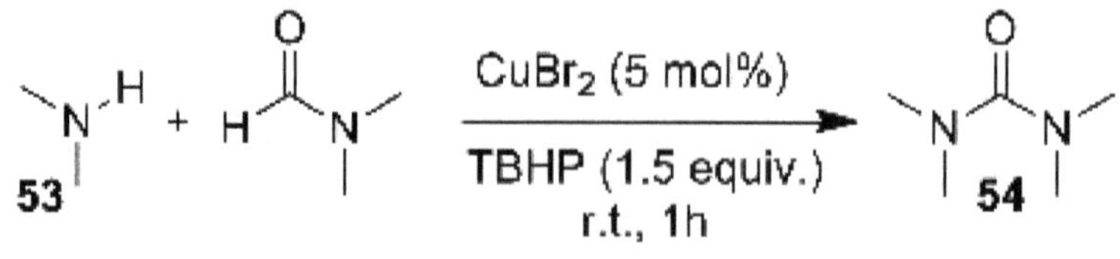

Figure 27.
Synthesis of chiral symmetrical urea derivatives from DMF.

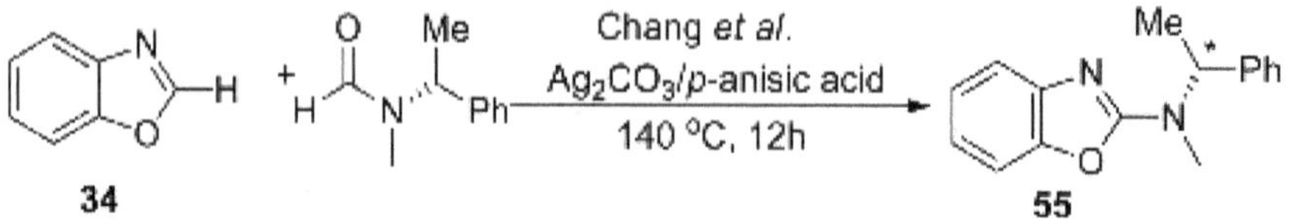

Figure 28.
Amidation of benzoxazole using Ag_2CO_3 catalyst.

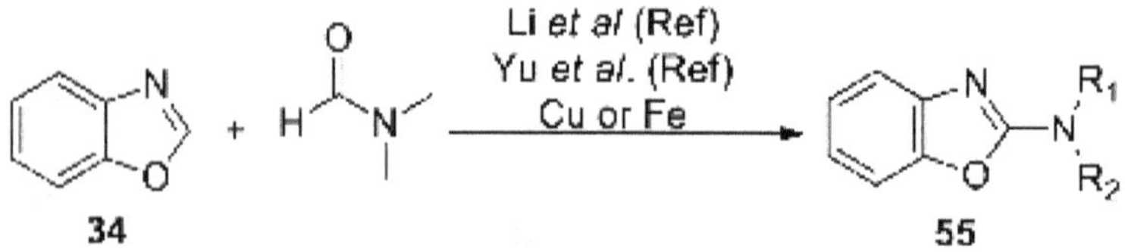

Figure 29.
Amidation of benzoxazole using Cu or Fe catalyst.

The importance of this green reaction is, it avoids the use of pre-functionalized substrates, atom economical [55].

3.1 Amination

Chang et al., reported that benzoxazoles **34** on treatment with *N*,*N*-dimethylformamide (DMF) using the Ag_2CO_3 as catalyst in the presence of an acid additive, 2-aminated benzoxazole **55** was obtained as a single product in moderate yield (**Figure 28**).

Interestingly, this method is also suitable for the optically active formamide, the desired product was obtained in better yield without racimization [56].

Li et al., gave a method for the synthesis of 2-aminoazole derivatives **58** in which construction of C-N bond of azoles **34** either by decarboxylative coupling with formamides as nitrogen source or by a direct C-H amination with secondary amines as nitrogen source by the use of inexpensive Cu catalyst, O_2 or air as oxidant is green and benzoic acid has its main role in the release of amine from amides by decarbonylation other than C-H activation [57].

Similarly, Yu et al., developed a decarbonylative coupling between azoles and formamides. The iron catalyzed direct C-H amination of azoles at C_2 took place in the presence of formamides and amines as nitrogen source (**Figure 29**). Easily accessible iron (II) salts acted as Lewis acid which activated the C_2 position of benzoxaoles **34** and oxidant and imidazole was used as an additive in the catalyst under air. This direct azole amination was catalyzed by inexpensive and environmentally benign reagents. The reaction was also carried with amines in the presence of acetonitrile [58].

Peng and coworkers developed a facile and efficient route for one pot synthesis of 2-acyl-4-(dimethylamino)-quinazoline **57** through direct amination of 2-aryl quinazoline-4(3*H*)ones **56** with DMF in which 4-toluene sulfonyl chloride acted as

C-OH bond activator (**Figure 30**). KO^{t}Bu was used as base which leads to the formation of tosylate which attacks DMF which in turn undergoes hydrolysis to give aminated product **59**. This reaction is inexpensive and uses easy to handle reagents [59].

Eycken et al. demonstrated a convenient microwave-assisted de-sulfitative dimethylamination of 5-chloro-3-(phenylsulfanyl)-2-pyrazinones **58** using DMF as a dimethylamine source and sodium carbonate as an essential (**Figure 31**). The solvent system used for this reaction is DMF:H_2O in 1:1 ratio and the corresponding de-sulfitative aminated product **59** was obtained in good yield. Finally, the utility of this methodology was also examined on oxazinone in place of pyrazinones under the optimized conditions and the desired products were formed in good yield [60].

Hongting et al. developed an efficient, atom-economic and eco-friendly approach for synthesizing enamines **61** by intermolecular hydroamination of activated alkynes (**Figure 32**). The reaction was carried out under solvent free condition using a catalyst at room temperature. Primary or secondary amines **53** were added to triple bonds **60** without generating any waste products. DMF pretreated

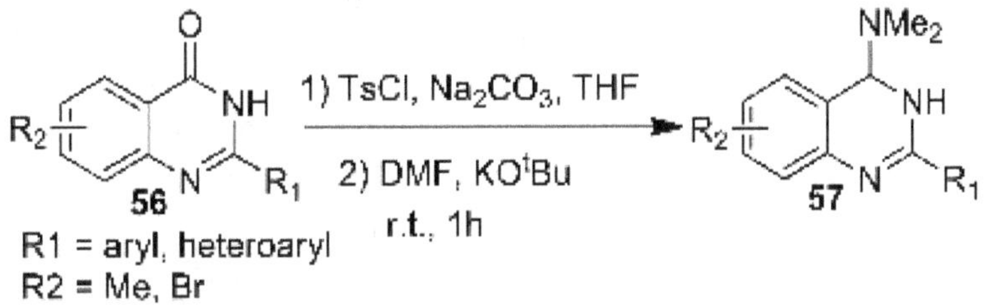

Figure 30.
Direct amination of 2-aryl quinazoline-4(3H)ones with DMF.

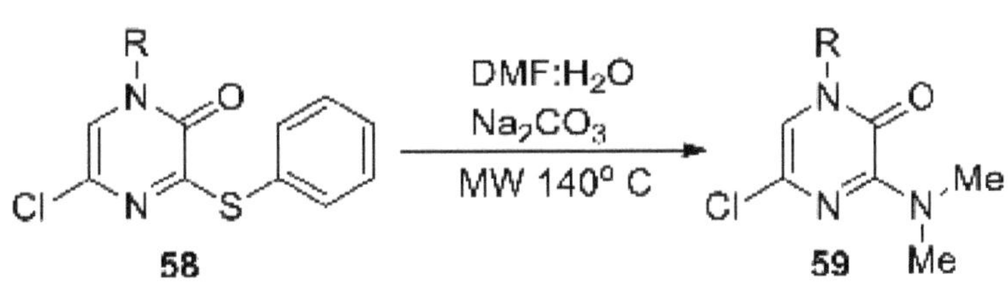

Figure 31.
De-sulfitative amination of 2(1H) pyrazinone.

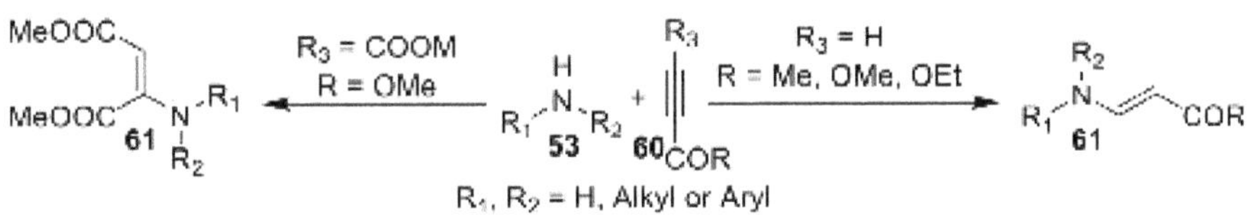

Figure 32.
Intermolecular hydroamination of activated alkynes.

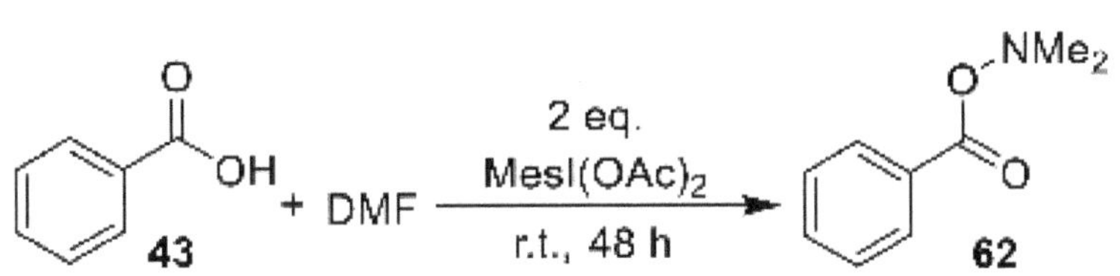

Figure 33.
Synthesis of O-aroyl-N,N-dimethyl hydroxyl amines.

with metal Na was used for synthesis of (E)-ethyl-3-(dimethylamino)acrylate and a new way for synthesis of quinolines was given [61].

Li et al., developed hypervalent iodine mediated reaction between carboxylic acids **43** and *N,N*-dimethylformamide which occur under mild conditions at room temperature to provide novel *O*-aroyl-*N,N*-dimethyl hydroxyl amines **62** in good yields (**Figure 33**), which are important electrophilic amination reagents. The process shows good functional group compatibility, air and moisture tolerance [62].

Liang and coworkers gave a simple and efficient one-pot multicomponent reaction of chalcones **63**, malononitrile **64** and DMF in the presence of NaOH for the synthesis of functionalized 4-oxobutanamides **65** (γ-ketoamides) from simple α,β-unsaturated enones (**Figure 34**). This reaction has a high atom economy, easily available starting materials, operational simplicity with mild conditions, broad substrate scope and good tolerance with diverse functional groups [63].

Xia and coworkers proposed a simple and green approach for the synthesis of sulfonamides through t-BuOK mediated direct S-N bond formation from sodium sulfinates **66** with formamides (**Figure 35**). This reaction undergoes in a metal-free conditions and formamides are used as amine source. It avoids pre-functionalized starting materials and forms an alternative method for the synthesis of sulfonamids **67** [64].

Gong et al., reported a base-promoted amination of aromatic halides **32** using a limited amount of *N,N*-dimethylformamide or amine as an amino source. Various aryl halides, including F, Cl, Br, and I, have been successfully aminated **68** in good to excellent yields (**Figure 36**) [65]. This protocol is valuable for industrial application due to the simplicity of operation, the unrestricted availability of amino sources and aromatic halides.

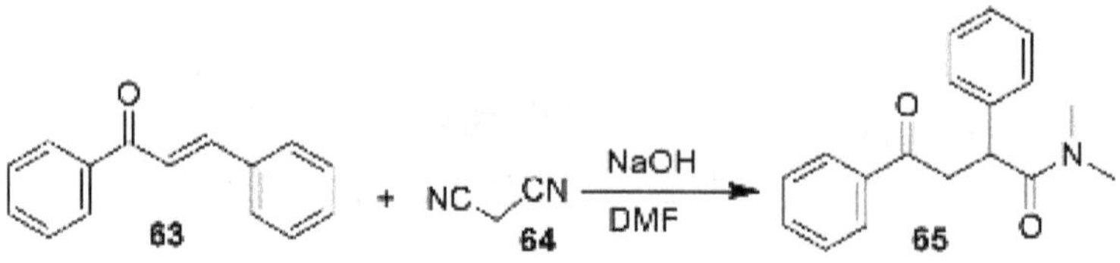

Figure 34.
Synthesis of γ-ketoamide.

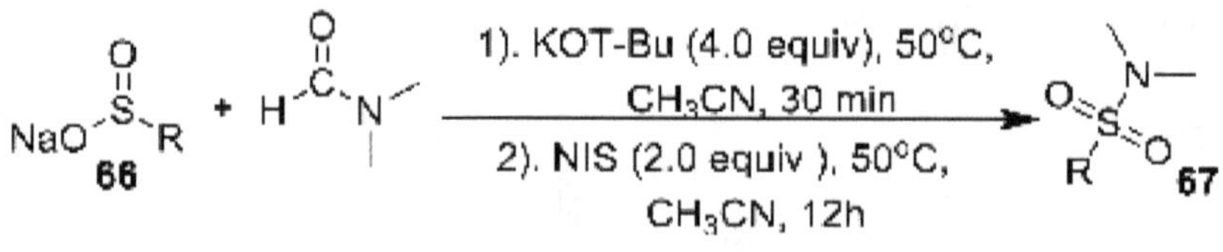

Figure 35.
Synthesis of sulfonamides using DMF as a amine source.

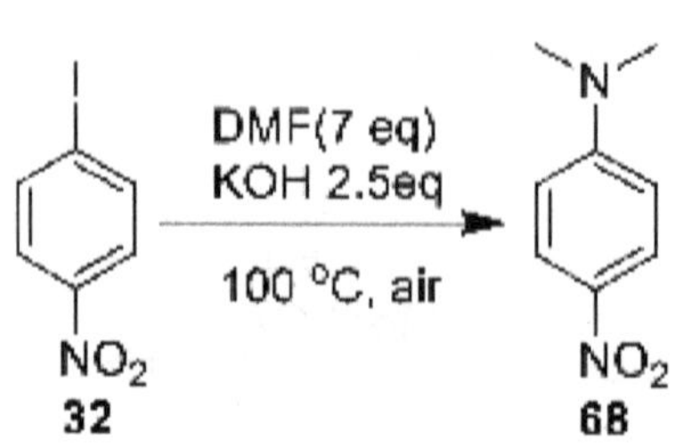

Figure 36.
A base-promoted amination of aromatic halides.

3.2 Methylenation

In recent past several methods were developed for using DMF as a methylene source.

Wang et al., developed a new method for the synthesis of vinylquinolines **70** from methyl quinolines **69** (**Figure 37**) using DMF as a methylene source. The synthesis was carried out *via* an iron-catalyzed sp^3 C-H functionalization and a subsequent C-N cleavage using TBHP as a radical initiator. This method is simple and effective for synthesis of large number of vinyl substituted quinoline derivatives in excellent yield. It also avoids the usage of organometallic compounds as reagents [66].

Qian Xu and coworkers developed an eco-friendly iron-catalyzed benzylic vinylation which transfers the carbon atom in *N*,*N*-dimethyl group from DMA or DMF to 2-methyl azaarenes **71** to generate 2-vinyl azaarenes **72** (**Figure 38**). The reaction of *N*,*N*-dimethyl amides as one carbon source proceeded *via* radical mechanism [67].

Miura et al., demonstrated an effective way for α-methylenation of benzyl pyridines **73** using copper catalyst. In the methylenation, *N*-methyl group of DMA was incorporated as the one-carbon source to produce α-styrylpyridine **74** derivatives (**Figure 39**), which are famous for their unique biological properties [68].

Li et al., developed an iron-catalyzed α-methylenation of aryl ketones **75** by using *N*,*N*-dimethylacetamides as a one-carbon source to form α, β-unsaturated carbonyl compounds (**Figure 40**). Potassium persulfate is used as oxidant and this method acts as an excellent synthetic method for synthesis of α, β-unsaturated carbonyl compounds **76** [69].

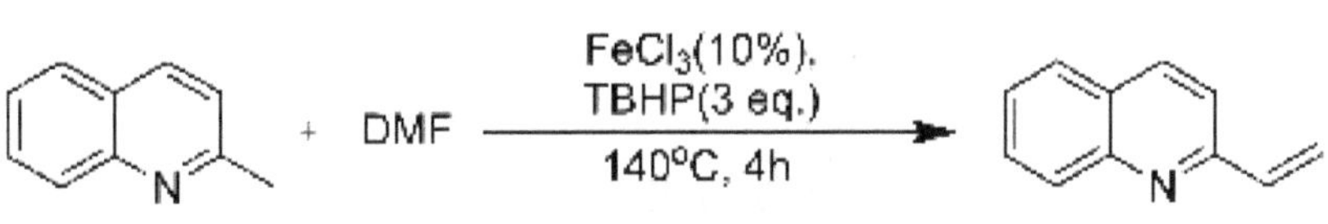

69 70

Figure 37.
Synthesis of vinyl quinolones using DMF with iron catalyst.

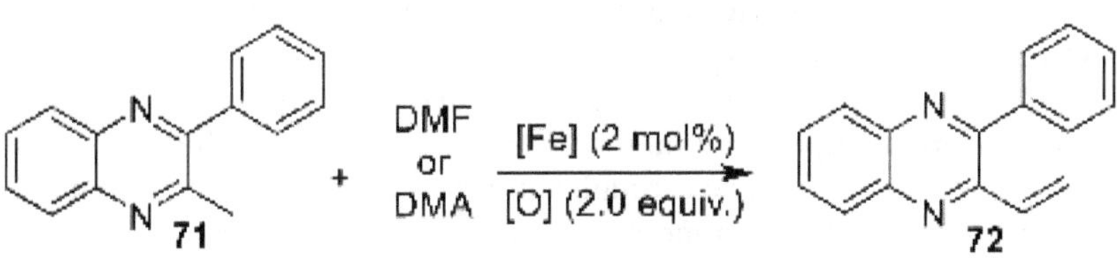

71 72

Figure 38.
DMA or DMF Synthesis of vinyl 2-vinylazaarenes.

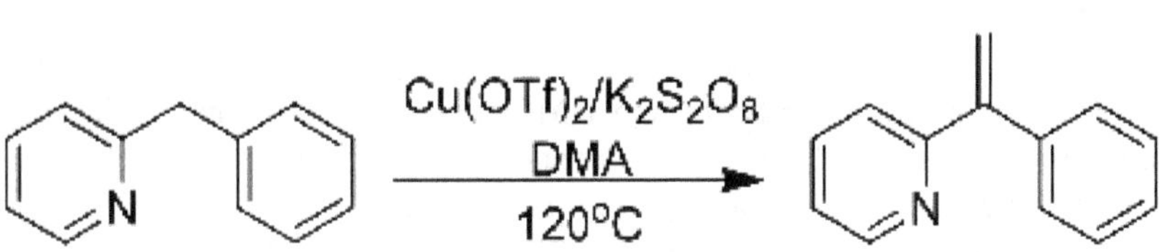

73 74

Figure 39.
α-methylenation of benzylpyridines using DMA.

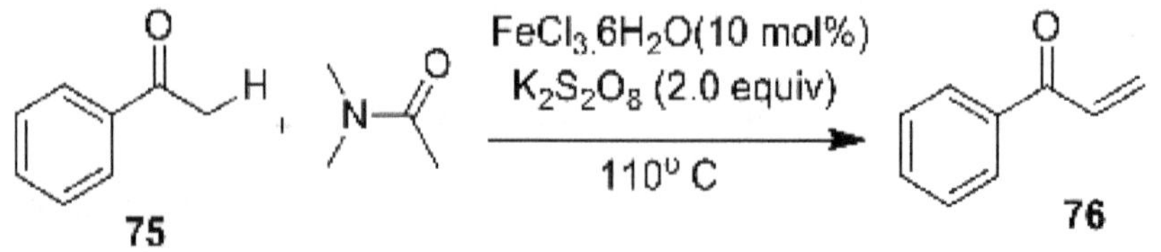

Figure 40.
α-methylenation of acetophenones.

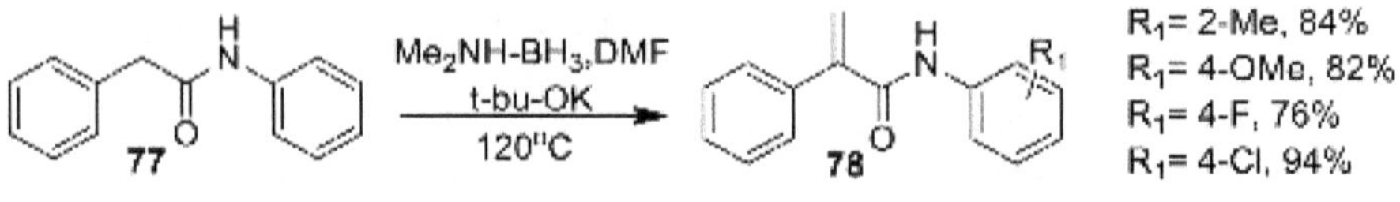

Figure 41.
α-methylenation of 2-arylacetamides with DMF.

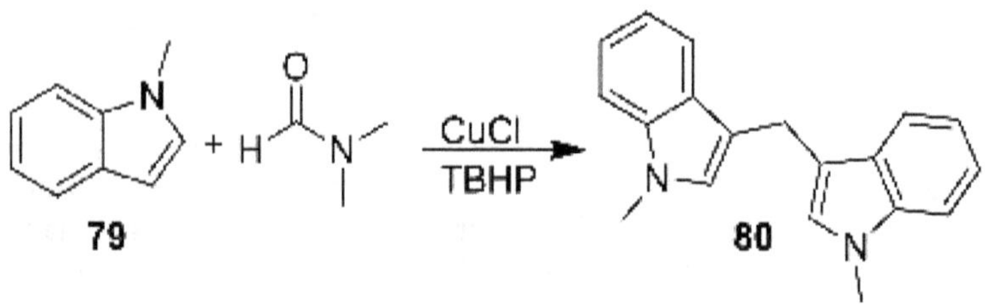

Figure 42.
Cu-catalyzed synthesis of diindolylmethane.

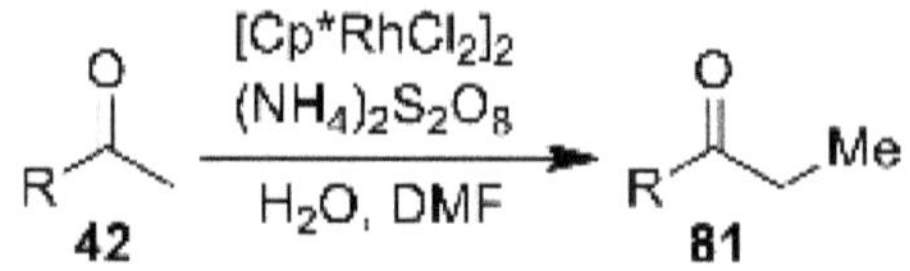

Figure 43.
Rh-catalyzed direct methylation and hydrogenation of ketones using DMF.

In 2019, Wang et al., reported a one-pot procedure for the synthesis of 3-indolyl-3-methyl oxindoles **78** *via* C(sp^3)-H methylenation of 2-arylacetamides **77** using DMF/Me_2NH-BH_3 as the methylene source (**Figure 41**) [70].

Liu and coworkers reported a method for the synthesis of diindolylmethane **80** and its derivatives which is done through copper catalyzed C-H activation of indole **79** where in DMF was used as a methylenating reagent. CuCl was mainly used as a catalyst which affords high regioselectivity and TBHP as oxidant. The reaction utilizes readily available copper catalyst and inexpensive DMF as carbon source and it has a broad scope of substrates with relatively mild reaction conditions (**Figure 42**) [71].

In 2014, Xue and co-workers developed methylation of ketones **42** with DMF, control experiment studies indicate that DMF plays dual functions as the source of carbon for methylation and source of hydrogen in the rhodium-catalyzed reduction of the methylene into a methyl group (**Figure 43**) [72].

A possible mechanism was proposed as shown in **Figure 44**. Initially, persulfate oxidizes DMF to give a reactive iminium intermediate. The intermediate **A** generated by attack of enolate is converted to intermediate **B** followed by C-N bond cleavage to generate unsaturated ketone intermediate **C** . Afterwards, the

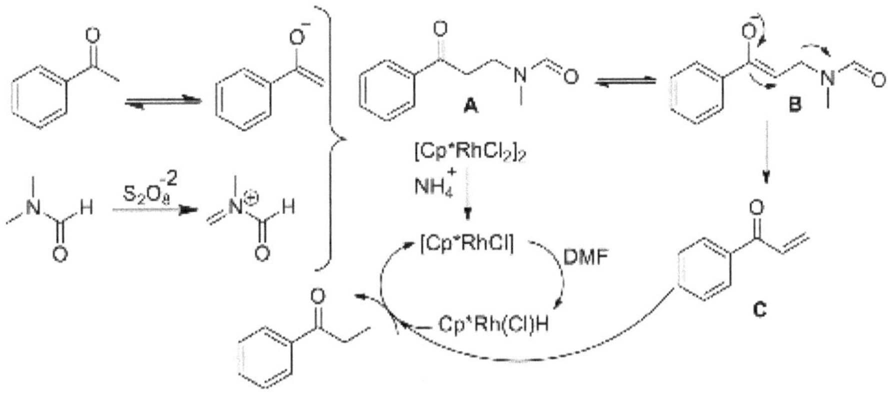

Figure 44.
A possible mechanism for methylation and hydrogenation of ketone.

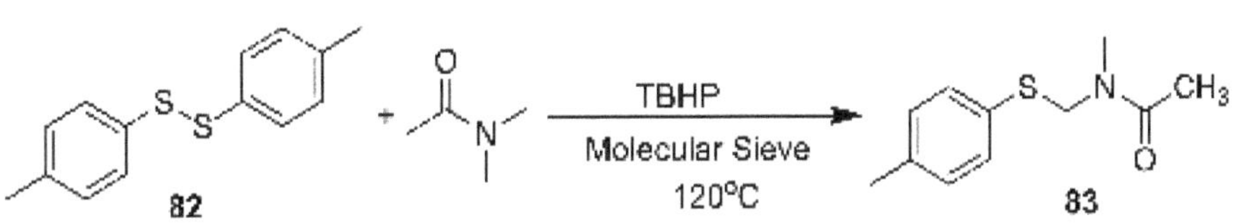

Figure 45.
Thiolation of sp³ C-H bond next to a nitrogen atom.

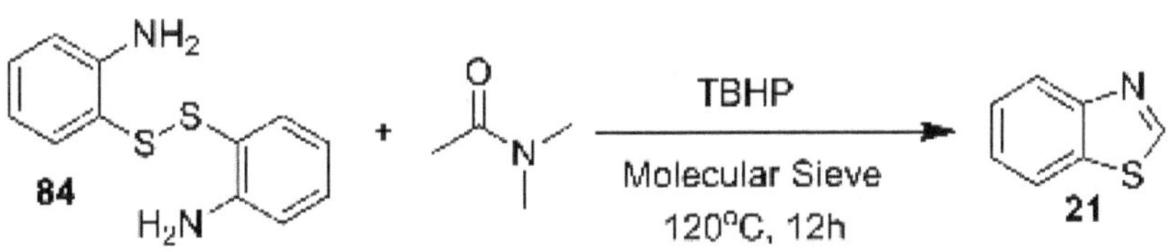

Figure 46.
TBHP-mediated synthesis of benzothiazoles.

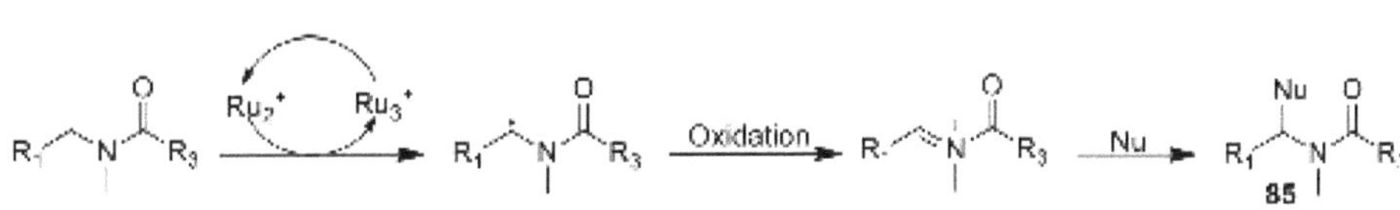

Figure 47.
FC amidoalkylation using alkyl amides.

intermediate **C** is reduced, which is probably generated by using DMF *via* dehydrogenation with the aid of $[Cp^*RhCl_2]_2$, which results in the formation of methylated product.

3.3 Amidoalkylation

Li et al., reported direct oxidative thiolation of sp^3 C-H bond next to a nitrogen atom **83** with disulfides **82** under metal free condition for the synthesis of several N, S containing compounds (**Figure 45**).

In this oxidative thiolation reaction, thiol group was successfully coupled with sp^3 C-H bond of *N,N*-dialkyl amides in the presence of TBHP/Molecular sieves through the formation of radical intermediate.

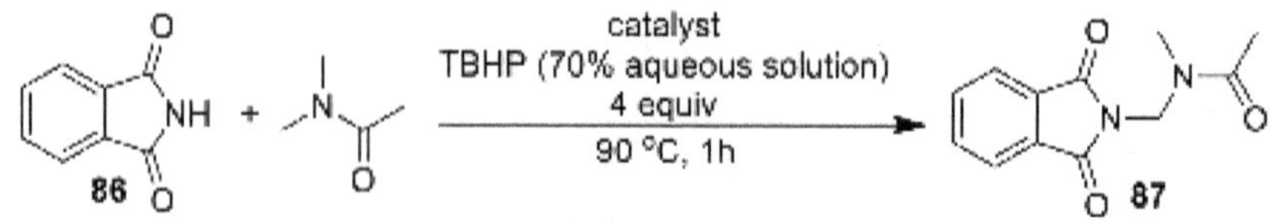

Figure 48.
Amidoalkylation under metal free condition using DMA.

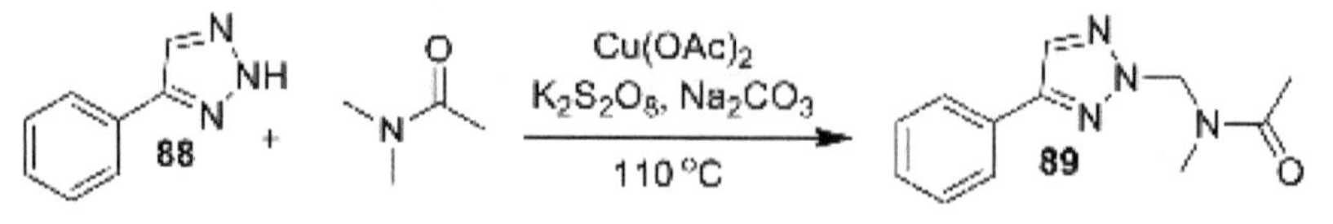

Figure 49.
Copper-catalyzed C-N bond formation of triazoles.

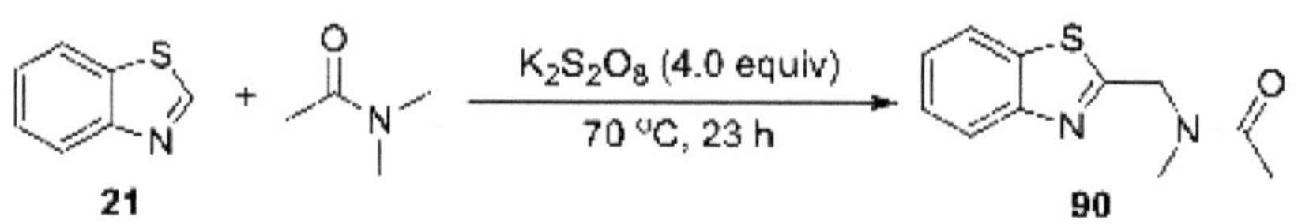

Figure 50.
Amidoalkylation of benzothiazoles with DMA.

It is noteworthy that various benzothiazole and a fipronil analogs could also be synthesized through this methodology (**Figure 46**) [73].

Stephenson et al., developed Friedel-Craft amidoalkylation of alcohols and electron rich arenes as potent nucleophile with alkyl amides **1b** *via* thermolysis and oxidative photocatalysis (**Figure 47**). The FC amidoalkylated product **85** was obtained by oxidation of *N,N*-dialkyl amides with the aid of persulfate and photocatalyst. On the other hand, persulfate at 55°C also afford amidoalkylated product.

In this method inexpensive and efficient persulfate was used as oxidant for the construction of C-O and C-C bonds. Most of the time, photo catalysis provided better selectivity and good yields for the Friedel-Crafts reactions as compared with the thermolytic reaction conditions [74].

Li et al., gave a transition metal-free method for amidation of sp^3 C-H bond in amides through cross dehydrogenative coupling process by using iodide anion as catalyst and TBHP as oxidant (**Figure 48**). It proceeds through free radical intermediate which is confirmed by TEMPO and the products has an potential bioactivity **87**. This is an efficient method for direct C-N bond formation because of its mild conditions and readily available reagents [75].

In 2017, Chen and coworkers demonstrated copper-catalyzed C-N bond formation of triazoles *via* cross dehydrogenative coupling (CDC) of *NH*-1,2,3-triazoles **88** with *N,N*-dialkylamides to construct *N*-amidoalkylated triazoles **89** (**Figure 49**). When the reaction was performed with 4-aryl-substituted *NH*-1,2,3-triazoles the desired N^2-substituted 1,2,3-triazoles was obtained and small amount of N^1 products were also observed. This method is useful for the synthesis of N^2-substituted 1,2,3-triazolesselectively [76].

Zhu and Co-Workers discovered a new methodology for the synthesis of 2-amidoalkylated benzothiazole and 3-amidoalkyl substituted indolinone derivatives using *N,N*-dialkylamides and potassium persulfate as an oxidant under metal free condition (**Figure 50**). The corresponding amidoalkylation products were formed selectively using simple *N,N*-dialkyl amides including formamides [77].

3.4 Cyanation

It is interesting to note that dialkylamides could undergo reaction to generate cycano group. In 2011 Ding et al., reported a novel and another kind of pathway to produce the aryl nitriles *through* the Pd-catalyzed cyanation of indoles **79** and benzofurans by functionalization of C-H bond using DMF as a source of CN and control experiments revealed that N and C of the cyano group are generated from DMF [78].

Similarly, in 2015, Chen and co-workers developed a selective copper-catalyzed C_3-cyanation of indole under an oxygen atmosphere with DMF as a safe CN source and as a solvent (**Figure 51**) [79].

Wang et al., demonstrated a copper catalyzed cyanation of indoles **82** using DMF as a single surrogate of CN (**Figure 52**). Electron rich arenes and aryl aldehydes can be transformed to acyl nitriles. Acyl aldehydes is the key intermediate for this transformation. The mechanism of this reaction involved C-H activation with

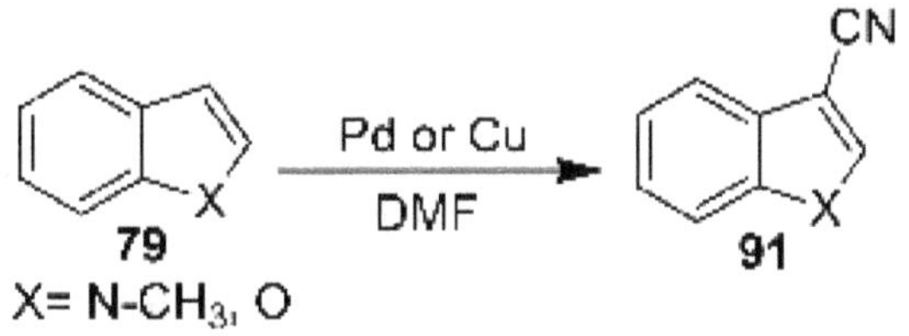

Figure 51.
Cyanation of indole and benzofuran.

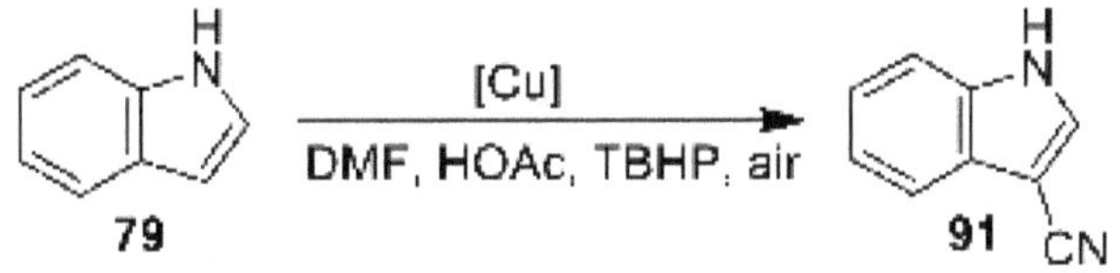

Figure 52.
Cyanation of indole with DMF.

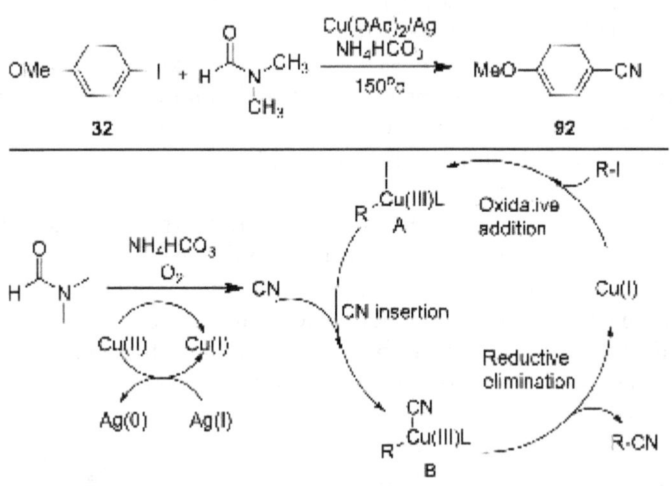

Figure 53.
Cyanation of arylhalides and plausible mechanism.

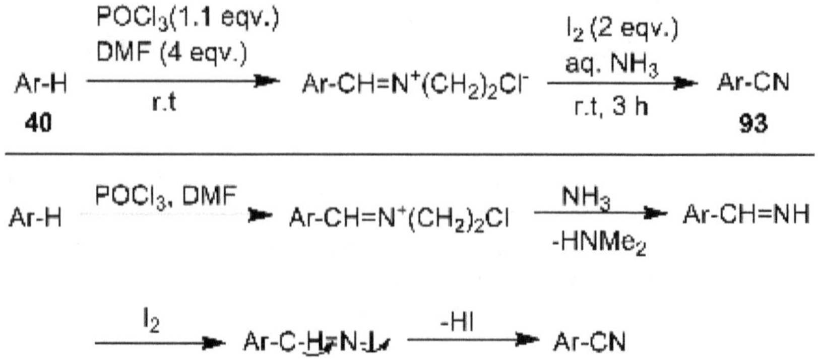

Figure 54.
Conversion of electron-rich aromatics into aromatic nitriles.

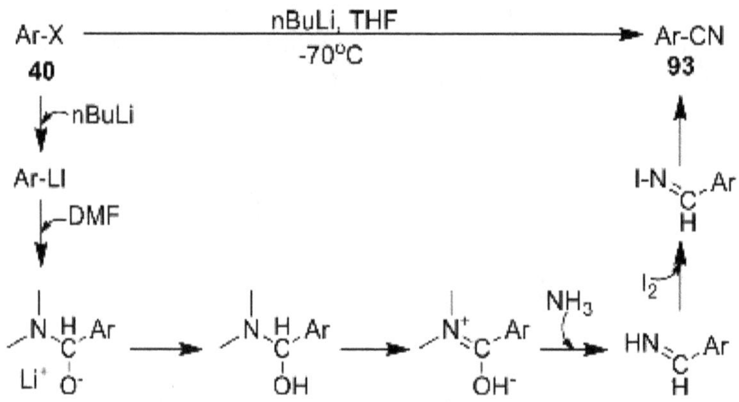

Figure 55.
Conversion of electron-rich aromatics into aromatic nitriles and plausible mechanism.

the help of copper catalyst then followed by carbonylation. 3-cyanoindoles have attracted much great extend owing to their importance in medicinal field especially in the preparation of therapeutic estrogen receptor ligand [80].

Chang et al., reported a new approach for the synthesis of Aryl nitriles **93**. Cyanation of aryl halides **32** catalyzed with copper acetate and Ag as an oxidant, in combination of ammonium bicarbonate as N source and DMF as a C source for cyanide functional group (**Figure 53**). With respect to the key roles of Cu(II) species in the *in-situ* formation of CN units and followed by cyanation of aryl halides, Ag_2CO_3 re-oxidizes the resultant Cu(I) species under copper-catalyzed oxidative conditions. This strategy is a practical and safe method and capable of providing nitriles in moderate to good yields [81].

Ushijima et al., reported the synthesis of aromatic nitriles **93** from electron-rich aromatics **40** under metal free one pot reaction condition. When the combination of molecular iodine in aqueous ammonia, with $POCl_3$ and DMF (**Figure 54**).

A possible mechanism for this reaction was given in **Figure 54**. When treated with ammonia, the iminium salt can be transformed into the aromatic imine. Then molecular iodine serves as an oxidizing agent and reacts with the aromatic imine to provide the corresponding aromatic *N*-iodoimine, which generates the aromatic nitrile through elimination in aqueous ammonia [82].

However, the need of highly electron-rich aromatics in the formation of aromatic *N,N*-dimethyl iminium salts limits the scope of this transformation. So, the authors should develop more convenient methods for this transformation.

Following this work, they reported a novel one-pot method for the preparation of aromatic nitriles from aryl bromides and arenes through the formation of aryl lithium and their DMF adducts (**Figure 55**) [83].

Followed by the treatment with molecular iodine in aqueous ammonia. Similarly, the same author reported synthesis of aryl nitriles from aryl bromides in the presence of Mg [84].

3.5 Formylation

Further, dialkylamides were also used as a formylation source. Wang et al., transformylated different amines, primary or secondary, aromatic or alkyl cyclic or linear, mono- or di-amine with DMF as formylation reagent to obtain corresponding formamides **95** with CeO_2 catalyst and the reaction does not require any homogeneous acidic or basic additives and it is tolerant to water.

The best part about the CeO_2 catalyst is the strong basicity and medium water-tolerant acidity (**Figure 56**) [85].

In 2017, Jagtap and coworkers reported highly efficient Ni(II) metal complex catalyzing *N*-formylation **96** and *N*-acyltion **97** of amines using *N,N*-dimethylformamide and *N,N*-dimethylacetamide as acyl source (CHO) in the presence of imidazole at a temperature of 150°C in a homogeneous medium (**Figure 57**). It has a broad substrate scope to aliphatic, aromatic and heterocyclic compounds.

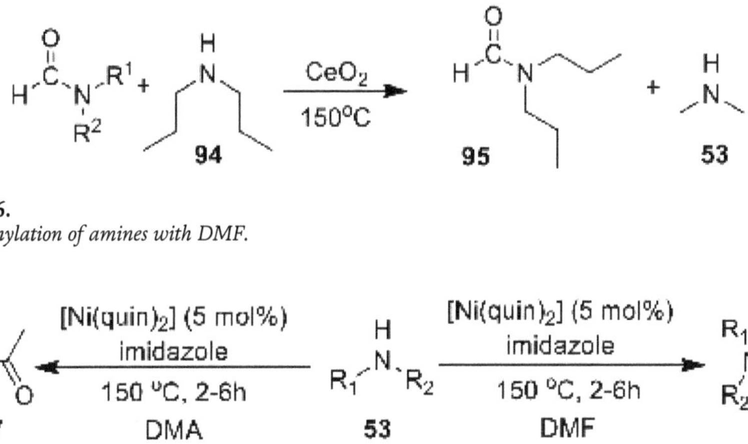

Figure 56.
Transformylation of amines with DMF.

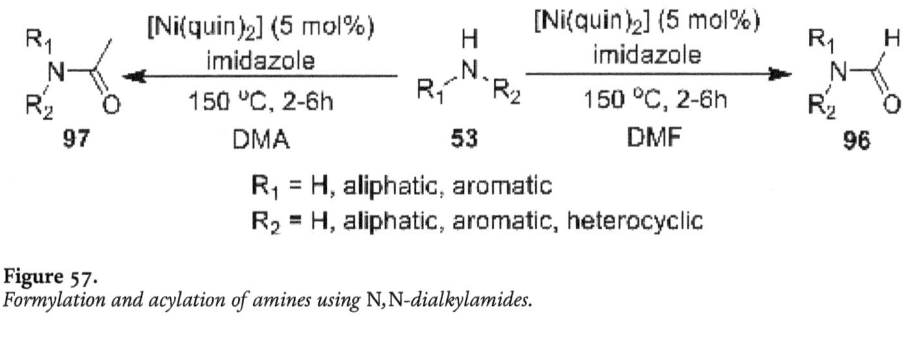

Figure 57.
Formylation and acylation of amines using N,N*-dialkylamides.*

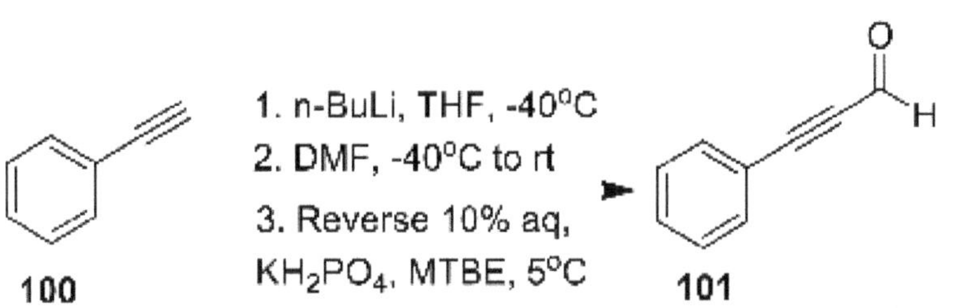

Figure 58.
Synthesis of α,β-acetylenic aldehydes.

The importance of this reactions are cost-effective, easily available starting material, high reactivity and inertness toward air and water [86].

Larsen et al., developed a convenient method for the synthesis of α,β-acetylenic aldehydes **101**, acetylides that are initially transformed to lithium acetylides with the aid of *n*-BuLi (**Figure 58**). The formylation of lithium acetylides was accomplished in the presence of DMF and followed by α-aminoalkoxide with 10% aqueous KH_2PO_4 to provide desired product with good yield [87].

Jeon and co-workers reported methyl benzoate **102** promoted *N*-formylation of different primary and secondary amines **38** employing DMF as a formylating agent under microwave irradiation (**Figure 59**). Key advantage of this methodology is selective *N*-formylation in the presence of a hydroxyl group [88].

3.6 Hydrogenation

Dialkylamides have ability to acts as hydrogen source and it has been used in several functional group transformations. It is advantageous to use hydrogen gas *in situ* generated from dialkylamides rather than handling easily flammable hydrogen gas.

Hua et al. reported triruthenium dodecacarbonyl $[Ru_3(CO)_{12}]$ catalyzed stereo divergent semi-hydrogenation of diaryl alkynes **104** with *N,N*-dimethylformamide/ water as hydrogen source for the synthesis of cis-**105** and trans **106**-stilbenes (**Figure 60**). When the HOAc was used excellent stereoslectivity was observed in favor of formation of *cis*-product. Surprisingly, the stereochemical preference changed to *trans*-isomer, with TFA as additive. This strategy is useful for the

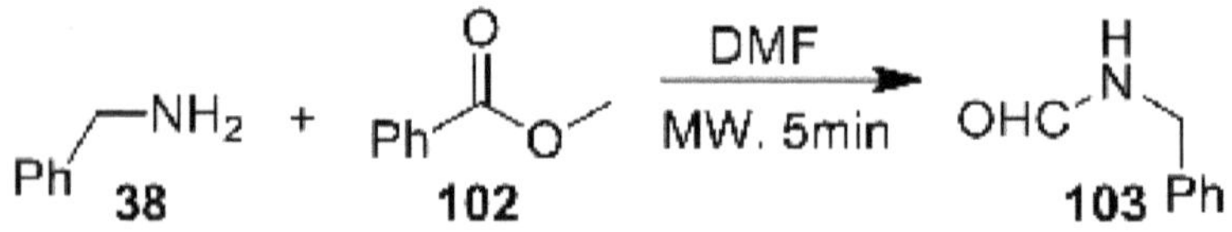

Figure 59.
N-formylation of various 1° and 2°.

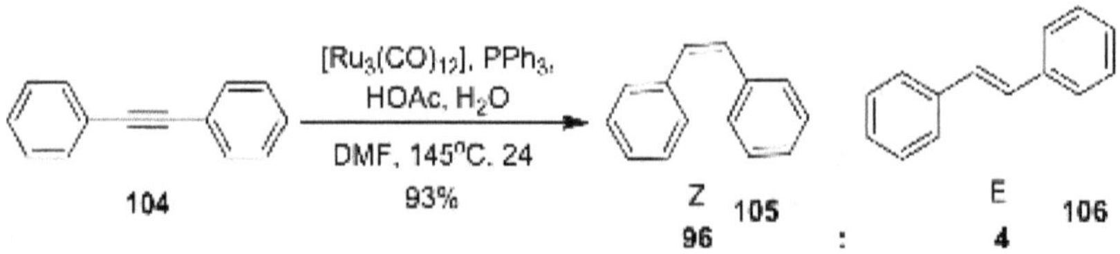

Figure 60.
Stereodivergent $[Ru_3(CO)_{12}]$ catalyzed semihydrogenation of diaryl alkynes.

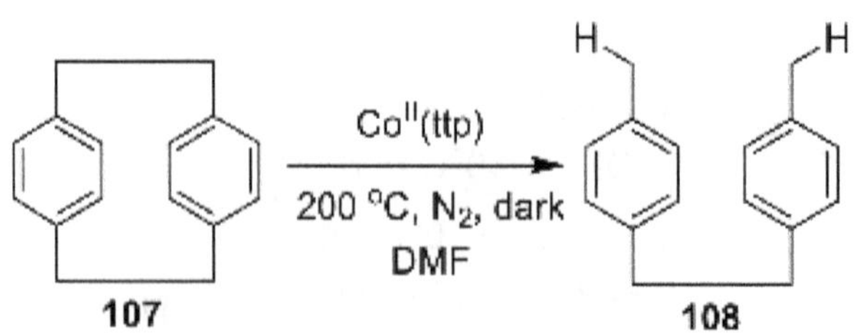

Figure 61.
DMF as hydrogenating reagent for benzylic positions.

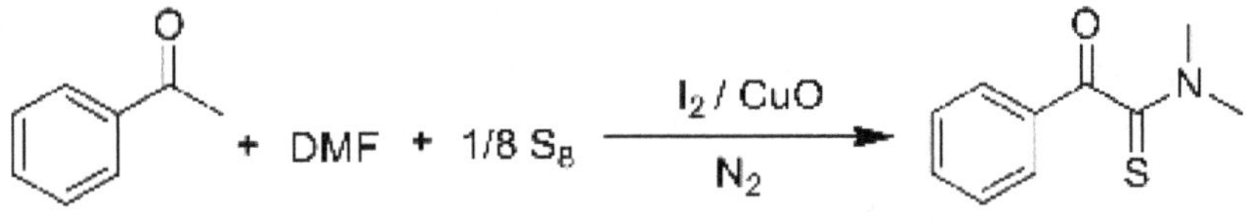

Figure 62.
Synthesis of α-arylketothioamides.

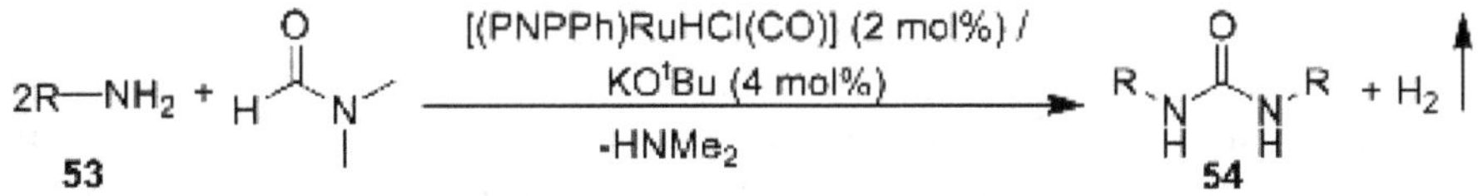

Figure 63.
Carbonylation of amines with DMF.

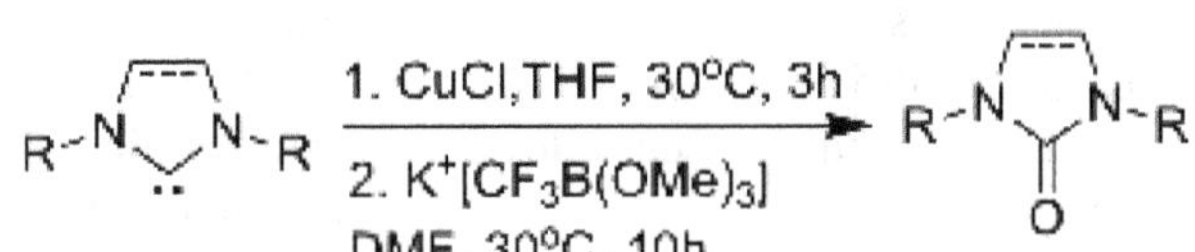

Figure 64.
Formation of complicated imidazolinones with DMF.

synthesis of analogs of natural products such as cis-combretastatin A-4 and trans-resveratrol [89].

Chan et al., reported a hydrogenation reaction catalyzed by cobalt porphyrins which hydrogenated C-C bond of [2.2] paracyclophane **107** (PCP) with DMF as solvent as well as hydrogen atom transfer agent (**Figure 61**). Metalloradical Co(II) porphyrins attacks the C-C sigma bond of PCP and the resultant benzyl radical abstracts a hydrogen atom from DMF to afford the hydrogenated product **108**. Results obtained from various control experiment revealed that the presence of benzyl radical intermediates in undergoing hydrogen atom transfer from DMF [90].

In 2017, Liu and coworkers synthesized α-arylketothioamides **110** *via* copper oxide and iodine mediated direct redox reaction from acetophenones **78**, elemental sulfur **109** and DMF under the nitrogen atmosphere (**Figure 62**). The elemental sulfur acts as a nucleophilic building block while DMF act as solvent and as the source of amino group (dimethylamine). This reaction tolerates a wide range of functional groups and proceeded in a redox efficient manner [91].

3.7 Carbonylation

Carbonylation is another important reaction in which the poisonous "CO" gas is generated from dialkylamides in the presence of suitable catalysts. Thus carbonylation reaction using dialkylamides is highly advantageous.

Gunanathan and coworkers developed a new mode of bond activation which is used effectively for the synthesis of simple and functionalized symmetrical and unsymmetrical urea derivatives from amines using DMF as CO source (**Figure 63**). Activation of N-H bond of amines by Ruthenium pincer complex and after that CO insertion from DMF with the liberation of hydrogen. Nucleophilicity of amines is essential for urea formation. The significance of this reaction occurs in an open condition, it avoids side products, doesn't require any pressure setup [92].

Furthermore, Chen and co-workers reported a unique and highly effective method for the formation of imidazolinones **112** from carbene complexes **111** through oxygen atom insertion reaction of NHC copper complexes in the presence of DMF as the source of oxygen (**Figure 64**) [93].

4. Conclusion

It is noteworthy that, the utilization of DMF as a precursor in heterocyclic synthesis was important development in the field of synthetic organic chemistry. With advent of new reagents, catalytic systems and need for development of efficient synthetic protocols it could be predicted that dialkyl amides will continue to find new applications in organic synthesis. So far dialkyl amides have been mainly utilized as a synthon through mono functionalization of one of the groups. Further, there is a lot of scope for its utilization as a difuctionalization, for example, alkyl group attached to carbonyl and nitrogen in DMA could be functionalized at both the ends simultaneously. Dialkyl amides due to low cost, ready availability and flexibility in reactivity, will continue to gain attention of synthetic chemists as a synthon, ligand, dehydrating agent and solvent. We appreciate all of the authors cited herein for their tremendous contributions that have developed this field. We hope that it is sufficiently impressive and thorough that it will increase the interest on organic chemistry and will initiate further developments in the applications of DMF/DMA beyond being just a polar solvent, because it can be used as substrates in several reactions such as formylation, amination, amidoalkylation, aminocarbonylation, amidation, and cyanation and it has been achieved under both metal-catalyzed and metal-free conditions. We believe this book chapter will make it easy for the synthetic chemists and invoke an idea about utility of dialkyl amides for some novel functional group transformations.

Acknowledgements

P.S thanks to UGC-RFSMS, New Delhi for the award of the fellowship for Ph.D.

Author details

Andivelu Ilangovan*, Sakthivel Pandaram and Tamilselvan Duraisamy
School of Chemistry, Bharathidasan University, Tiruchirappalli, Tamilnadu, India

*Address all correspondence to: ilangovanbdu@yahoo.com

References

[1] Muzart J. *N,N*-dimethylformamide: Much more than a solvent. Tetrahedron. 2009;**65**:8313-8323. DOI: 10.1016/j. tet.2009.06.091

[2] Dubey A, Upadhyay A, Kumar P. Pivaloyl chloride/DMF: A new reagent for conversion of alcohols to chlorides. Tetrahedron Letters. 2010;**51**: 744-746. DOI: 10.1016/j.tetlet. 2009.11.131

[3] Liu Y, He G, Chen K, Jin Y, Li Y, Zhu H. DMF-catalyzed direct and regioselective C-H functionalization: Electrophilic/nucleophilic 4-halogenation of 3-oxypyrazoles. European Journal of Organic Chemistry. 2011;**2011**:5323-5330. DOI: 10.1002/ejoc.201100571

[4] Rai A, Rai VK, Singh AK, Yadav LDS.[2 + 2] Annulation of aldimines with sulfonic acids: A novel one-pot *cis*-selective route to β-sultams. European Journal of Organic Chemistry. 2011; **2011**:4302-4306. DOI: 10.1002/ ejoc.201100628

[5] Gowda M, Pande S, Ramakrishna R, Prabhu K. Acylation of Grignard reagents mediated by *N*-methyl pyrrolidone: A remarkable selectivity for the synthesis of ketones. Organic & Biomolecular Chemistry. 2011;**9**: 5365-5368. DOI: 10.1039/C1OB05780D

[6] Liu X, Li C, Xu J, Lv J, Zhu M, Guo Y, et al. Surfactant-free synthesis and functionalization of highly fluorescent gold quantum dots. Journal of Physical Chemistry C. 2008;**112**:10778-10783. DOI: 10.1021/jp8028227

[7] Kawasaki H, Yamamoto H, Fujimori H, Arakawa R, Inada M, Iwasaki Y. Surfactant-free solution synthesis of fluorescent platinum subnanoclusters. Chemical Communications. 2010;**46**:3759-3761. DOI: 10.1039/ B925117K

[8] Hyotanishi M, Isomura Y, Yamamoto H, Kawasaki H, Obora Y. Surfactant-free synthesis of palladium nanoclusters for their use in catalytic cross-coupling reactions. Chemical Communications. 2011;**47**:5750-5752. DOI: 10.1039/ C1CC11487E

[9] Yao W, Gong WJ, Li HX, Li FL, Gao J, Lang JP. Synthesis of DMF-protected Au NPs with different size distributions and their catalytic performance in the Ullmann homocoupling of aryl iodides. Dalton Transactions. 2014;**43**:15752-15759. DOI: 10.1039/ C4DT01856G

[10] Azuma R, Nakamichi S, Kimura J, Yano H, Kawasaki H, Suzuki T, et al. Solution synthesis of *N,N*-dimethylformamide-stabilized iron-oxide nanoparticles as an efficient and recyclable catalyst for alkene hydrosilylation. ChemCatChem. 2018; **10**:2378-2382. DOI: 10.1002/ cctc.201800161

[11] Shanab FA, Sherif SM, Mousa SAS. Dimethylformamide dimethyl acetal as a building block in heterocyclic synthesis. Journal of Heterocyclic Chemistry. 2009;**46**(5):801-827. DOI: 10.1002/jhet.69

[12] Ding S, Jiao N. *N,N*-dimethylformamide: A multipurpose building block. Angewandte Chemie, International Edition. 2012;**51**:9226-9237. DOI: 10.1002/anie.201200859

[13] Batra A, Singh P, Singh KN. Cross dehydrogenative coupling (CDC) reactions of *N,N* disubstituted formamides, benzaldehydes and cycloalkanes. European Journal of Organic Chemistry. 2016:4927-4947. DOI: 10.1002/ejoc.201600401

[14] Bras JL, Muzart J. Recent uses of *N,N*-dimethylformamide and *N,N*- dimethylacetamide as reagents.

Molecules. 2018;**23**:1939. DOI: 10.3390/molecules23081939

[15] Heravi MM, Ghavidel M, Mohamadkhani L. Beyond a solvent: triple roles of dimethylformamide in organic chemistry. RSC Advances. 2018;**8**: 27832-27862. DOI: 10.1039/C8RA04985H

[16] Kodimuthali A, Mungara A, Prasunamba PL, Pal M. A simple synthesis of aminopyridines: Use of amides as amine source. Journal of the Brazilian Chemical Society. 2010;**21**: 1439-1445. DOI: 10.1590/S0103-50532010000800005

[17] Gu DW, Guo XX. Synthesis of *N*-arylcarboxamides by the efficient transamidation of DMF and derivatives with anilines. Tetrahedron. 2015;**71**: 9117-9122. DOI: 10.1016/j.tet.2015.10.008

[18] Chen C, Tan L, Zhou P. Approach for the synthesis of *N*-phenylamides from β-ketobutylanilides using dimethylformamide and dimethylacetamide as the acyl donors. Journal of Saudi Chemical Society. 2015; **19**:327-333

[19] Mondal S, Samanta S, Santra S, Bagdi AK, Hajra A. *N*,*N*-dimethylformamide as a methylenating reagent: Synthesis of heterodiarylmethanes via copper-catalyzed coupling between imidazo [1,2-a]pyridines and indoles/*N*,*N*-dimethylaniline. Advanced Synthesis and Catalysis. 2016;**358**:3633-3641. DOI: 10.1002/adsc.201600674

[20] Weng JQ, Xu WX, Dai XQ, Zhang JH, Liu XH. Alkylation reactions of benzothiazoles with *N*,*N*-dimethylamides catalyzed by the two-component system under visible light. Tetrahedron Letters. 2019;**60**:390-396. DOI: 10.1016/j.tetlet.2018.12.064

[21] Iranpoor N, Firouzabadi H, Rizi ZT, Erfan S. WCl_6/DMF as a new reagent system for the phosphine-free Pd(0)-catalyzed aminocarbonylation of aryl halides. RSC Advances. 2014;**4**: 43178-43182. DOI: 10.1039/C4RA04673K

[22] Venu B, Vishali B, Naresh G, Kumar VV, Sudhakar M, Kishore R, et al. C-H bond cyanation of arenes using *N*,*N*-dimethylformamide and NH_4HCO_3 as a CN source over a hydroxyapatite supported copper catalyst. Catalysis Science & Technology. 2016;**6**:8055-8062. DOI: 10.1039/C6CY01536K

[23] Kim J, Choi J, Shin K, Chang S. Copper-mediated sequential cyanation of aryl C-B and arene C-H bonds using ammonium iodide and DMF. Journal of the American Chemical Society. 2012;**134**:2528-2531. DOI: 10.1021/ja211389g

[24] Mata EG, Suarez AG. Regioselective acylation of benzodioxin derivatives employing $AlCl_3$-DMSO or $AlCl_3$-DMF reagent in the Friedel-Crafts reaction. Synthetic Communications. 1997;**27**: 1291-1300. DOI: 10.1080/00397919708003368

[25] Ahmed S, Boruah R. An efficient conversion for conjugated oximes into substituted pyridines under Vilsmeier conditions. Tetrahedron Letters. 1996; **37**:8231-8232. DOI: 10.1016/0040-4039(96)01909-0

[26] Zhao MN, Hui RR, Ren ZH, Wang YY, Guan ZH. Ruthenium-catalyzed cyclization of ketoxime acetates with DMF for synthesis of symmetrical pyridines. Organic Letters. 2014;**16**:3082-3085. DOI: 10.1021/ol501183z

[27] Weng Y, Zhou H, Sun C, Xie Y, Su W. Copper-catalyzed cyclization for access to 6H-chromeno[4,3-b]quinolin-6-ones employing DMF as the carbon source. The Journal of Organic Chemistry. 2017;**82**:9047-9053. DOI: 10.1021/acs.joc.7b01515

[28] Bai B, Tang L, Huanga H, Deng GJ. Synthesis of 2,4-diarylsubstituted-pyridines through a Ru-catalyzed four component reaction. Organic & Biomolecular Chemistry. 2015;**13**:4404-4407. DOI: 10.1039/05c5ob00162e

[29] Guo W, Liao J, Liu D, Li J, Ji F, Wu F, et al. A four-component reaction strategy for pyrimidine carboxamide synthesis. Angewandte Chemie. 2016; **128**:1-6. DOI: 10.1002/ange.201608433

[30] Lv Y, Li Y, Xiong T, Pu W, Zhang H, Sun K, et al. Copper-catalyzed annulation of amidines for quinazoline synthesis. Chemical Communications. 2013;**49**:6439-6644. DOI: 10.1039/c3cc43129k

[31] Zheng LY, Guo W, Fan XL. Metal-free, TBHP-mediated, [3+ +2+ +1]-type intermolecular cycloaddition reaction: Synthesis of pyrimidines from amidines, ketones, and DMF through C(sp3)C-H activation. Asian Journal of Organic Chemistry. 2017;**6**:837-840. DOI: 10.1002/ajoc.201700105

[32] Rao DN, Rasheed SK, Das P. Palladium/silver synergistic catalysis in direct aerobic carbonylation of C(sp2)-H bonds using DMF as a carbon source: Synthesis of pyrido-fused quinazolinones and phenanthridinones. Organic Letters. 2016;**18**:3142-3145. DOI: 10.1021/acs.orglett.6b01292

[33] Chen J, Feng JB, Natte K, Wu X. Palladium-catalyzed carbonylative cyclization of arenes by C-H bond activation with DMF as the carbonyl source. Chemistry - A European Journal. 2015;**21**:16370-16373. DOI: 10.1002/chem.201503314

[34] Zhang Q, Song C, Huang H, Zhang K, Chang J. Cesium carbonate promoted cascade reaction involving DMF as a reactant for the synthesis of dihydropyrrolizino[3,2-b]indol-10ones. Organic Chemistry Frontiers. 2018;**5**: 80-87. DOI: 10.1039/C7QO00771J

[35] Wang JB, Li YL, Deng J. Metal-free activation of DMF by dioxygen: A cascade multiple-bond-formation reaction to synthesize 3-acylindoles from 2-alkenylanilines. Advanced Synthesis and Catalysis. 2017;**359**:3460. DOI: 10.1002/adsc.201700584

[36] Gao X, Yu B, Mei Q, Yang Z, Zhao Y, Zhang H, et al. Atmospheric CO_2 promoted synthesis of N-containing heterocycles over $B(C_6F_5)_3$ catalyst. New Journal of Chemistry. 2016;**40**:8282-8287. DOI: 10.1039/ C6NJ01721E

[37] Rasal KB, Yadav GD. One-pot synthesis of benzimidazole using DMF as a multitasking reagent in presence $CuFe_2O_4$ as catalyst. Catalysis Today. 2018;**309**:51-60

[38] Yoshida H, Ito Y, Ohshita J. Three-component coupling using arynes and DMF: Straightforward access to coumarins via ortho-quinone methides. Chemical Communications. 2011;**47**: 8512-8514. DOI: 10.1039/c1cc11955a

[39] Mori S, Shibuya M, Yamamoto Y. Ruthenium-catalyzed hydrocarbamoylative cyclization of 1,6-diynes with formamides. Chemistry Letters. 2017;**46**:2. DOI: 10.1246/ cl.160961

[40] Hosoi K, Nozaki K, Hiyama T. Carbon monoxide free aminocarbonylation of aryl and alkenyl iodides using DMF as an amide source. Organic Letters. 2002;**4**:2849-2851. DOI: 10.1021/ol026236k

[41] Schnyder A, Beller M, Mehltretter G, Nsenda T, Studer M, Indolese AF. Synthesis of primary aromatic amides by aminocarbonylation of aryl halides using formamide as an ammonia synthon. The Journal of Organic Chemistry. 2001;**66**(12): 4311-4315. DOI: 10.1021/jo015577t

[42] Ju J, Jeong M, Moon J, Jung HM, Lee S. Aminocarbonylation of aryl

halides using a nickel phosphite catalytic system. Organic Letters. 2007;**9**(22): 4615-4618. DOI: 10.1021/ol702058e

[43] He T, Li H, Li P, Wang L. Direct amidation of azoles with formamides *via* metal-free C-H activation in the presence of *tert*-butyl perbenzoate. Chemical Communications. 2011; **47**:8946-8948. DOI: 10.1039/C1CC13086B

[44] Xu K, Hu Y, Zhang S, Zha Z, Wang Z. Direct amidation of alcohols with N-substituted formamides under transition-metal-free conditions. Chemistry - A European Journal. 2012; **18**:9793-9797. DOI: 10.1002/chem.201201203

[45] Gao L, Tang H, Wang Z. Oxidative coupling of methylamine with an aminyl radical: Direct amidation catalyzed by I_2/TBHP with HCl. Chemical Communications. 2014;**50**:4085-4088. DOI: 10.1039/c4cc00621f

[46] Fan W, Shi D, Feng B. TBAI-catalyzed synthesis of α-ketoamides via sp^3 C-H radical/radical cross-coupling and domino aerobic oxidation. Tetrahedron Letters. 2015;**56**: 4638-4641. DOI: 10.1016/j.tetlet.2015.06.021

[47] Mai WP, Wang HH, Li ZC, Yuan JW, Xiao YM, Yang LR, et al. $n$$Bu_4$NI-catalyzed direct synthesis of a-ketoamides from aryl methyl ketones with dialkylformamides in water using TBHP as oxidant. Chemical Communications. 2012;**48**:10117-10119. DOI: 10.1039/C2CC35279F

[48] Bi X, Li J, Shi E, Wang H, Gao R, Xiao J. Ru-catalyzed direct amidation of carboxylic acids with *N*-substituted formamides. Tetrahedron. 2016;**72**: 8210-8214. DOI: 10.1016/j.tet.2016.10.043

[49] Bannwart L, Abele S, Tortoioli S. Metal-free amidation of acids with formamides and T3P. Synthesis. 2016; **48**(13):2069-2078. DOI: 10.1055/s-0035-1561427

[50] Mete TB, Singh A, Bhat RG. Transition-metal-free synthesis of primary to tertiary carboxamides: A quick access to prodrug-pyrazinecarboxamide. Tetrahedron Letters. 2017;**58**:4709-4712

[51] Ali W, Rout SK, Guin S, Modi A, Banerjee A, Pater BK. Copper-catalyzed cross dehydrogenative coupling of *N*,*N*-disubstituted formamides and phenols: A direct access to carbamates. Advanced Synthesis and Catalysis. 2015;**357**: 515-522. DOI: 10.1002/adsc.201400659

[52] Phan NTS, Nguyen TT, Vu PHL. A copper metal-organic framework as an efficient and recyclable catalyst for the oxidative cross-dehydrogenative coupling of phenols and formamides. ChemCatChem. 2013;**5**:3068-3077. DOI: 10.1002/cctc.201300400

[53] Yuan YG, Guo SR, Xiang JN. Cu $(OAc)_2$-catalyzed thiolation of acyl C-H bonds with thiols using TBHP as an oxidant. Synlett. 2013;**24**(4):443-448. DOI: 10.1055/s-0032-1318188

[54] Singh P, Batra A, Singh P, Kaur A, Singh KN. Oxidative C-Se coupling of formamides and diselenides by using aqueous *tert*-butyl hydroperoxide: A convenient synthesis of selenocarbamates. European Journal of Organic Chemistry. 2013:7688-7692. DOI: 10.1002/ejoc.201301248

[55] Kumar GS, Kumar RA, Kumar PS, Reddy NV, Kumar KV, Kantam ML, et al. Copper catalyzed oxidative coupling of amines with formamides: A new approach for the synthesis of unsymmetrical urea derivatives. Chemical Communications. 2013;**49**: 6686-6688. DOI: 10.1039/C3CC42381F

[56] Cho S, Kim J, Lee S, Chang S. Silver-mediated direct amination of

benzoxazoles: Tuning the amino group source from formamides to parent amines. Angewandte Chemie International Edition. 2009;**48**: 9127-9130. DOI: 10.1002/anie.200903957

[57] Li Y, Xie Y, Zhang R, Jin K, Wang X, Duan C. Copper-catalyzed direct oxidative C-H amination of benzoxazoles with formamides or secondary amines under mild conditions. The Journal of Organic Chemistry. 2011;**76**:5444-5449. DOI: 10.1021/jo200447x

[58] Wang J, Hou JT, Wen J, Zhang J, Yu XQ. Iron-catalyzed direct amination of azoles using formamides or amines as nitrogen sources in air. Chemical Communications. 2011;**47**:3652-3654. DOI: 10.1039/c0cc05811d

[59] Chen X, Yang Q, Zhou Y, Deng Z, Mao X, Peng Y. Synthesis of 4-(dimethylamino) quinazoline via direct amination of quinazolin-4(3*H*)-one using *N*,*N*-dimethylformamide as a nitrogen source at room temperature. Synthesis. 2015;**47**(14): 2055-2062. DOI: 10.1055/s-0034-1380550

[60] Sharma A, Mehta VP, Eycken EVD. A convenient microwave-assisted desulfitative dimethylamination of the 2 (1*H*)-pyrazinone scaffold using *N*,*N*-dimethylformamide. Tetrahedron. 2008;**64**:2605-2610. DOI: 10.1016/j.tet.2008.01.030

[61] Ruijie Z, Hongting S, Bo R, Yan F, Hao W, Yehua S, et al. An efficient and green approach to synthesizing enamines by intermolecular hydroamination of activated alkynes. Chemical Research in Chinese Universities. 2015;**31**(2):212-217. DOI: 10.1007/s40242-015-4388-8

[62] Zhang C, Yue Q, Xiao Z, Wang X, Zhang Q, Li D. Synthesis of *O*-aroyl-*N*,*N*-dimethylhydroxylamines through hypervalent iodine-mediated amination of carboxylic acids with *N*,*N*-dimethylformamide. Synthesis. 2017;**49**(18):4303-4308. DOI: 10.1055/s-0036-1588460

[63] Wei E, Liu B, Lin S, Liang F. Multicomponent reaction of chalcones, malononitrile and DMF leading to γ-ketoamides. Organic & Biomolecular Chemistry. 2014;**12**:6389-6392. DOI: 10.1039/C4OB00971A

[64] Bao XD, Rong X, Liu Z, Gu Y, Liang G, Xia Q. Potassium *tert*-butoxide-mediated metal-free synthesis of sulfonamides from sodium sulfinates and *N*,*N*-disubstituted formamides. Tetrahedron Letters. 2018;**50**:2853-2858. DOI: 10.1016/j.tetlet.2018.06.031

[65] Yang C, Zhang F, Deng GJ, Gon H. Amination of aromatic halides and exploration of the reactivity sequence of aromatic halides. The Journal of Organic Chemistry. 2019;**84**(1):181-190. DOI: 10.1021/acs.joc.8b02588

[66] Li Y, Guo F, Zha Z, Wang Z. Iron-catalyzed synthesis of 2-vinylquinolines *via* sp3 C-H functionalization and subsequent CN cleavage. Chemistry, An Asian Journal. 2013;**8**:534-537. DOI: 10.1002/asia.201201039

[67] Lou SJ, Xu DQ, Shen DF, Wang YF, Liua YK, Xu ZY. Highly efficient vinylaromatics generation *via* iron-catalyzed sp^3 C-H bond functionalization CDC reaction: A novel approach to preparing substituted benzo [α]phenazines. Chemical Communications. 2012;**48**:11993-11995. DOI: 10.1039/C2CC36708D

[68] Liu J, Yi H, Zhang X, Liu C, Liu R, Zhang G, et al. Copper-catalysed oxidative Csp^3-H methylenation to terminal olefins using DMF. Chemical Communications. 2014;**50**:7636-7638. DOI: 10.1039/C4CC02275K

[69] Li YM, Lou SJ, Zhou QH, Zhu LW, Zhu LF, Li L. Iron-catalyzed α-methylenation of ketones with *N*,*N*-dimethylacetamide: An approach for α,β-unsaturated carbonyl compounds. European Journal of Organic Chemistry. 2015;**2015**:3044-3047. DOI: 10.1002/ejoc.201500189

[70] Liu Y, Wang CL, Xia HM, Wang Z, Wang YF. Direct Csp^3-H methylenation of 2-arylacetamides using DMF/Me_2NH-BH_3 as the methylene source. Organic & Biomolecular Chemistry. 2019;**17**:6153-6157. DOI: 10.1039/C9OB00875F

[71] Pu F, Li Y, Song YH, Xiao J, Liu ZW, Wang C, et al. Copper-catalyzed coupling of indoles with dimethylformamide as a methylenating reagent. Advanced Synthesis and Catalysis. 2016;**358**:539-542. DOI: 10.1002/adsc.201500874

[72] Li Y, Xue D, Lu W, Wang C, Liu ZT, Xiao J. DMF as carbon source: Rh-catalyzed α-methylation of ketones. Organic Letters. 2014;**16**:66-69. DOI: 10.1021/ol403040g

[73] Tang RY, Xie YX, Xie YL, Xiang JN, Li JH. TBHP-mediated oxidative thiolation of an sp^3 C-H bond adjacent to a nitrogen atom in an amide. Chemical Communications. 2011; **47**:12867-12869. DOI: 10.1039/c1cc15397h

[74] Dai C, Meschini F, Narayanam JMR, Stephenson CRJ. Friedel-Crafts amidoalkylation *via* thermolysis and oxidative photocatalysis. The Journal of Organic Chemistry. 2012;**77**:4425-4431. DOI: 10.1021/jo300162c

[75] Lao ZQ, Zhong WH, Lou QH, Li ZJ, Meng XB. KI-catalyzed imidation of sp^3 C-H bond adjacent to amide nitrogen atom. Organic & Biomolecular Chemistry. 2012;**10**:7869. DOI: 10.1039/c2ob26430g

[76] Deng X, Lei X, Nie G, Jia L, Li Y, Chen Y. Copper-catalyzed cross-dehydrogenative N_2-coupling of NH-1,2,3-triazoles with *N*,*N*-dialkylamides: N-amidoalkylation of NH-1,2,3-triazoles. The Journal of Organic Chemistry. 2017;**82**:6163-6171. DOI: 10.1021/acs.joc.7b00752

[77] Wang J, Li J, Huang J, Zhu Q. Transition metal-free amidoalkylation of benzothiazoles and amidoalkylarylation of activated alkenes with *N*,*N*-dialkylamides. The Journal of Organic Chemistry. 2016;**81**: 3017-3022. DOI: 10.1021/acs. joc.6b00096

[78] Ding S, Jiao N. Direct transformation of *N*,*N*-dimethylformamide to CN: Pd-catalyzed cyanation of heteroarenes via C-H functionalization. Journal of the American Chemical Society. 2011;**133**: 12374-12377. DOI: 10.1021/ja204063z

[79] Xiao J, Li Q, Chen T, Han LB. Copper-mediated selective aerobic oxidative C_3-cyanation of indoles with DMF. Tetrahedron Letters. 2015;**56**: 5937-5940. DOI: 10.1016/j. tetlet.2015.09.044

[80] Zhang L, Lu P, Wang Y. Copper-mediated cyanation of indoles and electron-rich arenes using DMF as a single surrogate. Organic & Biomolecular Chemistry. 2015;**13**:8322. DOI: 10.1039/c5ob01244a

[81] Pawara AB, Chang S. Catalytic cyanation of aryl iodides using DMF and ammonium bicarbonate as the combined source of cyanide: A dual role of copper catalysts. Chemical Communications. 2014;**50**:448. DOI: 10.1039/c3cc47926a

[82] Ushijima S, Togo H. Metal-free one-pot conversion of electron-rich aromatics into aromatic nitriles. Synlett. 2010;7:1067-1070. DOI: 10.1055/s-0029-1219575

[83] Ushijima S, Togo H. One-pot conversion of aromatic bromides and aromatics into aromatic nitriles. Synlett. 2010;**10**:1562-1566. DOI: 10.1055/s- 0029-1219935

[84] Ishii G, Moriyama K, Togo H. Transformation of aromatic bromides into aromatic nitriles via formations of Grignard reagents and their DMF adducts. Tetrahedron Letters. 2011;**52**: 2404-2406. DOI: 10.1016/j.tetlet.2011.02.110

[85] Wang Y, Wang F, Zhang C, Zhang J, Li M, Xu J. Transformylating amine with DMF to formamide over CeO_2 catalyst. Chemical Communications. 2014;**50**:2438. DOI: 10.1039/c3cc48400a

[86] Sonawane RB, Rasal NK, Jagtap SV. Nickel-(II)-catalyzed N-formylation and N-acylation of amines. Organic Letters. 2017;**19**:2078-2081. DOI: 10.1021/acs.orglett.7b00660

[87] Journet M, Cai D, Dimichele LM, Larsen RD. Highly efficient synthesis of α,β-acetylenic aldehydes from terminal alkynes using DMF as the formylating reagent. Tetrahedron Letters. 1998;**39**: 6427-6428. DOI: 10.1016/S0040-4039 (98)01352-5

[88] Yang D, Jeon HB. Convenient N-formylation of amines in dimethylformamide with methyl benzoate under microwave irradiation. Bulletin of the Korean Chemical Society. 2010;**31**(5):1424-1426. DOI: 10.5012/36 bkcs.2010.31.5.1424

[89] Li J, Hua R. Stereodivergent ruthenium-catalyzed transfer semihydrogenation of diaryl alkynes. Chemistry - A European Journal. 2011; **17**:8462-8465. DOI: 10.1002/chem.201003662

[90] Tam CM, To CT, Chan KS. Carbon-carbon σ-bond transfer hydrogenation with DMF catalyzed by cobalt porphyrins. Organometallics. 2016;**35**: 2174-2177. DOI: 10.1021/acs.organomet.6b00434

[91] Liu W, Chen C, Zhou P. Concise access to α-arylketothioamides by redox reaction between acetophenones, elemental sulfur and DMF. ChemistrySelect. 2017;**2**:5532. DOI: 10.1002/slct.201700866

[92] Krishnakumar V, Chatterjee B, Gunanathan C. Ruthenium-catalyzed urea synthesis by N-H activation of amines. Inorganic Chemistry. 2017;**56**: 7278-7284. DOI: 10.1021/acs.inorgchem.7b00962

[93] Zeng W, Wang E, Qiu R, Sohail M, Wu S, Chen FX. Oxygen-atom insertion of NHC-copper complex: The source of oxygen from *N,N*-dimethylformamide. Journal of Organometallic Chemistry. 2013;**743**:4448. DOI: 10.1016/j.jorganchem.2013.06.017

9

Approaches to the Total Synthesis of Puupehenone-Type Marine Natural Products

Yan-Chao Wu, Yun-Fei Cheng and Hui-Jing Li

Abstract

Puupehenones have been isolated from the marine sponge *Chondrosia chucalla*, which belong to a growing family of natural products with more than 100 members. These marine natural products have attracted increasing attention mainly due to their wide variety of biological activities such as antitumor, antiviral, and anti-HIV, and thus offer promising opportunities for new drug development. This chapter covers the approaches to the total synthesis of puupehenone-type marine natural products including puupehenol, puupehenone, puupehedione, and halopuupehenones. The routes begin with the construction of their basic skeletons, followed by the modification of their C- and D-rings. The contents are divided into two sections in terms of the key strategies employed to construct the basic skeleton. One is the convergent synthesis route with two synthons coupled by nucleophilic or electrophilic reaction, and the other is the linear synthesis route with polyene series cyclization as a key reaction.

Keywords: total synthesis, marine natural product, puupehenones, convergent synthesis, linear synthesis

1. Introduction

In recent years, the synthesis and application of marine natural products have become the focus of a much greater research effort, which is due in large part to the increased recognition of marine organisms as a rich source of novel compounds with biological applications [1–4]. The puupehenone-type marine natural products obtained from deep sea sponge have played a very important role in health care and prevention of diseases [5–14].

As shown in **Figure 1**, the most representative of this natural product family includes puppehenone, halopuupehenones, puupehedione, puupehenol, 15-cyanopuupehenol, 15-oxopuupehenol, and bispuupehenonen. Structurally, puupehenones are tetracyclic compounds consisting of a bicyclic sesquiterpene A- and B-rings and a shikimic acid/O-benzoquinone/O-phenol D-ring connected by tetrahydropyran/dihydropyran C-ring. In addition, the chiral center of the C-8 of this series of natural products listed in the figure is 8S, which is also the structural specificity of them.

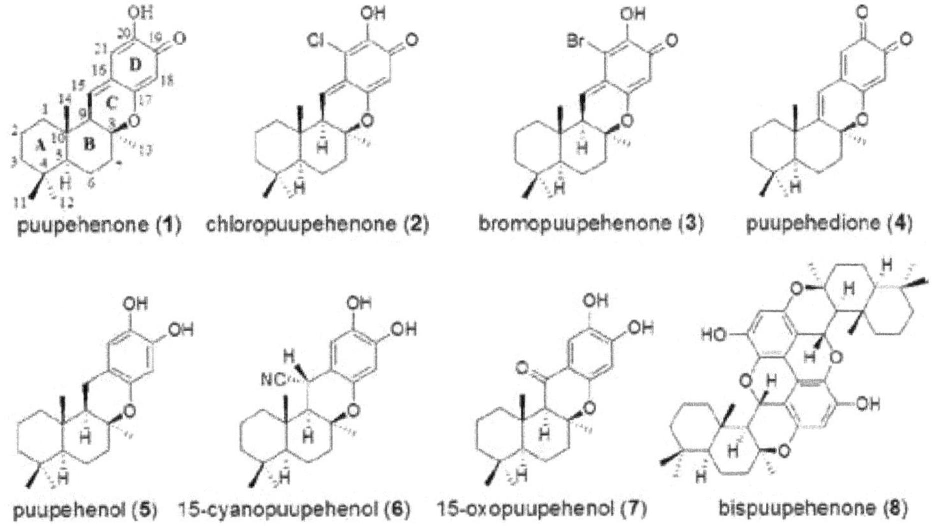

Figure 1.
Representatives of puupehenone-type natural products.

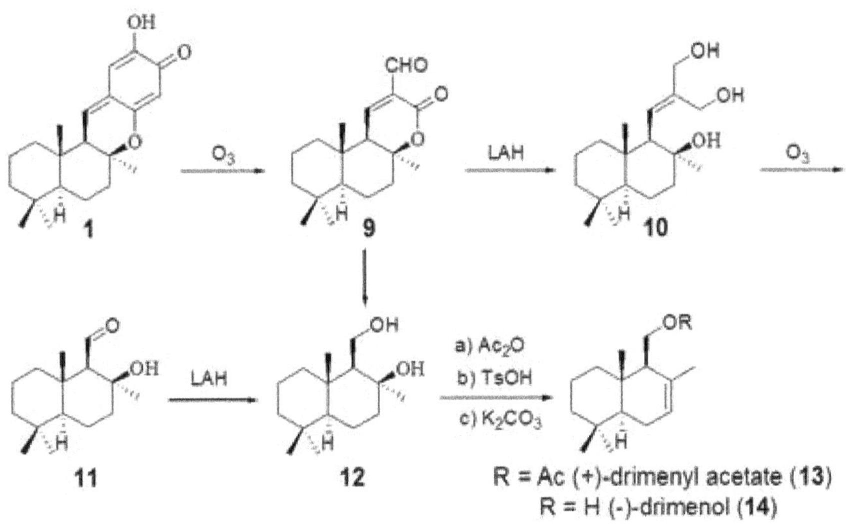

Figure 2.
The confirmation of the absolute configuration of puupehenone by chemical decomposition [18].

2. Isolation and biological activities

The natural product puupehenone was first isolated from the Hawaiian sponge *Chondrosia chucalla* by Schauer group in 1979 [15]. Subsequently, it was obtained from sponges such as *Heteronema*, *Hyrtios*, and *Strongylophora* sp. [14, 16, 17]. At that time, the assignment of an absolute stereochemistry to puupehenone was not permitted by spectroscopic analysis or degradative studies. As shown in **Figure 2**, it was not until 1996 that Capon group [18] used chemical decomposition, ozone oxidative decomposition, and lithium aluminum hydride reduction to finally decompose the natural product into the known structure (+)-drimenyl acetate (**13**) and (−)-drimenol (**14**), and since then the absolute configuration of puupehenone has been determined.

Studies show that puupehenone-type marine natural products have antitumor [5–8], anti-HIV [9], anticancer [10], antiviral [11], antimalaria [12], antimite [9, 13], immunomodulation [14], and other important physiological activities. In view of their important biological activities, such natural products have been favored by organic synthetic chemists since their separation.

3. Total synthesis of puupehenone-type marine natural products

Compound supply and appropriate structural analysis are two main barriers to develop a natural product into drug [19–31]. Chemical synthesis of marine natural products could provide the technological base for preparing enough materials for further research of bioactivity [19]. Thus, the total synthesis of puupehenones has been widely researched and published in excellent literature.

In the present chapter, approaches to the total synthesis of puupehenone-type marine natural products have been reviewed. In general, the strategies employed in the total synthesis of puupehenones are as follows:

- Convergent synthesis route with two synthons coupled by nucleophilic or electrophilic reaction.
- Linear synthesis route with polyene series cyclization as a key reaction.

3.1 Convergent synthesis route

Barrero group has been working on the study of total synthesis of puupehenone-type natural products, and has obtained great achievements [32–35]. In 1997, Barrero and coworkers reported the first enantiospecific synthesis of puupehenol and puupehenone in 32 and 22% yield, respectively [33]. As shown in **Figure 3**, acetoxyaldehyde **17** and aromatic synthon **18** were prepared from commercially available sclareol **15** and veratraldehyde **16** in high yields through a series of

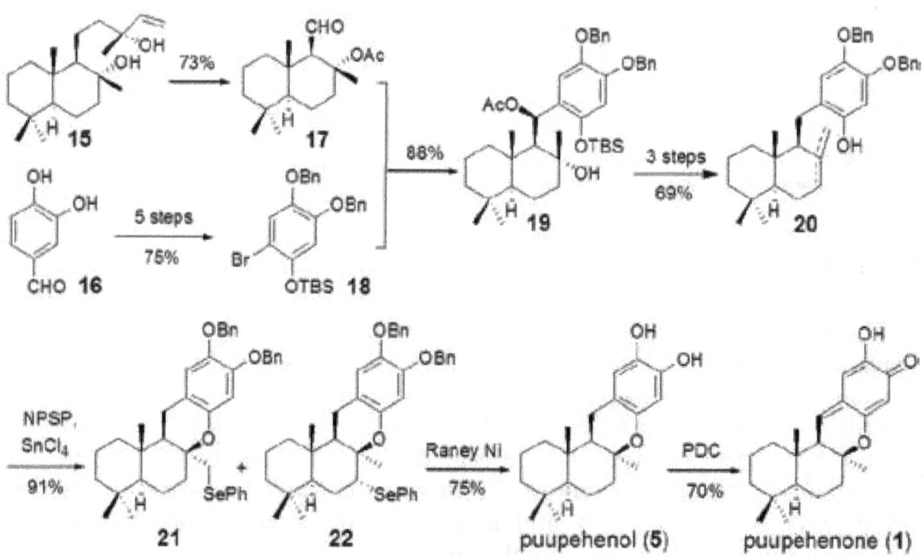

Figure 3.
Barrero's stereoselective synthesis of puupehenol and puupehenone [33].

transformations. The acetoxy alcohol **19** was completed by condensation of **17** with the aryllithium derived from **16**, and after three steps compound **19** gave the phenolic derivatives **20**. Finally, complete diastereoselectivity was achieved by organoselenium-induced cyclization. The treatment of **20** with NPSP(N-phenylselenophthalimide) and $SnCl_4$ obtained a mixture of the selenium derivatives **21** and **22**. Treatment with Raney Ni allowed both deprotection of the phenylselenyl group and removal of the benzyl ethers, producing puupehenol (**5**) as the only product, which was easily oxidized to (+)-puupehenone (**1**) in the presence of pyridinium dichromate (PDC).

Besides the above-mentioned research work, in 1999, Barrero group applied a base-mediated cyclization via 8,9-epoxy derivative to achieve the first asymmetric synthesis of puupehedione in 17% overall yield [35]. As shown in **Figure 4**, Sclareol **15** and veratraldehyde **16** were employed as the starting materials to obtain synthons **23** and **18**, which were accordingly converted to the key skeleton **24** in two steps. The treatment of **24** in the presence of mCPBA gave epoxydes **25**, and finally alcohol **26** was obtained in high yield when 8a, 9a-epoxyde **25** was treated with KOH in methanol. The subsequent two-step routine transformations, involving dehydration of alcohol **26** and oxidation, gave the target compound puupehedione.

In 2001, Maiti group reported the total synthesis of 8-epi-puupehedione with angiogenesis inhibitory activity [36]. As shown in **Figure 5**, commercially available carvone (**27**) and sesamol (**28**) were converted into tosylhydrazone **29** and aromatic synthon **30** in eight and three steps, respectively. Exposure of the vinyl lithium species, produced by the addition of tosylhydrazone **29** to an excess of n-BuLi, to **30** afforded the diene **31**. Then, the cleavage of the O-allyl ether of compound **31** with a catalytic amount of $RhCl_3{\cdot}3H_2O$ in refluxing EtOH resulted in spontaneous cyclization [37], affording a mixture of the puupehedione (**4**) and 8-epi-puupehedione (**32**).

In 2002, Quideau and coworkers completed asymmetric total synthesis of puupehenone in 10 steps starting from commercially available (+)-sclareolide [38]. The main feature of this synthesis strategy is an intramolecular attack of the terpenoid-derived C-8 oxygen function onto an oxidatively activated 1,2-

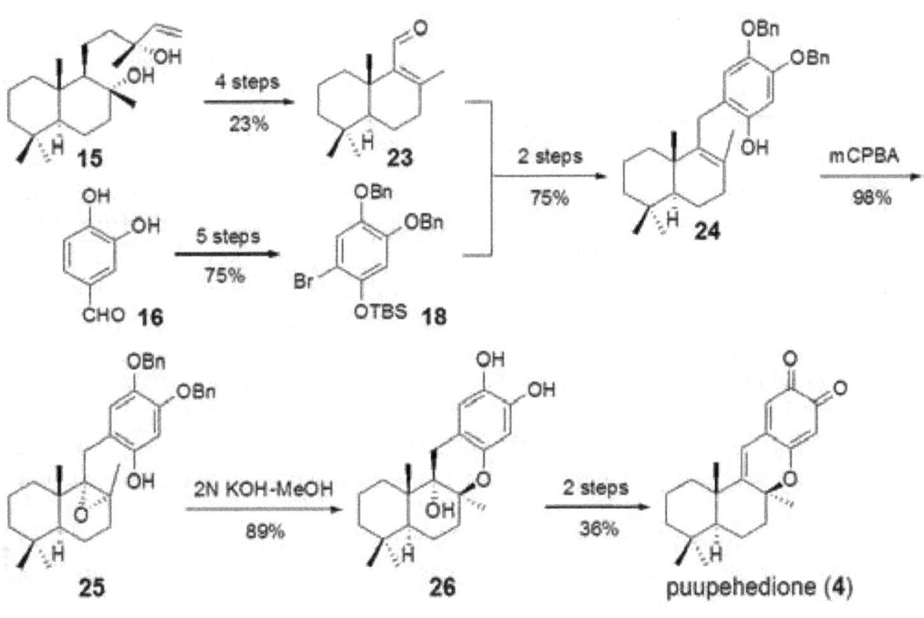

Figure 4.
Barrero's asymmetric synthesis of puupehedione [35].

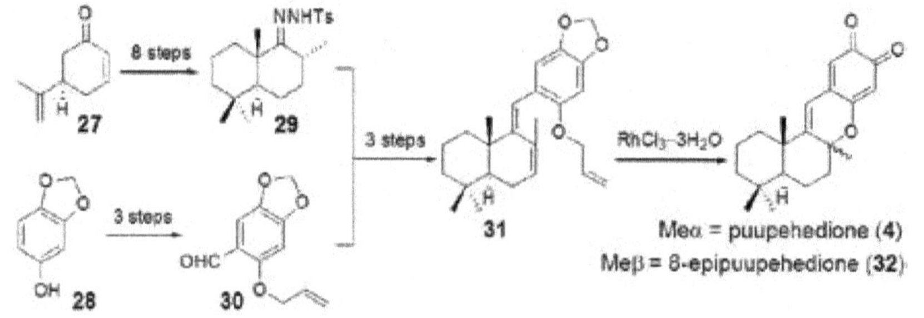

Figure 5.
Maiti's $RhCl_3$ catalyzed cyclization synthesis of 8-epi-puupehedione [36].

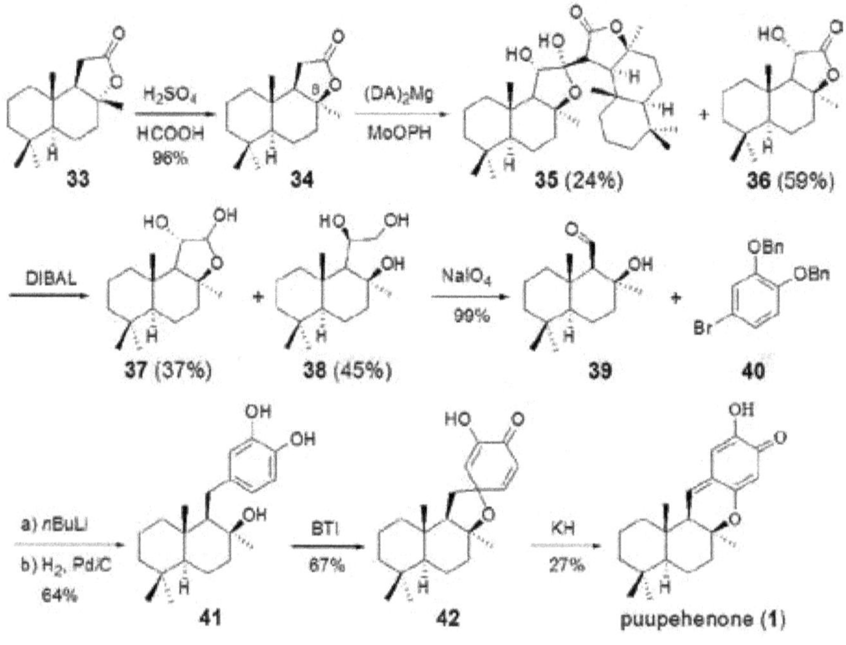

Figure 6.
Quideau's asymmetric synthesis of puupehenone [38].

dihydroxyphenyl unit to construct the heterocycle. As shown in **Figure 6**, the first step in their synthesis is inversion of the configuration at C-8 to construct a C-8 chiral center via simple acid treatment before coupling two key synthons. Subsequent treatment with $(DA)_2Mg$ and MoOPH afforded **35** and **36**, which were converted into **39** after hydride reduction with DIBAL and oxidation with $NaIO_4$. Then, coupling of aldehyde **15** with bromide **40** was achieved via a standard halogen-metal exchange protocol. Then, the key skeleton catechol **41** was obtained in good yield by a subsequent hydrogenolysis to remove both the benzyl protective groups. Finally, key oxidative activation of the catechol unit toward intramolecular attack by the drimane 8-oxygen and rearrangement with KH accomplished total synthesis of puupehenone.

In 2005, Alvarez-Manzaneda group reported a new strategy toward puupehenone-related natural products based on the palladium(II)-mediated diastereoselective cyclization of a drimenylphenol [39] to complete the first enantiospecific synthesis of 15-oxopuupehenol, together with improved syntheses of 15-cyanopuupehenone, puupehenone and puupehedione. As shown in **Figure 7**,

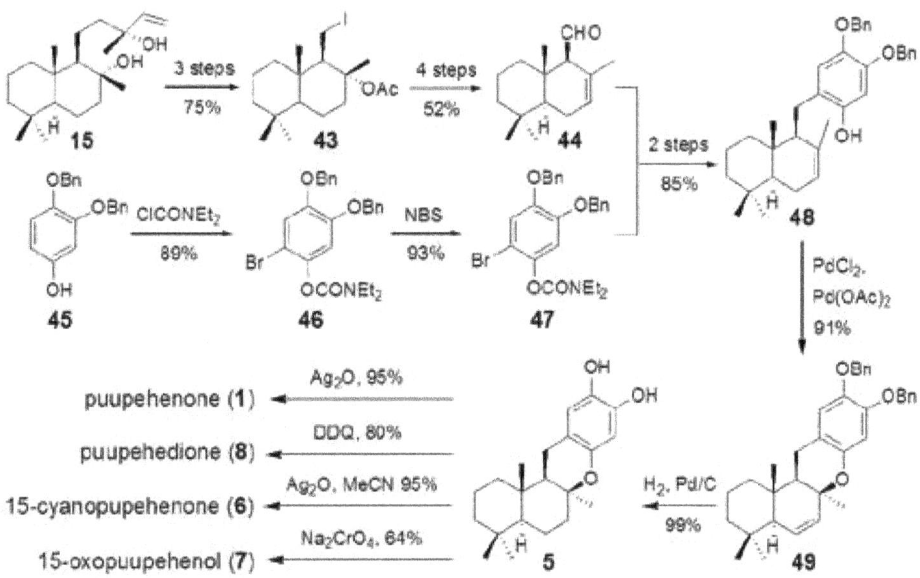

Figure 7.
Synthesis of several puupehenone-type natural products by palladium-catalyzed cyclization [39].

the drimane synthon **44** is easily prepared from sclareol (**15**) in seven steps. According to the procedure reported by Barrero [40], the drimane precursor **43** was prepared over three steps from **15** in 75% overall yield. Treating **43** with t-BuOK in a mixed solvent of DMSO-H_2O, followed by oxidative hydroboration, dehydration, and oxidation, afforded synthon **44** in 52% yield over four steps. The new synthon **47** from the 3,4-bis(benzyloxybenzyloxy)phenol (**45**), in a two-step sequence in 83% overall yield. Then, the key skeleton **48** was obtained by the coupling of **44** and **47**. Alvarez-Manzaneda and coworkers realized that catalytic $PdCl_2$ and $Pd(OAc)_2$ allowed to obtain the desired C8α-Me epimer with complete diastereoselectivity by inducing cyclization, yielding the most satisfactory compounds. Thus, puupehenol (**5**) was achieved by catalytic hydrogenation of **49**, which was obtained in high yield via palladium(II) catalysis of compound **48**. Finally, puupehenol (**5**) can be transformed into 15-oxopuupehenol (**7**) and the other puupehenone-related natural products.

Continuing their research into the total synthesis of this type of natural product, in 2007, Alvarez-Manzaneda group reported a new synthetic route toward puupehenone-related natural products starting from sclareol oxide (**50**) [41]. As shown in **Figure 8**, the key structure **53** was constructed by the coupling of two synthons **51** and **52**, based on a Diels-Alder cycloaddition approach. They employed sclareol oxide (**50**) as starting material to afford **51** over four steps which was treated with dienophile R-chloroacrylonitrile to afford compound **53** utilizing Diels-Alder cycloaddition. Treatment of **53** with DBU in benzene and DDQ in dioxane at room temperature led to aromatic nitrile **54**. Then, ent-chromazonarol (**55**) was obtained over three steps in 63% yield. The oxidation of phenol **55** to the appropriate ortho-quinone precursor of target compound **32** was then addressed.

In 2009, Manzaneda group [42] reported an enantiospecific route toward puupehenone and other related metabolites based on the cationic-resin-promoted Friedel-Crafts alkylation of alkoxyarenes with an α,β-unsaturated ketone **57**. As shown in **Figure 9**, Manzaneda and coworkers developed a very efficient synthesis of compound **57** which is a key synthon employed in the total synthesis of puupehenones, starting from commercially available sclareol (**15**) in 60% yield.

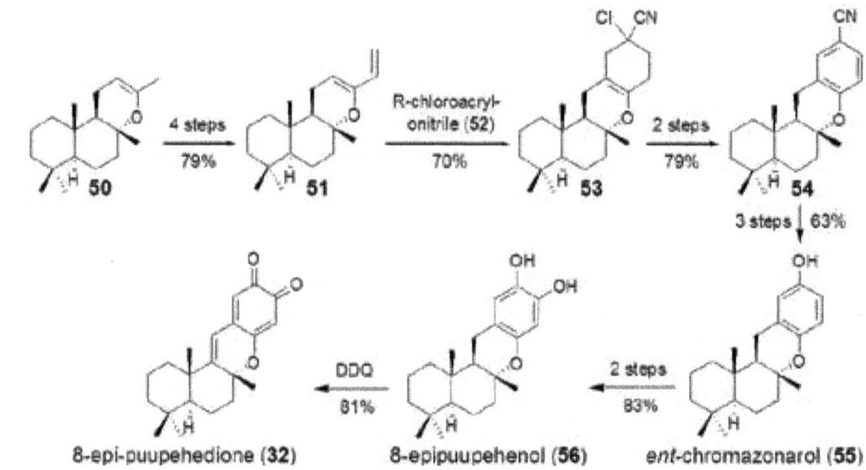

Figure 8.
Synthesis of 8-epi-puupehenone-type compound by Diels-Alder cyclization [41].

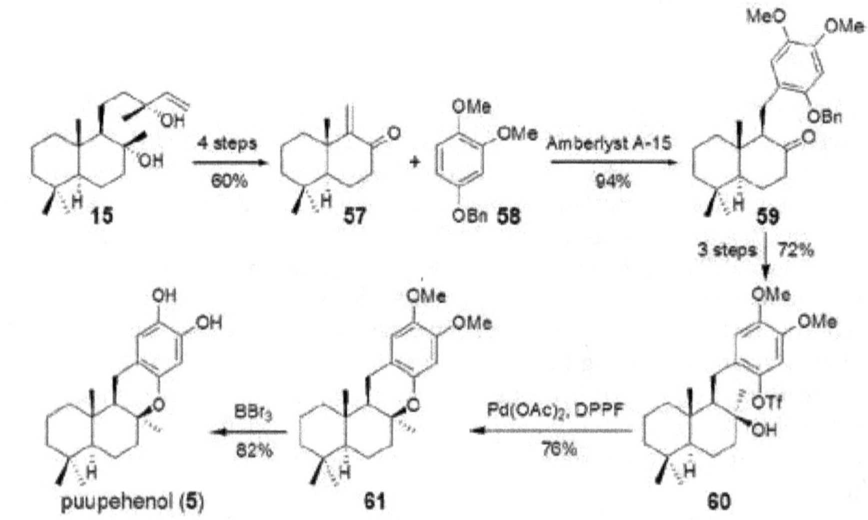

Figure 9.
Synthesis of puupehenol by Friedel-Crafts coupling reaction [42].

Then, the key intermediate ketone **59** was obtained in high yield and with complete diastereoselectivity by treatment of **57** with protected phenol **58** under the condition of Amberlyst A-15. Alternatively, treatment of ketone **59** with MeMgBr, further cleavage of the benzyl ether and protection of hydroxyl gave triflate **60** in 72% yield, which was a perfect intermediate for synthesizing puupehenone-type derivatives. Finally, puupehenol (**5**) was achieved in 82% yield by the deprotection of tetracyclic compound 61 obtained by the cyclization of triflate **60** with $Pd(OAc)_2$, DPPF (1,1-bis(diphenylphosphanyl) ferrocene), and sodium tertbutoxide in toluene.

In 2012, Baran group [43] described a scalable, divergent synthesis of bioactive meroterpenoids via borono-sclareolide (**63**) of which the preparation requires the excision of carbon monoxide from **33** and incorporation of BOH in its place

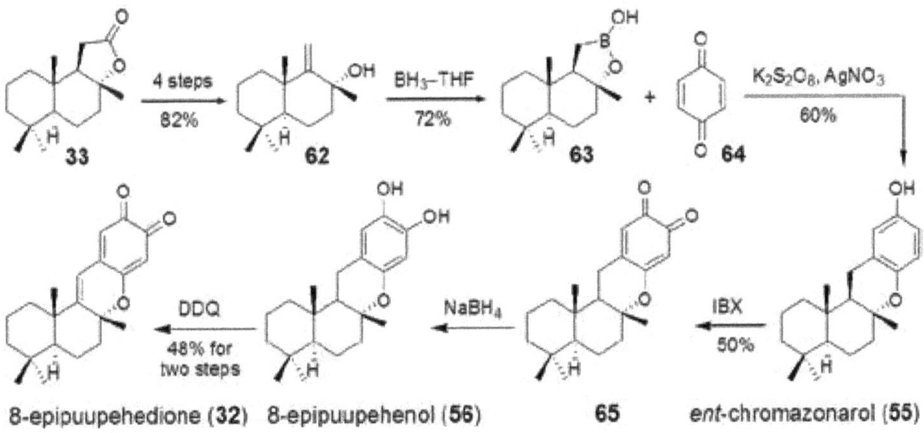

Figure 10.
Baran's synthesis of puupehenone-type natural products [43].

(**Figure 10**). Thus, compound **63** was accessed from **33** in 59% yield over five steps including DIBAL-mediated reduction of **33**, PIDA/I_2-mediated C—C bond cleavage, dehydroiodination, hydrolysis (AgF in pyridine followed by K_2CO_3 in methanol), and hydroboration with BH_3. This strategy constitutes the most efficient synthesis and highest yielding of **63** by far. Then, the key skeleton **55** was synthesized by treating **63** with an excess of 1,4-benzoquinone under the condition of $K_2S_2O_8$ and $AgNO_3$ in $PhCF_3/H_2O$ at 60°C. By following an oxidation-reduction-oxidation procedure, compound **55** was converted into 8-epi-puupehedione (**32**) in 24% yield.

The generation of boron-sclareolide **63** in such a direct manner enables total synthesis of puupehenone-type compounds to be more succinct than those previously established. However, the synthesis of C8α-Me boron-sclareolide is problematic, probably due to its lower stability than its C8α-Me epimer.

In 2017, Wu and his coworkers developed a hemiacetalization/dehydroxylation/hydroxylation/retro-hemiacetalization tandem reaction as the key step to synthesize puupehenone-type marine natural products [44], and this novel synthetic strategy is superior to other reported routes in terms of synthetic steps, purification of the intermediates, and overall yield.

As shown in **Figure 11**, the key synthon β-hydroxyl aldehyde **39** was accomplished starting from commercially available sclareolide (**33**) over four steps with an markedly higher overall yield (66%) including the stereospecific 8-episclareolide with H_2SO_4 in HCO_2H, α-hydroxylation, reduction with LiH_4Al, and in situ lactol-oxidation/ester-hydrolysis. The key skeleton **67** was constructed by the coupling of aldehyde **39** and ketone **66**. Treatment of **66** with LDA in THF at −78°C in the presence of **39** gave **67** in 67% yield. The following hemiacetalization/dehydroxylation/hydroxylation/retro-hemi-acetalization of **67** permitted to produce enone **68** as the only product in 92% yield, which can be converted into α-hydroxylated product **69** in 19% yield and natural product puupehenone (**1**) in 38% yield when treated with KHMDS and subsequent reaction with $P(OMe)_3$. Besides, natural products puupehenol (**5**) and puupehedione (**4**) were also achieved in good yield. Reduction of one with $NaBH_4$ gave puupehenol (**5**) in 92% yield and oxidation of **5** with DDQ afforded puupehedione (**4**) in 71% yield.

It is worth mentioning that the preparation strategy of the key intermediates **67** can be employed for the total synthesis of haterumadienone- and puupehenone-type natural products without using protecting groups.

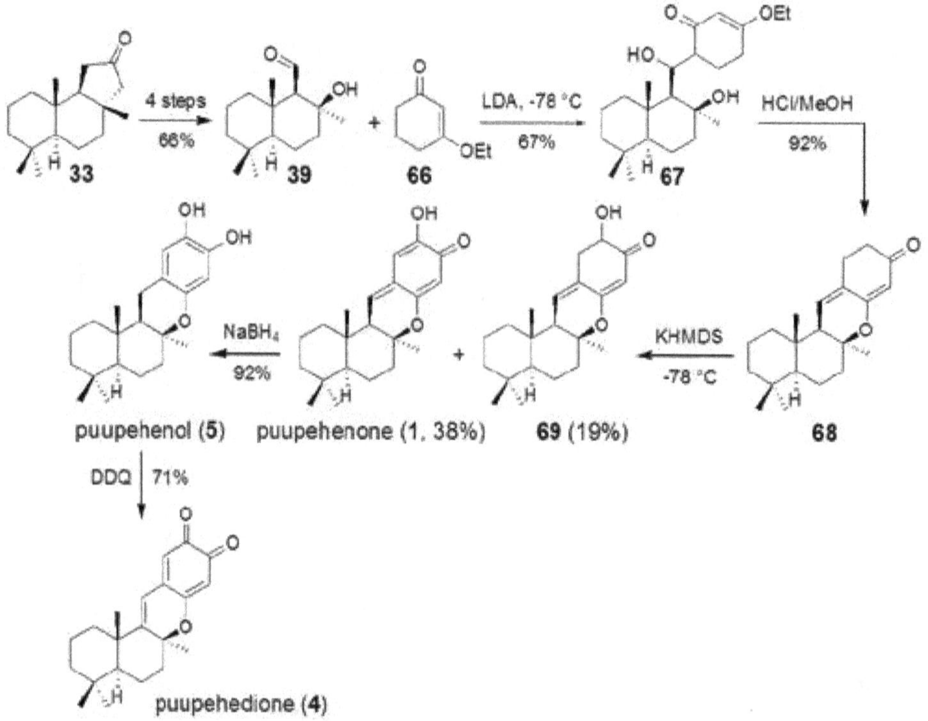

Figure 11.
Wu's synthesis of puupehenone-type natural products [44].

In the same year, Wu's group reported an enantiospecific semisynthesis of puupehedione commencing from sclareolide (**33**) in only seven steps with an overall yield of 25% [45].

The key drimanal trimethoxystyrene skeleton **71** and **72** were constructed by the palladium-catalyzed cross-coupling reaction of an aryl-iodine and a drimanal hydrazine (**70**) which was obtained from commercially available sclareolide over five steps. Treatment of compound **70** and aryl iodine in the presence of $Pd(PPh_3)_4$ and K_2CO_3 in toluene at 110°C afforded key skeletons **71** and **72** in 40 and 45% yields, respectively. Exposure of the mixture of drimanal trimethoxystyrenes **71** and **72** with Pb/C produced compound **73** in 62% yield. Then, the p-benzoquinone (**74**) can be prepared by treating **73** with CAN (ceric ammonium nitrate) in 84% yield. Treatment of **74** with pTsOH at room temperature produced compound **75** by intramolecular oxa-Stork-Danheiser transposition. Finally, puupehenone (**1**) was achieved over nine steps in 26% overall yield by exposing the resulting product **75** with K_2CO_3 in an enolization process. Besides, natural product puupehenol (**5**) can be obtained by reduction of **75** in presence of $NaBH_4$ in EtOH at room temperature (**Figure 12**).

Interestingly, natural product puupehedione (**4**) can be accomplished as the sole diastereoisomer in 47% yield when the mixture of **71** and **72** was treated with CAN at room temperature.

In 2018, Wu and his coworkers reported the divergent synthesis of (+)-8-epi-puupehedione [46].

Figure 13 shows the synthesis of 8-epi-puupehedione based on the Lewis acid catalyzed cyclization with sclareolide as starting material. Drimanal hydrazone **75** was obtained over four steps, as mentioned above. Then, the key skeleton was obtained by cross-coupling reaction of aryl iodide and drimanal hydrazone **75**, yielding intermediates **76** and **77** in 32 and 54% yields, respectively. Allylic product **78** was

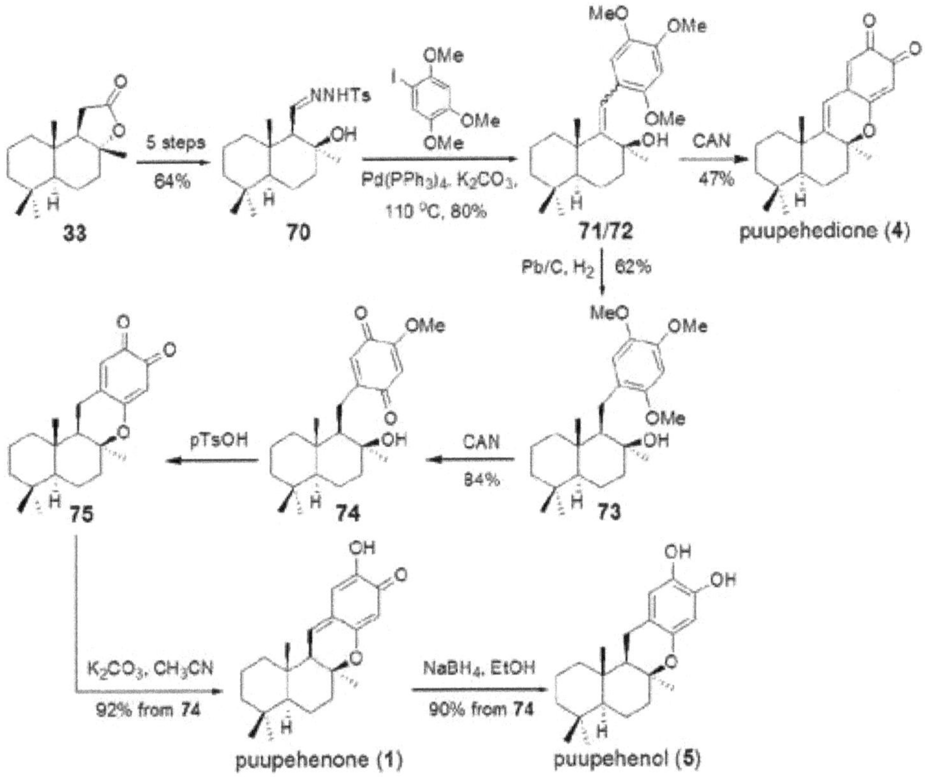

Figure 12.
Wu's synthesis of puupehenone-type natural products [45].

prepared in 91% yield by reduction of compounds **76** and **77** with TFA (trifluoroacetic acid) in the presence of Et_3SiH. Exposure of product **78** to CAN produced compound **80** as the major product in 48% yield, together with byproduct **79** in 9% yield. Then, the cyclization product 8-epi-19-methoxy puupehenol (**82**) was synthesized in 87% yield from compound **80** over two steps including treating **80** with $Na_2S_2O_4$ in the presence of tetrabutylammonium bromide (TBAB) and treating **81** with $BF_3{\cdot}Et_2O$. Exposure of **82** to CAN afforded **83** in 77% yield. Finally, 8-epi-puupehedione (**32**) was completed in 48% overall yield by reducing **83** with $NaBH_4$ and subsequent treatment with DDQ (2,3-dichloro-5,6-dicyano-1,4-benzoquinone).

Figure 14 shows another synthesis route of 8-epi-puupehedione (**32**) based on the tandem cyclization. Compound **84** was prepared in 62% yield by a ring opening reaction starting from 8-epi-19-methoxy puupehenol (**82**) by treatment with DDQ. Then, compound **84** was converted into **83** in 92% yield via an intramolecular oxa-Stork-Danheiser transposition reaction when it was treated with pTsOH. Reduction of **83** with $NaBH_4$ gave 8-epi-puupehenol (**56**), which can be transformed into 8-epi-puupehedione (**32**) by oxidation in the presence of DDQ.

Figure 15 shows an alternative synthesis of (+)-8-epi-puupehedione (**32**) based on the 6π electrocyclic reaction. Compound **87** was achieved in 86% yield when **80** was reacted with base in MeOH. Then, treatment of **87** with DDQ in a mixed solvent of CH_2Cl_2 and H_2O (10:1, v/v) obtained 8-epi-puupehedione (**32**) in 65% yield.

In 2018, Li's group developed an efficient synthesis of 8-epi-puupehenol [47] and central to this strategy is the Barton decarboxylative coupling, comprising a one-pot radical decarboxylation and quinone.

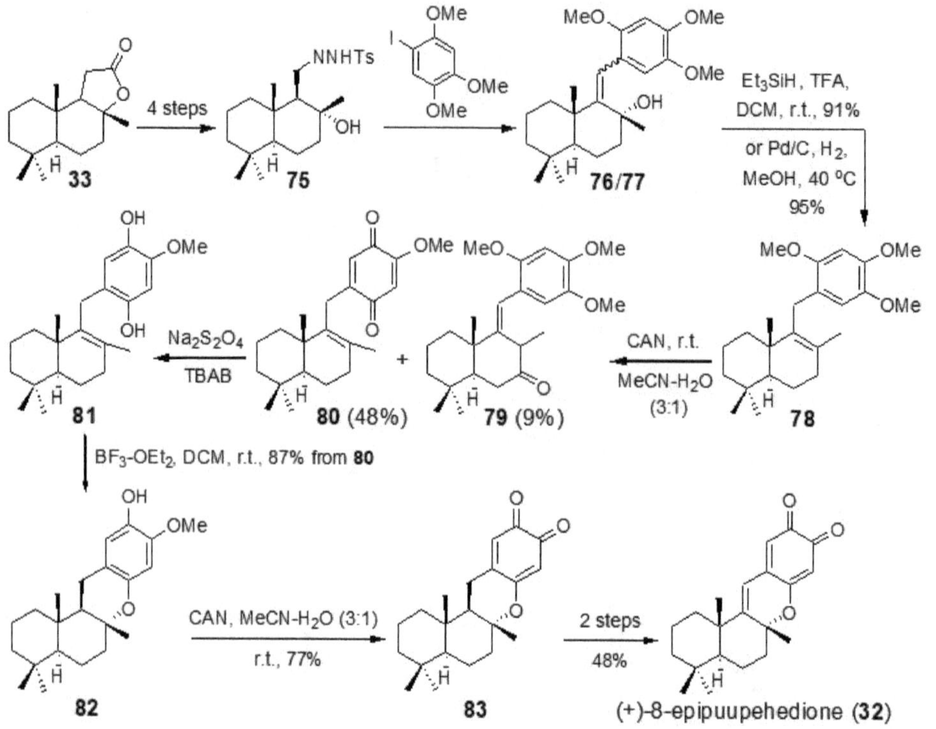

Figure 13.
Wu's synthesis of 8-epi-puupehedione based on the Lewis acid catalyzed cyclization [46].

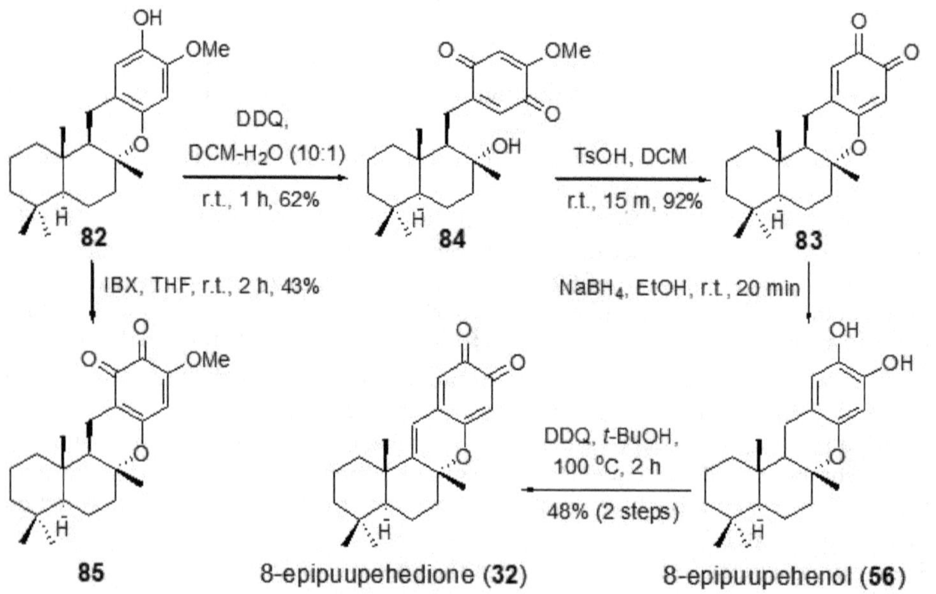

Figure 14.
Wu's synthesis of 8-epi-puupehedione based on the tandem cyclization [46].

As shown in **Figure 16**, the 8-O-acetylhomodrimanic acid (**89**) was obtained by oxidative degradation of sclareol (**15**) with potassium permanganate and Ac_2O, and then the key intermediate thiohydroxamic ester **90** was achieved from the coupling of

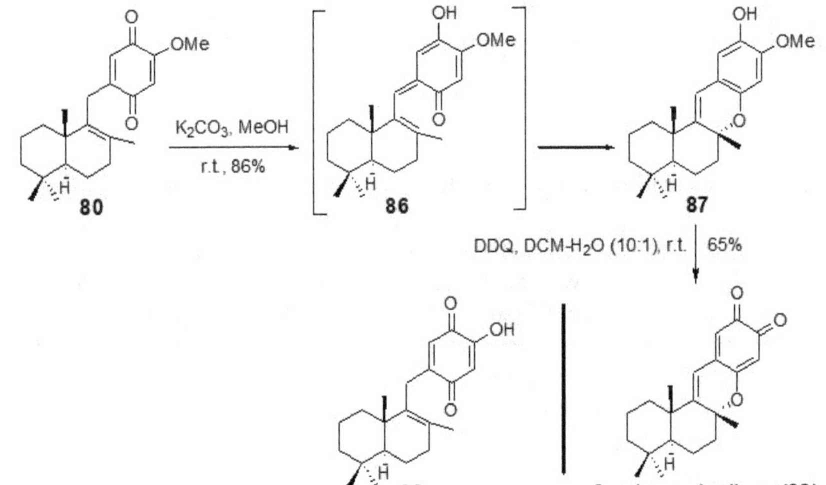

Figure 15.
Wu's synthesis of 8-epi-puupehedione based on 6π-electrocyclic reaction [46].

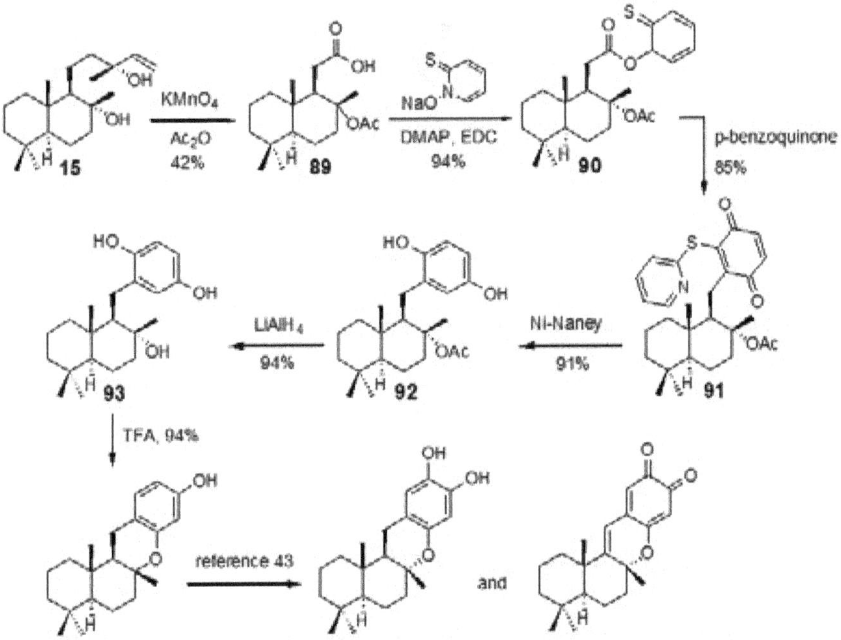

Figure 16.
Li's formal synthesis of 8-epi-puupehenol and 8-epi-puupehedione [47].

8-O-acetylhomodrimanic acid (**89**) with 2-mercaptopyridine N-oxide under Steglichesterification conditions. Treatment of Barton ester [48, 49] 90 with 250 W light in the presence of the electron-deficient benzoquinone gave pyridylthioquinone meroterpenoid 91 in 85% yield which was converted into acetate **92** in 91% yield when it was treated with Raney-nickel in EtOH at room temperature. To a solution of compound **92** in anhydrous THF added $LiAlH_4$ gave **93** in 93% yield which was treated with TFA (trifluoroacetic acid) to obtain **94** in excellent yield. Finally, synthesis of

8-epi-puupehenol (**56**) and 8-epi-puupehedione (**32**) was accomplished via IBX oxidation, followed by redox manipulation, according to the published literature [43].

3.2 Linear synthesis route

In 2004, Yamamoto group [50] developed a liner synthesis route of 8-epi-puupehenone (**32**) employing a new artificial cyclase **97**. Utilizing this cyclase, polycyclic terpenoids bearing a chroman skeleton can be obtained effectively.

8-epi-puupehenone **32** was achieved in 57% overall yield from **95** over four steps. Firstly, treatment of **95** with (R)-catalyst **97** through the enantio- and diastereoselective cyclization gave compound **96** in 62% yield. Then, **96** was transformed into 8-epi-puupehenone **32** through treatment of **96** with DDQ in 1,4-dioxane followed by hydrosilylative acetal cleavage employing Et_3SiH and $B(C_6F_5)_3$ and DDQ oxidation (**Figure 17**).

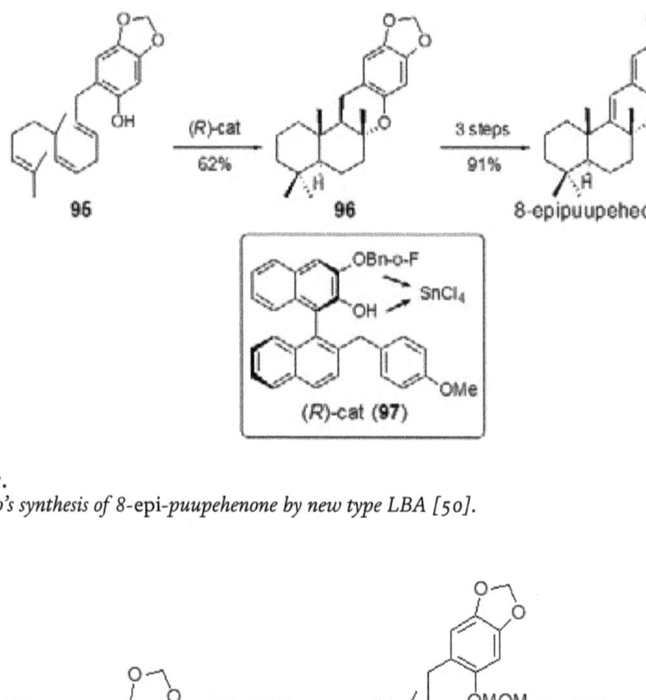

Figure 17.
Yamamoto's synthesis of 8-epi-puupehenone by new type LBA [50].

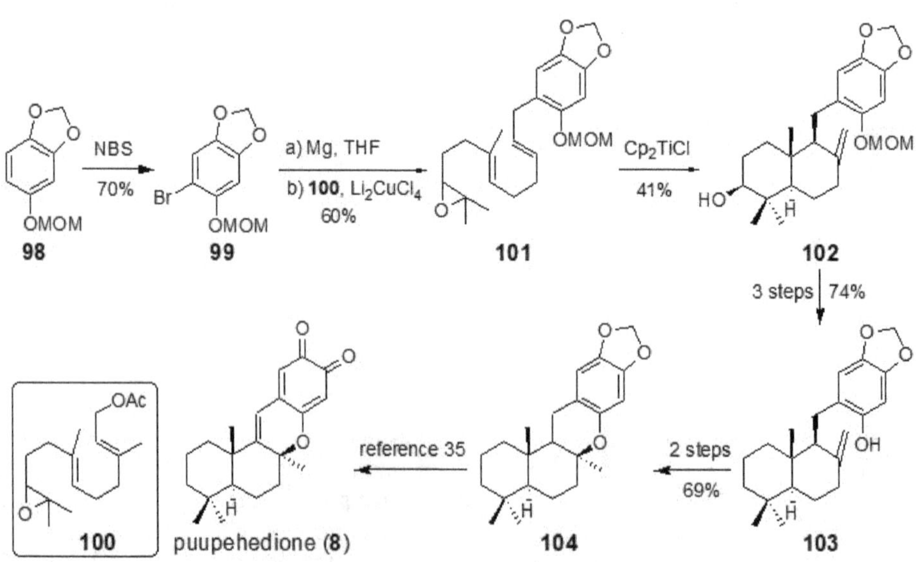

Figure 18.
Gansäuer's formal synthesis of puupehedione [51].

In 2006, Gansäuer and coworkers reported a highly stereoselective and catalytic synthesis strategy for the marine natural product puupehedione (**8**) [51].

As shown in **Figure 18**, compound **98** was converted into cyclization precursor **101** over two steps in 42% yield. Bromination of **98** with NBS (N-bromosuccinimide) gave compound **99** in 70% yield and treatment of **100** with Grignard reagent derived from **99** in the presence of Li_2CuCl_4 via copper-catalyzed allylic substitution reaction. Then, the bicyclic alcohol **102** was obtained in 41% yield by Cp_2TiCl-catalyzed epoxypolyene cyclization of **101**. The desired building unit **103** was achieved over three steps from compound **102** including deoxygenation of **102** by a Barton-McCombie reaction and high yielding cleavage of protecting group. Treating **103** with N-(phenylseleno) phthalimide and reduction with Bu_3SnH obtained compound **104**. Then, puupehedione (**8**) was completed according to the literature published by Barrero [35].

4. Conclusions

Undoubtedly, puupehenone-type marine natural products play a vital role in new drug development. Thus, the total synthesis of puupehenones has become a research hotspot for organic chemists [52].

Recent accomplishments made in total syntheses of puupehenone-type marine natural products are highlighted as above in terms of the employed synthetic strategy. The main routes to synthesize puupehenones include Diels-Alder cycloaddition reaction, coupling of the aldehydes with halogenated aromatic synthon, Friede-Crafts coupling reaction, hemiacetalization/dehydroxylation/hydroxylation/retro-hemiacetalization tandem reaction, and linear synthesis routes. Advances in total synthesis above offer new strategies for the chemical optimization of biologically active puupehenones.

Acknowledgements

This work was supported by the Natural Science Foundation of Shandong (ZR2019MB009), the Fundamental Research Funds for the Central Universities (HIT.NSRIF.201701), the Science and Technology Development Project of Weihai (2012DXGJ02, 2015DXGJ04), the Natural Science Foundation of China (21672046, 21372054), and the Found from the Huancui District of Weihai City.

Author details

Yan-Chao Wu*, Yun-Fei Cheng and Hui-Jing Li
School of Marine Science and Technology, Harbin Institute of Technology, Weihai, P. R. China

*Address all correspondence to: ycwu@iccas.ac.cn

References

[1] Butler MS. Natural products to drugs: Natural product-derived compounds in clinical trials. Natural Product Reports. 2008;**25**:475-516. DOI: 10.1039/B514294F

[2] Newman DJ, Cragg GM, Snader KM. Natural products as sources of new drugs over the period 1981–2002. Journal of Natural Products. 2003;**66**: 1022-1037. DOI: 10.1021/np030096l

[3] Gerwick WH, Fenner AM. Drug discovery from marine microbes. Microbial Ecology. 2013;**65**:800-806. DOI: 10.1007/s00248-012-0169-9

[4] Jin L, Quan C, Hou X. Potential pharmacological resources: Natural bioactive compounds from marine-derived fungi. Marine Drugs. 2016;**14**: 76. DOI: 10.3390/md14040076

[5] Kohmoto S, McConnell OJ, Wright A. Puupehenone, a cytotoxic metabolite from a deep water marine sponge, *Stronglyophora hartman*. Journal of Natural Products. 1987;**50**:336-336. DOI: 10.1021/np50050a064

[6] Pina IC, Sanders ML, Crews P. Puupehenone congeners from an indo-Pacific *Hyrtios* sponge. Journal of Natural Products. 2003;**66**:2-6. DOI: 10.1021/np020279s

[7] Longley RE, McConnell OJ, Essich E. Evaluation of marine sponge metabolites for cytotoxicity and signal-transduction activity. Journal of Natural Products. 1993;**56**:915-920. DOI: 10.1021/np50096a015

[8] Sova VV, Fedoreev SA. Metabolites from sponges as beta-1,3-gluconase inhibitors. Khimiya Prirodnykh Soedinenii. 1990;**4**:497-500

[9] El Sayed KA, Bartyzel P, Shen XY. Marine natural products as antituberculosis agents. Tetrahedron. 2000;**56**:949-953. DOI: 10.1016/S0040-4020(99)01093-5

[10] Castro ME, González-Iriarte M, Barrero AF. Study of puupehenone and related compounds as inhibitors of angiogenesis. International Journal of Cancer. 2004;**110**:31-38. DOI: 10.1002/ijc.20068

[11] John FD. Marine natural products. Natural Product Reports. 1998;**15**: 113-158. DOI: 10.1039/A815113Y

[12] Hamann MT, Scheuer PJ. Cyanopuupehenol, an antiviral metabolite of a sponge of the order Verongida. Tetrahedron Letters. 1991; **32**:5671-5672. DOI: 10.1016/S0040-4039(00)93525-1

[13] Kraus GA, Nguyen T, Bae J. Synthesis and antitubercular activity of tricyclic analogs of puupehenone. Tetrahedron. 2004;**60**:4223-4225. DOI: 10.1016/j.tet.2004.03.043

[14] Nasu SS, Yeung BK, Hamann MT. Puupehenone-related metabolites from two Hawaiian sponges, *Hyrtios* spp. The Journal of Organic Chemistry. 1995;**60**: 7290-7292. DOI: 10.1021/jo00127a039

[15] Ravi BN, Perzanowski HP, Ross RA. Recent research in marine natural products: The puupehenones. Pure and Applied Chemistry. 1979;**51**:1893-1900. DOI: 10.1351/pac197951091893

[16] Bourguet-Kondracki M-L, Debitus C, Guyot M. Dipuupehedione, a cytotoxic new red dimer from a new Caledonian marine sponge *Hyrtios* sp. Tetrahedron Letters. 1996;**37**:3861-3864. DOI: 10.1016/0040-4039(96)00700-9

[17] Bourguet-Kondracki M-L, Lacombe F, Guyot M. Methanol adduct of puupehenone, a biologically active derivative from the marine sponge

Hyrtios species. Journal of Natural Products. 1999;**62**:1304-1305. DOI: 10.1021/np9900829

[18] Urban S, Capon RJ. Absolute stereochemistry of puupehenone and related metabolites. Journal of Natural Products. 1996;**59**:900-901. DOI: 10.1021/np9603838

[19] Suyama TL, Gerwick WH, McPhail KL. Survey of marine nature product structure revisions: A synergy of spectroscopy and chemical synthesis. Bioorganic & Medicinal Chemistry. 2011;**19**:6675-6701. DOI: 10.1016/j. bmc.2011.07.017

[20] Baran PS, Maimone TJ, Richter JM. Total synthesis of marine natural products without using protecting groups. Nature. 2007;**446**:404-408. DOI: 10.1038/nature05569

[21] Hanessian S. Structure-based synthesis: From natural products to drug prototypes. Pure and Applied Chemistry. 2009;**81**:1085-1091. DOI: 10.1351/PAC-CON-08-07-12

[22] Hashimoto S. Natural product chemistry for drug discovery. The Journal of Anibiotics. 2011;**64**:697-701. DOI: 10.1038/ja.2011.74

[23] Morris JC, Phillips AJ. Marine natural products: Synthetic aspects. Natural Product Reports. 2011;**28**: 269-289. DOI: 10.1039/C0NP00066C

[24] Morris JC, Phillips AJ. Marine natural products: Synthetic aspects. Natural Product Reports. 2010;**27**: 1186-1203. DOI: 10.1039/B919366A

[25] Carter GT. Natural products and pharma 2011: Strategic changes spur new opportunities. Natural Product Reports. 2011;**28**:1783-1789. DOI: 10.1039/C1NP00033K

[26] Henkel T, Brunne RM, Reichel F. Statistical investigation into the structural complementarity of natural products and synthetic compounds. Angewandte Chemie (International Ed. in English). 1999;**38**:643-647. DOI: 10.1002/(SICI)1521-3773(19990301)38: 5<643::AID-ANIE643>3.0.CO;2-G

[27] Capon RJ. Marine natural products chemistry: Past, present, and future. Australian Journal of Chemistry. 2010; **63**:851-854. DOI: 10.1071/ch10204

[28] Morris JC, Phillips AJ. Marine natural products: Synthetic aspects. Natural Product Reports. 2009;**26**: 245-265

[29] Morris JC, Phillips AJ. Marine natural products: Synthetic aspects. Natural Product Reports. 2008;**25**: 95-117. DOI: 10.1039/B701533J

[30] Morris JC, Nicholas GM, Phillips AJ. Marine natural products: Synthetic aspects. Natural Product Reports. 2007; **24**:87-108. DOI: 10.1039/B602832M

[31] Nicholas GM, Phillips AJ. Marine natural products: Synthetic aspects. Natural Product Reports. 2006;**23**:79-99. DOI: 10.1039/B501014B

[32] Barrero AF, Manzaneda EA, Altarejos J. Synthesis of biologically active drimanes and homodrimanes from (−)-sclareol. Tetrahedron. 1995;**51**: 7435-7450. DOI: 10.1016/0040-4020 (95)00370-N

[33] Barrero AF, Alvarez-Manzaneda EJ, Chahboun R. Enantiospecific synthesis of (+)-puupehenone from (−)-sclareol and protocatechualdehyde. Tetrahedron Letters. 1997;**38**:2325-2328. DOI: 10.1016/S0040-4039(97)00305-5

[34] Barrero AF, Alvarez-Manzaneda EJ, Chahboun R. Synthesis of wiedendiol-A and wiedendiol-B from labdane diterpenes. Tetrahedron. 1998;**54**: 5635-5650. DOI: 10.1016/S0040-4020 (98)00235-X

[35] Barrero AF, Alvarez-Manzaneda EJ, Chahboun R. Synthesis and antitumor activity of puupehedione and related compounds. Tetrahedron. 1999;**55**: 15181-15208. DOI: 10.1016/S0040-4020 (99)00992-8

[36] Maiti S, Sengupta S, Giri C. Enantiospecific synthesis of 8-epipuupehedione from (R)-(−)-carvone. Tetrahedron Letters. 2001;**42**: 2389-2391. DOI: 10.1016/S0040-4039 (01)00153-8

[37] Martin SF, Garrison PJ. General methods for alkaloid synthesis. Total synthesis of racemic lycoramine. The Journal of Organic Chemistry. 1982;**47**: 1513-1518. DOI: 10.1021/jo00347a029

[38] Quideau S, Lebon M, Lamidey A-M. Enantiospecific synthesis of the antituberculosis marine sponge metabolite (+)-puupehenone. The arenol oxidative activation route. Organic Letters. 2002;**4**:3975-3978. DOI: 10.1021/ol026855t

[39] Alvarez-Manzaneda EJ, Chahboun R, Barranco Pérez I, et al. First enantiospecific synthesis of the antitumor marine sponge metabolite (−)-15-oxopuupehenol from (−)-sclareol. Organic Letters. 2005;**7**: 1477-1480. DOI: 10.1021/ol047332j

[40] Barrero AF, Alvarez-Manzaneda EJ, Chahboun R. New routes toward drimanes and nor-drimanes from (−)-sclareol. Synlett. 2000;**2000**:1561-1564. DOI: 10.1055/s-2000-7924

[41] Alvarez-Manzaneda EJ, Chahboun R, Cabrera E. Diels–Alder cycloaddition approach to puupehenone-related metabolites: Synthesis of the potent angiogenesis inhibitor 8-epipuupehedione. The Journal of Organic Chemistry. 2007;**72**:3332-3339. DOI: 10.1021/jo0626663

[42] Alvarez-Manzaneda E, Chahboun R, Cabrera E. A convenient enantiospecific route towards bioactive merosesquiterpenes by cationic-resin-promoted Friedel–Crafts alkylation with A,B-enones. European Journal of Organic Chemistry. 2009;**2009**: 1139-1143. DOI: 10.1002/ejoc.200801174

[43] Dixon DD, Lockner JW, Zhou Q. Scalable, divergent synthesis of meroterpenoids via "borono-sclareolide". Journal of the American Chemical Society. 2012;**134**:8432-8435. DOI: 10.1021/ja303937y

[44] Wang HS, Li HJ, Wu YC. Protecting-group-free synthesis of haterumadienone- and puupehenone-type marine natural products. Green Chemistry. 2017;**19**:2140-2144. DOI: 10.1039/c7gc00704c

[45] Wang HS, Li HJ, Wu YC. Enantiospecific semisynthesis of puupehedione-type marine natural products. The Journal of Organic Chemistry. 2017;**82**:12914-12919. DOI: 10.1021/acs.joc.7b02413

[46] Wang HS, Li HJ, Wu YC. Divergent synthesis of bioactive meroterpenoids via palladium-catalyzed tandem carbene migratory insertion. European Journal of Organic Chemistry. 2018;**2018**: 915-925. DOI: 10.1002/ejoc.201800026

[47] Li SK, Zhang SS, Wang X. Expediently scalable synthesis and antifungal exploration of (+)-yahazunol and related meroterpenoids. Journal of Natural Products. 2018;**81**:2010-2017. DOI: 10.1021/acs.jnatprod.8b00310

[48] Ling T, Xiang AX, Theodorakis EA. Enantioselective total synthesis of avarol and avarone. Angewandte Chemie (International Ed. in English). 1999;**38**: 3089-3091. DOI: 10.1002/(SICI)1521-3773(19991018)38:20<3089::AID-ANIE3089>3.0.CO;2-W

[49] Marcos IS, Conde A, Moro RF. Synthesis of quinone/hydroquinone

sesquiterpenes. Tetrahedron. 2010;**66**: 8280-8290. DOI: 10.1016/j.tet. 2010.08.038

[50] Ishibashi H, Ishihara K, Yamamoto H. A new artificial cyclase for polyprenoids: enantioselective total synthesis of (−)-chromazonarol, (+)-8-epi-puupehedione, and (−)-11′-deoxytaondiol methyl ether. Journal of the American Chemical Society. 2004; **126**:11122-11123. DOI: 10.1021/ja0472026

[51] Gansäuer A, Rosales A, Justicia J. Catalytic epoxypolyene cyclization via radicals: Highly diastereoselective formal synthesis of puupehedione and 8-epi-puupehedione. Synlett. 2006; **2006**:927-929. DOI: 10.1055/s-2006-933139

[52] Shen B. A new golden age of natural products drug discovery. Cell. 2015;**163**: 1297-1300. DOI: 10.1016/j.cell. 2015.11.031

10

Design and Strategic Synthesis of Some β-Carboline-Based Novel Natural Products of Biological Importance

Tejpal Singh Chundawat

Abstract

β-Carboline compounds and their derivatives have attracted strong interest in medicinal chemistry due to their biological and pharmacological properties. Many bioactive β-carboline-based natural products have been found to be an important source of drugs and drug leads. β-Carboline has major players in natural products chemistry, which plays an important role in drug discovery. β-Carboline represents the core unit of several natural products, alkaloids, and bioactive compounds. The unusual and complex molecular architectures of natural products pose significant challenges to organic chemists and are a source of inspiration for the development of new organic reactions and innovative synthetic strategies. However, in many cases, β-carboline natural products are isolated in only minute quantities, and their constant supply from natural sources is problematic or virtually impossible. In addition, chemoselective derivatization of natural products themselves is usually quite difficult because of their sensitive and elaborate molecular structures, and access to their structural analogs is severely restricted in many cases. Since chemical synthesis is expected to be the only way to overcome these shortcomings, β-carboline natural products are rewarding synthetic targets for organic chemists. This chapter assimilates the reports pertaining to the synthetic applications of some β-carbolines for the synthesis of substituted and fused β-carbolines.

Keywords: β-carboline, scaffold, organic synthesis, natural product, biological importance

1. Introduction

The need for efficient and practical synthesis of biologically active molecules remains one of the greatest intellectual challenges with which chemists are faced in the twenty-first century.

Organic synthesis is a compound-creating activity often focused on biologically active molecules and occupies a central role in any pharmaceutical development endeavor. The field of organic synthesis has made phenomenal advances in the past 50 years, yet chemists still struggle to design synthetic routes that will enable them to obtain sufficient quantities of complex molecules for biological and medicinal studies. The diversity of β-carboline compounds offers a great advantage for being developed into new drugs because of their unique and complex structures, developed through old and underexplored species evolution.

The drive to develop methodology allowing improved access to such compounds has arisen after the demonstration of the useful physical and chemical properties possessed by this class of compounds such as improved lipophilicity and decrease in oxidative metabolism.

Naturally occurring compounds have always played a vital role in medicine and, in particular, β-carboline has progressively become real players in recent drug discovery. The β-carboline moiety represents core structure of several natural compounds and pharmaceutical agents. Compounds containing this subunit are pervasively present in plants, marine organisms, insects, mammalian including human tissues and body fluids in the form of alkaloids or hormones [1–7]. Several β-carboline-based compounds of natural or synthetic origin are ascribed with different pharmacological properties [8] which include antimalarial [9, 10], antineoplastic [11, 12], anticonvulsive [13], hypnotic and anxiolytic [14], antiviral [15], antimicrobial [16], as well as topoisomerase-II inhibitors [17, 18] and cGMP inhibi-tors [19] (**Figure 1**).

Further, the significance of β-carboline-based compound is underscored by the way that two of the β-carboline-based mixes Tadalafil and Abecarnil (**Figure 2**) are clinically utilized for erectile brokenness and CNS issue, individually [20–22].

Many bioactive β-carboline-based natural products have been found to be an important source of drugs and drug leads. Most of the natural products of interest to the pharmaceutical industry are secondary metabolites and several such β-carbolines, derived from marine invertebrates, have been in clinical trials as experimental anti-cancer drugs. The significant favorable position offered by utilizing these metabolites as valuable formats, is that they are as such exceedingly dynamic and specific. Being created ordinarily to secure a specific living being, they have been exposed to evolutive pressure for a few a huge number of years and have been chosen to achieve ideal action and to perform particular capacities.

Synthesis of medicinally important β-carboline-based natural products is challengeous in synthetic organic chemistry. Current research activities while primarily with the academic laboratories, have generated convincing evidence that these natural products have an exceedingly bright future in discovery of life saving drugs

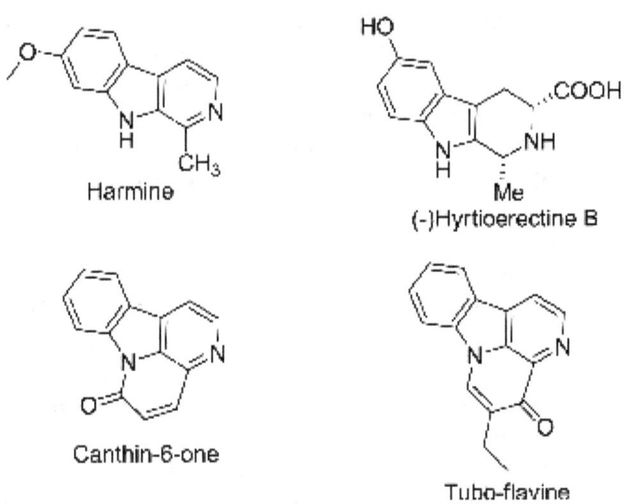

Figure 1.
Bioactive β-carboline-based compounds.

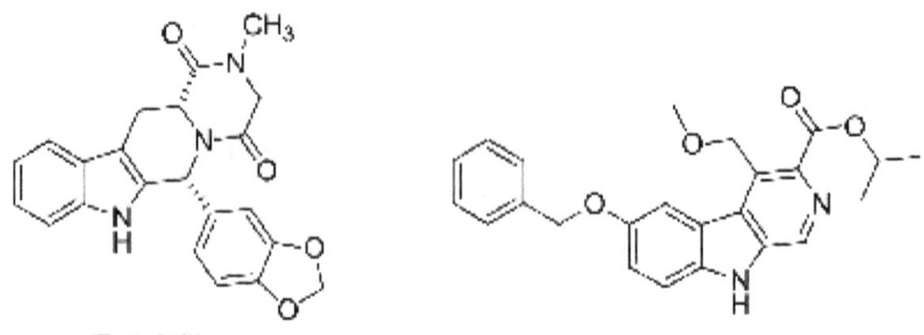

Figure 2.
β-Carboline-based drugs.

[23] included antibacterial, analgesic, anti-inflammatory, antimalarial, anticancer, antiparasitic and antiviral agents [24]. Although large numbers of novel β-carboline compounds have been isolated from plants, marine organisms, insects, mammalian including human tissues.

Furthermore, huge numbers of these substances have articulated natural action, without a doubt, not many have been advertised as pharmaceutical products. Some of the compounds have also been valuable as "lead" compounds, which have led to derivatives of them being marketed [25, 26].

In addition, the biological diversity of many of the β-carboline compounds still partially unknown. A considerable lot of them have indicated fascinating bioactivities both in vitro and in vivo measures, although just couple of molecules have been up to this point brought into facilities and onto the pharmaceutical market. Be that as it may, precedents are realized where cutting-edge clinical or preclinical preliminaries, did by utilizing common β-carboline items have prompted promising outcomes in the investigation of new prescriptions a variety of diseases including cancer and infective pathologies. Synthetic organic chemistry is able to produce sufficient amounts for a broad biological application and to provide access to synthetic analogs for structure-activity relationships (SAR) studies.

In particular, alkaloids establish one of the biggest classes of natural products and are synthesized by terrestrial and marine organisms on every transformative dimension and a standout amongst the most encouraging being indole alkaloids. Indole alkaloids, their action, synthesis, and potential use in medicines have been as of now inspected in a few articles [27–29]. Marine indole alkaloids speak to a rich gathering of characteristic natural compounds and can possibly turned out to be new medicinal chemistry leads for different psychiatric disorders, just as to give better bits of knowledge into the comprehension of serotonin receptor work. These atoms are sensible synthetic targets, which further improve their incentive as conceivable medicinal chemistry studies; be that as it may, hardly any, have been set up as a feature of manufactured or therapeutic science thinks about intended to produce advanced leads.

In this class, β-carbolines that consist of a pyridine ring that is fused to an indole skeleton and biological activity of their derivatives is also well established [30].

Also, substance blend might be utilized to illuminate normal procedures at the atomic dimension through biomimetic approaches, to affirm the structures of natural compounds which are typically settled depending just on spectral information, or to develop new synthetic methods for tackling the challenge of the complex chemical templates designed by nature. Significant endeavors are identified with the structure of particles that in nature are created by metabolic changes happening with high return and rate, and furthermore with high regio-, diastereo- and enantio-particularity.

2. Synthesis of β-carboline

During the last two decades, β-carboline-based natural products have been the focus of many investigations [31]. The β-carboline is a core-unit of several natural compounds and pharmaceutical agents. Compounds containing this core-unit are pervasively present in plants, marine animals, insects, mammalian including human tissues and body fluids in the form of alkaloids or hormones. Several β-carboline-based compounds of natural or synthetic origin are ascribed with different pharmacological properties which include antimalarial, antineoplastic, anticonvulsive, hypnotic and anxiolytic, antiviral, antimicrobial, as well as topoisomerase-II inhibitors and cGMP inhibitors.

The Pictet-Spengler reaction since its discovery in 1911 has been the key step of the synthetic strategies formulated for obtaining either substituted or fused β-carbolines [32]. The utility of Pictet-Spengler reaction is immense as it allows the option to either construct the tetrahydro-β-carboline (THBC) core first with appropriate substitution which could be extended after cyclization or to install the different substitutions which undergo cascade reactions during cyclization to afford the new THBC derivatives. These THBCs can then be oxidized to generate the desired β-carboline-derivative. However due to major significance associated with this heterocyclic moiety, alternate strategies for generating new β-carbolines are desired. In this context one of the possible strategies could be generation of a β-carboline core that bears a functional group at a suitable position that could be synthetically designed for producing substituted or fused β-carbolines. The presence of an electrophilic site in the form of formyl group in close proximity of the indole NH which is a nucleophilic site makes it an attractive template for the synthesis of substituted and 1–9 annulated β-carbolines. Alternatively, intramolecular cyclization could also be achieved with the N-2 to generate 1–2 annulated β-carbolines.

The synthesis of 1-formyl-9*H*-β-carboline was firstly reported by Gatta and Misiti [33] while carrying out the studies toward SeO_2 mediated oxidation of variously substituted THBCs. During the synthesis of carboline he unexpectedly obtained the 1-formyl-9*H*-β-carboline instead of the expected 1-methyl,1-phenyl-3-(methoxycarbonyl)-1,4-dihydro-4-oxo-β-carboline when the reaction of the diastereomeric mixture of 1-methyl,1-phenyl THBC was carried out with SeO_2 in dioxane. Probably the reaction was preceded through the oxidation of the benzylic moiety affording the benzaldehyde, followed by the aromatization of C-ring and finally the oxidation of the C-1-methyl to the formyl group (**Figure 3**).

Later Gatta and co-workers [34] reported an improved synthesis of methyl 1-formyl-9-*H*-pyrido [3,4-*b*] indole-3-carboxylate from 1-methyl-3-methoxycarbonyl-β-carboline via oxidation with SeO_2 in dioxane. These workers further reported the application of 1-formyl-9*H*-β-carboline for the synthesis of canthin-6-one [35]. They extended the synthetic utility of for the generation of pyrimido-[3,4,5-*lm*]-pyrido-[3,4-*b*]-indole derivatives in the synthesis of different derivatives of this carboline moiety.

Suzuki et al. [36] reported the total synthesis of various naturally occurring 4,8-dioxygenated β-carboline alkaloids (**Figure 4**). The synthetic route involved two methodologies (i) an improved Fischer indolization for affording 7-oxygenated indole via protecting the phenolic group with a tosyl group and (ii) construction of a 4-methoxy-β-carboline skeleton by the C-3 selective cyclization of the C-2 substituent of the indole. Then, 4-methoxy-β-carboline was converted into 1-nitrile derivative with diethylphosphoryl cyanide (DEPC) via *N*-oxide by a modified Reissert-Henze reaction.

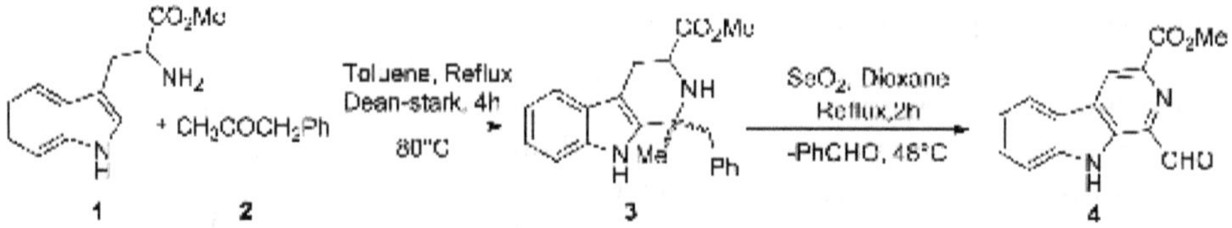

Figure 3.
Synthesis of 1-formyl-9H-β-carboline.

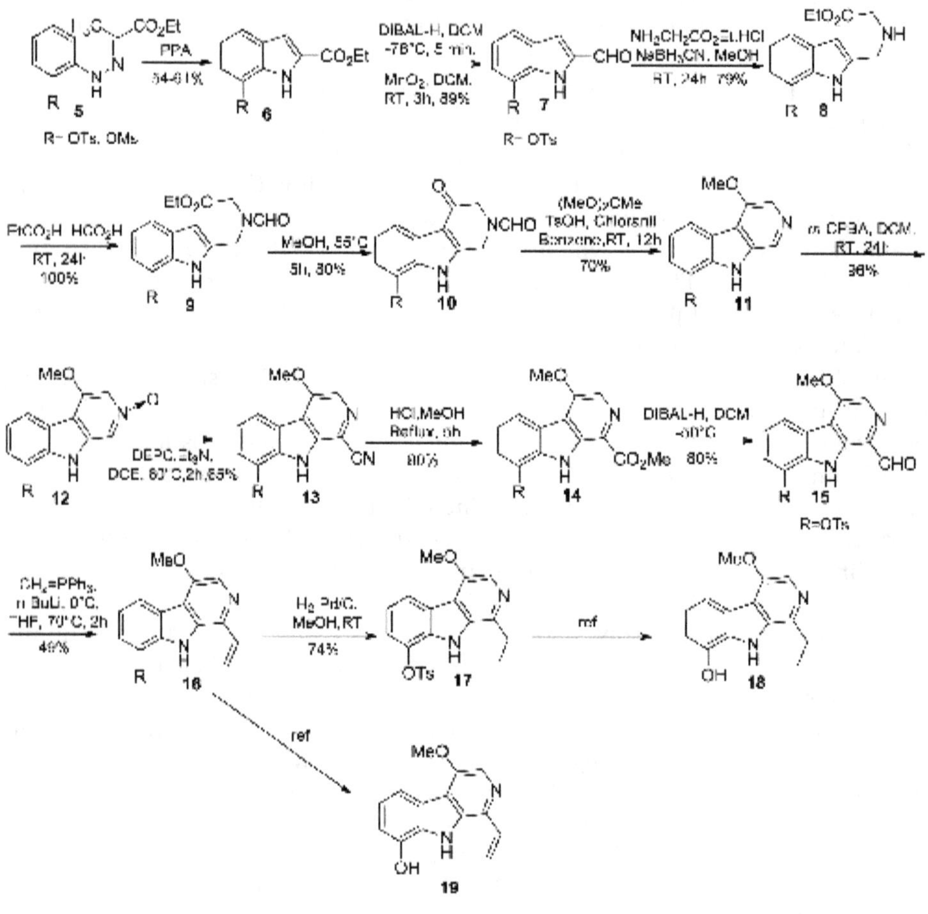

Figure 4.
Total synthesis of naturally occurring 4,8-dioxygenated β-carboline alkaloids.

Takasu et al. [37] also reported the synthesis of different β-carboline-based compounds including the natural products Kumujancine, MVC (4-methoxy vinyl β-carboline), Creatine and their corresponding salts. They followed the synthetic strategies which involved the Pictet-Spengler reaction of tryptamine hydrochloride with ethyl glyoxylate in ethanol, followed by acylation with acetyl chloride which furnished THBC in 44% yields (**Figure 5**).

Condie and Bergman [38] reported the condensation of 1-formyl-9*H*-β-carboline with ethyl azidoacetate which produced a non-isolable intermediate which immediately underwent intramolecular cyclization via the attack of nitrogen of indole subunit at the ester functionality. The resulting 5-azido-canthin-6-one was further transformed to 5-aminocanthin-6-one via catalytic reduction (**Figure 6**).

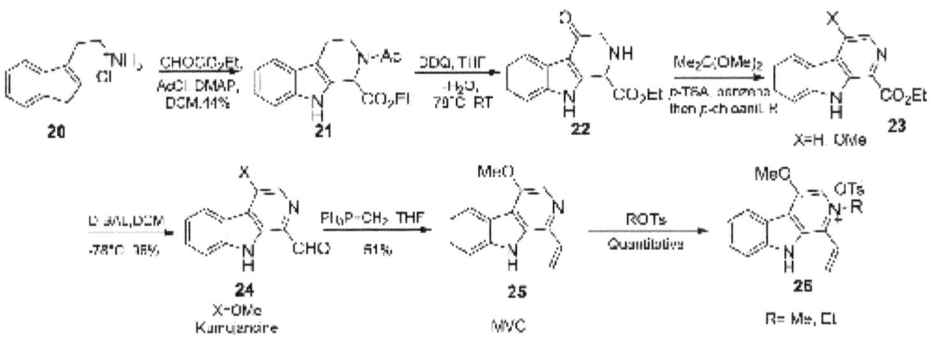

Figure 5.
Synthesis of tosyl salt of β-carboline-based compounds.

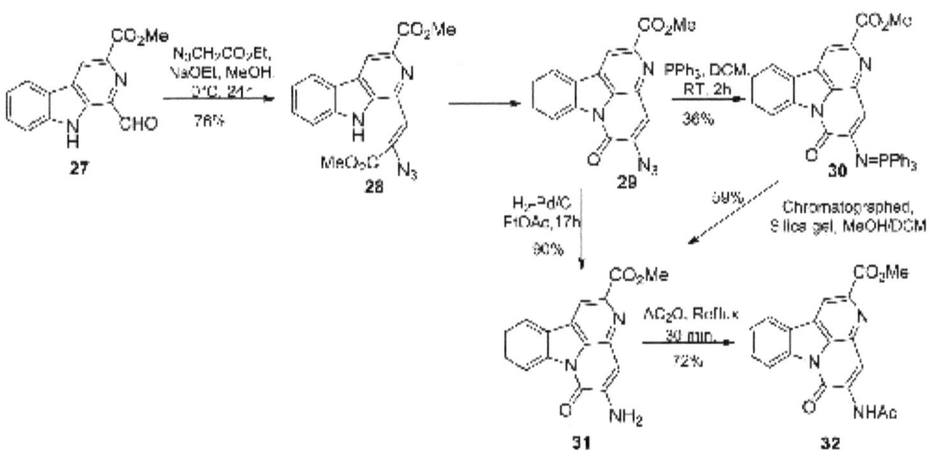

Figure 6.
Synthesis of 5-aminocanthin-6-one via intramolecular cyclization process.

Suzuki et al. [39] reported the synthesis of canthin-6-one derivative from 1-formyl-9*H*-β-carboline and its 4-methoxy derivative. In addition many researchers are continuous trying to do more research in this field. Because the β-carboline gives more interest to natural product chemist and it is a huge scope for researchers (**Figure 7**).

The Morita-Baylis-Hillman (MBH) reaction have also been used by Singh et al. [40] for 1-formyl-9*H*-β-carbolines (**38**) with various activated alkenes led to the formation of expected MBH product (**40**) as well as unnatural canthin-6-one derivatives (**41**). It was discovered that exclusive formation of either product **40** or **41** could be achieved by modulating the amount of DABCO used in the reaction as well as the reaction time (**Figure 8**).

In an extension of this study, they disclosed the potential of substituted 1-formyl-9*H*-β-carboline for achieving the synthesis of indolizinoindole derivatives as depicted in **Figure 9**. The N-alkylated derivatives (**42**) were subjected to MBH reaction with various acrylates and cycloalkenones in the presence of DABCO or DMAP to afford the MBH adducts (**43**) which were transformed into indolizinoindole derivatives **45** (R1 = CO_2Me) via reaction with PBr_3. The reaction was preceded through the formation of allyl bromide **44**.

A Claisen rearrangement have also been used for the synthesis of different β-carbolines by using of allyl alcohol in the presence of *p*-toluenesulfonic acid, which upon heating at 200°C for 30 min resulted the final product in 84% yield [41].

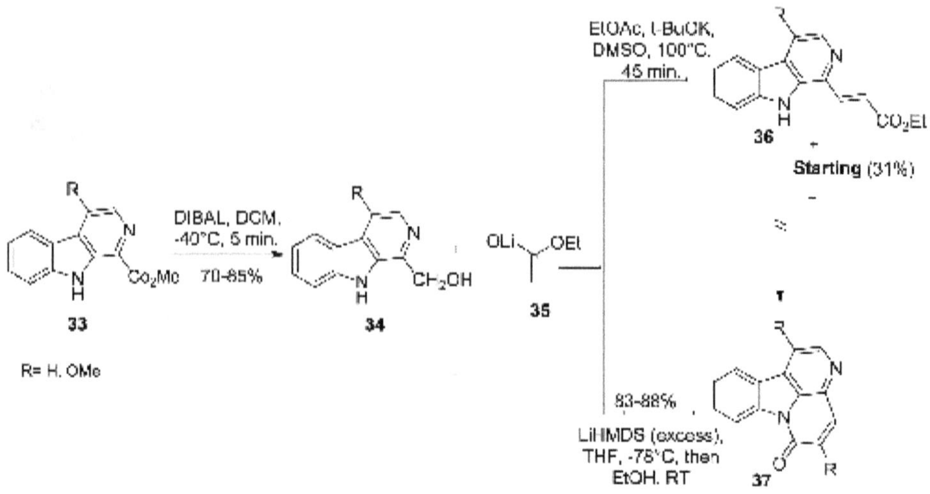

R= H, OMe

Figure 7.
Synthesis of 5-aminocanthin-6-one.

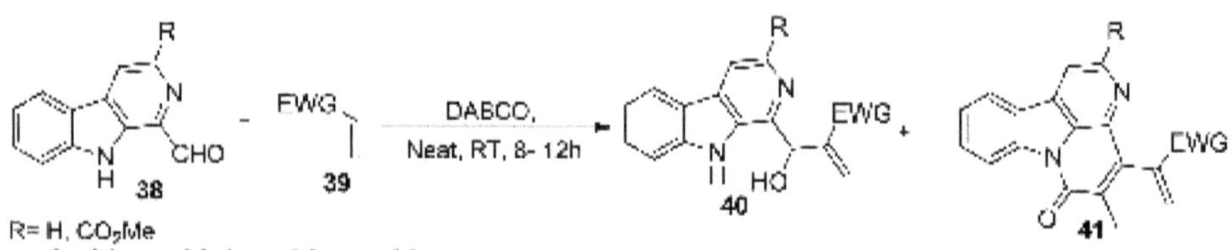

R= H, CO_2Me
EWG= CO_2Me, $CO_2C_2H_5$, CO_2nBu, CO_2tBu

Figure 8.
*Morita-Baylis-Hillman reaction of 1-formyl-9*H*-β-carbolines.*

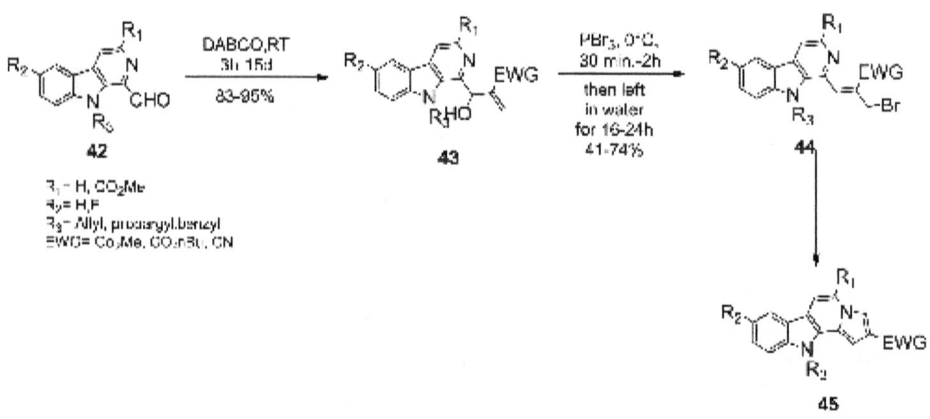

Figure 9.
*Synthesis of indolizinoindole derivatives of 1-formyl-9*H*-β-carbolines.*

Alternatively, 4-amino-β-carboline synthesized by Fischer indole synthesis reaction when the hydrazine was used as a reactant, which is postulated to occur via initial hydrazone formation, followed by isomerization and loss of ammonia (**Figure 10**).

Another oxidant for changing over tetrahydro-β-carbolines to the completely fragrant framework is elemental sulfur, which is usually utilized when utilization of palladium or platinum is not feasible. For example, in Still's synthesis of

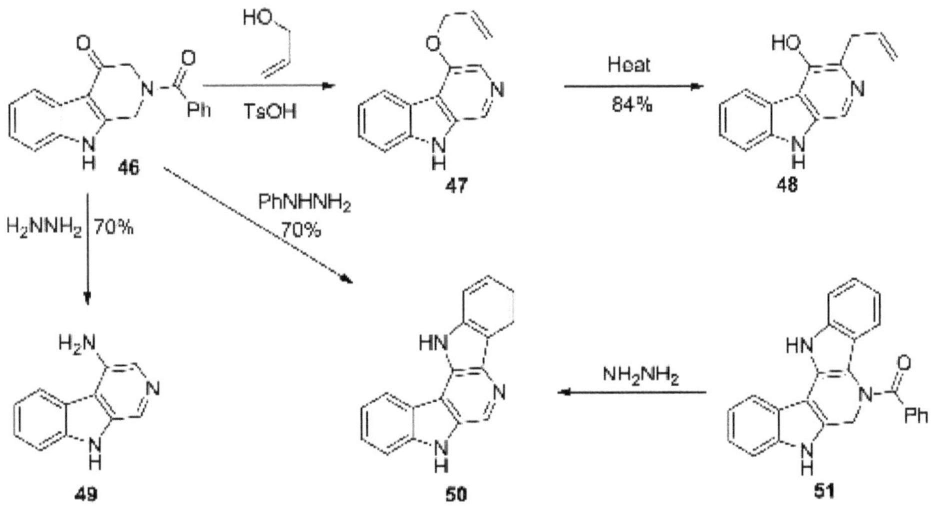

Figure 10.
Claisen rearrangement for the synthesis of different β-carbolines.

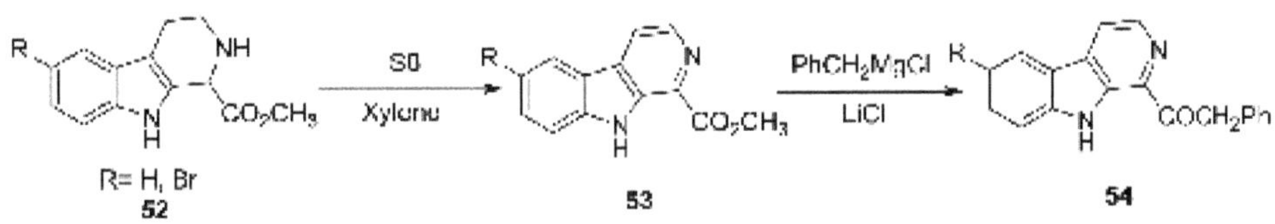

Figure 11.
Oxidation of tetrahydro-β-carbolines.

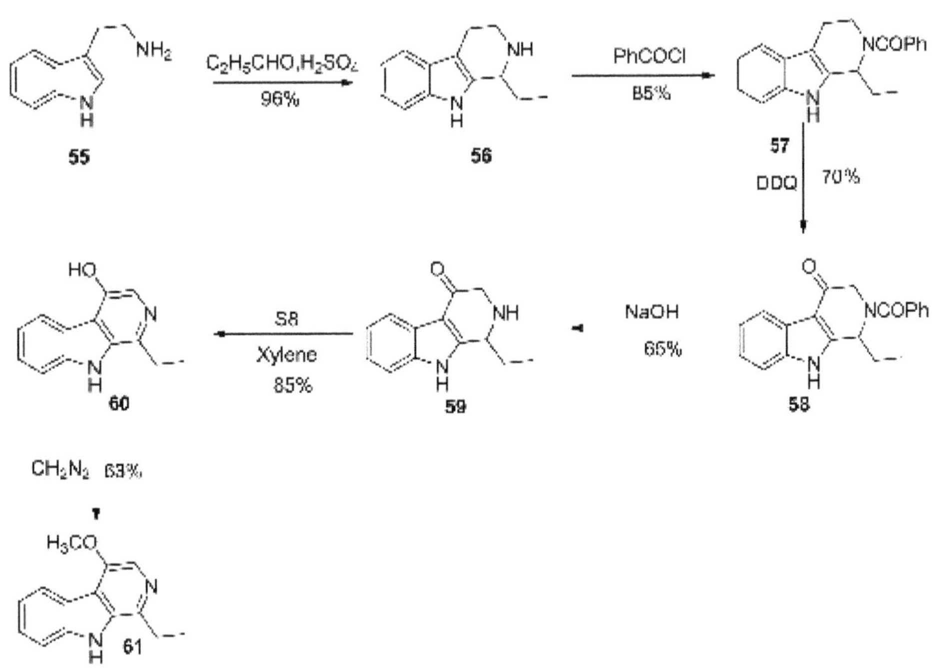

Figure 12.
Synthesis of 4-alkoxy-β-carbolines.

eudistomins **52,** aromatic esters **53** were produced by heating **52** with sulfur in xylenes at reflux condition [42] (**Figure 11**).

For the synthesis of 4-alkoxy-β-carbolines **61,** Oxidation of tetrahydro-β-carbolines **57** with 2,3-dichloro-5,6-dicyano-1,4-benzoquinone (DDQ) has also found one of the best way of synthetic method [43] (**Figure 12**).

3. Conclusion

Using different reaction conditions and reports, it is evident from the past years in medicinal chemistry filed that a wide range of synthetic methods have been reported for the generation of β-carboline moiety and its analogs. However with the new strategy developed for the synthesis of the β-carboline substrate this chapter demonstrated the extensive utility of this prototype design and synthesis of new β-carboline analogs. We believe that this substrate has great potential in medicinal chemistry division and would be more beneficial for pharmaceutical industry.

Author details

Tejpal Singh Chundawat
Department of Applied Sciences, The NorthCap University, Gurugram, Haryana, India

*Address all correspondence to: chundawatchem@yahoo.co.in;
tejpal@ncuindia.edu

References

[1] Saxton JE. Alkaloids of the aspidospermine group. In: Cordell GA, editor. The Alkaloids: Chemistry and Biology. Vol. 51. San Diego, CA: Academic Press; 1998. pp. 2-197

[2] Mansoor TA, Ramalhete C, Molnar J, Mulhovo S, Ferreira MJU. Tabernines A –C. β-Carbolines from the leaves of *Tabernaemontana elegans*. Journal of Natural Products. 2009;**72**:1147-1150

[3] Cao R, Peng W, Wang Z, Xu A. Beta-Carboline alkaloids: Biochemical and pharmacological functions. Current Medicinal Chemistry. 2007;**14**: 479-500

[4] Higuchi K, Kawasaki T. Simple indole alkaloids and those with a nonrearranged monoterpenoid unit. Natural Product Reports. 2007;**24**: 843-868

[5] Kawasaki T, Higuchi K. Simple indole alkaloids and those with a nonrearranged monoterpenoid unit. Natural Product Reports. 2005;**22**:761-793

[6] Carbrera GM, Seldes AMA. A β-carboline alkaloid from the soft coral *Lignopsis spongiosum*. Journal of Natural Products. 1999;**62**:759-760

[7] Gonzalez-Gomez A, Domınguez G, Perez-Castells J. Novel chemistry of β-carbolines. Expedient synthesis of polycyclic scaffolds Tetrahedron. 2009;**65**:3378-3391

[8] Wu Y, Zhao M, Wang C, Peng S. Synthesis and thrombolytic activity of pseudopeptides related to fibrinogen fragment. Bioorganic & Medicinal Chemistry Letters. 2002;**12**:2331-2335

[9] Schwikkard S, Heerden RV. Antimalarial activity of plant metabolites. Natural Product Reports. 2002;**19**:675-692

[10] Steele JCP, Veitch NC, Kite GC, Simmonds MSJ, Warhurst DC. Indole and carboline alkaloids from *Geissospermum sericeum*. Journal of Natural Products. 2002;**65**:85-88

[11] Takasu K, Shimogama T, Saiin C, Kim HS, Wataya Y, Ihara M. π-Delocalized β-carbolinium cations as potential antimalarials Bioorganic & Medicinal Chemistry Letters. 2004;**14**:1689-1694

[12] Boursereau Y, Coldham I. Synthesis and biological studies of 1-amino β-carbolines. Bioorganic & Medicinal Chemistry Letters. 2004;**14**:5841-5844

[13] Schlecker W, Huth A, Ottow E, Mulzer J. Regioselective metalation of 9-methoxymethyl-b-carboline-3-carboxamides with amidomagnesium chlorides. Synthesis. 1995:1225-1227

[14] Ozawa M, Nakada Y, Sugimachi K, Yabuuchi F, Akai T, Mizuta E, et al. Japanese Journal of Pharmacology. 1994;**64**:179-187

[15] Molina P, Fresnda PM, Gareia-Zafra S. An iminophosphorane-mediated efficient synthesis of alkaloid Eudistomin U of marine origin. Tetrahedron Letters. 1995;**36**:3581-3582

[16] Molina P, Fresnda PM, Gareia-Zafra S, Almendros PI. Iminophosphorane-mediated syntheses of the fscaplasyn alkaloid of marine origin and nitramarine. Tetrahedron Letters. 1994;**35**:8851-8854

[17] Deveau AM, Labroli MA, Dieckhaus CM. The synthesis of amino-acid functionalized beta-carbolines as topoisomerase II inhibitors. Bioorganic & Medicinal Chemistry Letters. 2001;**11**:1251-1255

[18] Batch A, Dodd RH. Ortho-Directed Metalation of 3-Carboxy-β-carbolines: Use of the SmI2-cleavable

9-N-(N',N'-dimethylsulfamoyl) blocking group for the preparation of 9-N-deprotected 4-amino derivatives via azide introduction or a palladium-catalyzed cross-coupling reaction. The Journal of Organic Chemistry. 1998;**63**:872-877

[19] Maw GN, Allerton CMN, Gbekor E, Million W. Design synthesis and biological activity of β-caroline based type-5 phosphodiesterase inhibhitors. Bioorganic & Medicinal Chemistry Letters. 2003;**13**(8):1425-1428

[20] Sorbera LA, Martin L, Leeson PA, Castaner J. Treatment of erectile dysfunction—Treatment of female sexual dysfunction—Phosphodiesterase 5 inhibitor. Drugs of the Future. 2001;**26**:15-19

[21] Daugan A, Grondin P, Ruault C, de Gouville A-CLM, Coste H, Kirilovsky J, et al. The discovery of tadalafil: A novel and highly selective PDE5 inhibitor. 2:2,3,6,7,12,12a-hexahydropyrazino [1',2':1,6]pyrido[3,4-b]indole-1,4-dione analogues Journal of Medicinal Chemistry. 2003;**46**:4525-4532

[22] Maw GN, Allerton CM, Gbekor E, Million WA. Design synthesis and biological activity of β-caroline based type-5 phosphodiesterase inhibhitors. Bioorganic & Medicinal Chemistry Letters. 2003;**13**:1425-1428

[23] Colwell R. Biotechnology in the marine sciences. In: Colwell S, Pariser, editors. Biotechnology in the Marine Sciences. New York: John Wiley; 1984

[24] Blunden G. Biologically active compounds from marine organisms. Phytotherapy Research. 2001;**15**:89-94

[25] Proksch P, Edrada-Ebel R, Ebel R. Drugs from the sea opportunities and obstacles. Marine Drugs. 2003;**1**:5-17

[26] O'Hagan D, Harper DB. Fluorine-containing natural products. Journal of Fluorine Chemistry. 1999;**100**:127-133

[27] Kochanowska-Karamyan AJ, Hamann MT. Marine indole alkaloids: Potential new drug leads for the control of depression and anxiety. Chemical Reviews. 2010;**110**:4489-4497

[28] Hill RA. Annual Reports on the Progress of Chemistry, Section B: Organic Chemistry. 2012;**108**:131-146

[29] Frederich M, Tits M, Angenot L. Potential antimalarial activity of indole alkaloids. Transactions of the Royal Society of Tropical Medicine and Hygiene. 2008;**102**:11-19

[30] Haefner B. Drugs from the deep. Marine natural products as drug candidates. Drug Discovery Today. 2003;**8**:536-544

[31] Sorriente A. Manoalide. Current Medical Chemistry. 1999;**6**:415-431

[32] Royer J, Bonin M, Micouin L. Chiral heterocycles by iminium ion cyclization. Chemical Reviews. 2004;**104**:2311-2352

[33] Gatta F, Misiti D. Selenium dioxide oxidation of tetrahydro-β-carboline derivatives. Journal of Heterocyclic Chemistry. 1987;**24**:1183-1187

[34] Bennasar M-L, Roca T, Monerris M. Total synthesis of the proposed structures of indole alkaloids lyaline and lyadine. The Journal of Organic Chemistry. 2004;**69**:752-756

[35] Benson SC, Li JH, Snyder JK. Indole as a dienophile in inverse electron demand Diels-Alder reactions. 3. intramolecular reactions with 1,2,4-triazines to access the canthine skeleton. The Journal of Organic Chemistry. 1992;**57**:5285-5287

[36] Suzuki H, Unemoto M, Hagiwara M, Ohyama T, Yokoyama Y, Murakami Y, et al. Synthetic studies on indoles and related compounds. Part 46.1. First total syntheses of 4,8-dioxygenated β-carboline alkaloids.

Journal of the Chemical Society, Perkin Transactions I. 1999:1717-1723

[37] Takasu K, Shimogama T, Saiin C, Kim H-S, Wataya Y, Ihara M. Pi-delocalized beta-carbolinium cations as potential antimalarials. Bioorganic & Medicinal Chemistry Letters. 2004;**14**:1689-1692

[38] Condie GC, Bergman J. Synthesis of some fused β-carbolines including the first example of the pyrrolo[3,2-c]-β-carboline system. Journal of Heterocyclic Chemistry. 2004;**41**:531-540

[39] Suzuki H, Adachi M, Ebihara Y, Gyoutoku H, Furuya H, Murakami Y, et al. Total synthesis of 1-methoxycanthin-6-one: An efficient one-pot synthesis of the canthin-6-one skeleton from β-carboline-1-carbaldehyde. Synthesis;**2005**:28-32

[40] Singh V, Hutait S, Batra S. One-step synthesis of the canthin-6-one framework by an unprecedented cascade cyclization reaction. European Journal of Organic Chemistry. 2009:6211-6216

[41] Fukada N, Trudell ML, Johnson B, Cook JM. Synthetic studies in the β-carboline area new entry into 4-substituted and 3,4-disubstituted β-carbolines. Tetrahedron Letters. 1985;**26**:2139

[42] Still IWJ, McNulty J. The synthesis of eudistomins S and T: β-Carbolines from the tunicate *Eudistoma olivaceum.* Journal of Heterocycles. 1989;**29**:2057

[43] Cain M, Mantei R, Cook JM. Dichlorodicyanoquinone oxidations in the indole area. Synthesis of crenatine. The Journal of Organic Chemistry. 1982;**47**:4933-4966

11

Hydrolase-Catalyzed Promiscuous Reactions and Applications in Organic Synthesis

Yun Wang and Na Wang

Abstract

The potential of biocatalysis becomes increasingly recognized as an efficient and green tool for modern organic synthesis. Biocatalytic promiscuity, a new frontier extended the use of enzymes in organic synthesis, has attracted much attention and expanded rapidly in the past decade. It focuses on the enzyme catalytic activities with unnatural substrates and alternative chemical transformations. Exploiting enzyme catalytic unconventional reactions might lead to improvements in existing catalysts and provide novel synthesis pathways that are currently not available. Among these enzymes, hydrolase (such as lipase, protease, acylase) undoubtedly has received special attention since they display remarkable activities for some unexpected reactions such as aldol reaction and other novel carbon-carbon and carbon-heteroatom bond-forming reactions. This chapter introduces the recent progress in hydrolase catalytic unconventional reactions and application in organic synthesis. Some important examples of hydrolase catalytic unconventional reactions in addition reactions are reviewed, highlighting the catalytic promiscuity of hydrolases focuses on aldol reaction, Michael addition, and multicomponent reactions.

Keywords: enzyme, biocatalysis, promiscuity, hydrolases, lipase, aldol reactions, Michael addition, multicomponent reactions

1. Introduction

Biocatalysis is the application of enzymes for chemical transformations of organic compounds. Enzymes as biocatalysts have many advantages [1_3]: (1) enzymes are very efficient catalysts. Typically the rates of enzyme-mediated processes are accelerated, compared to those of the corresponding nonenzymatic reaction, by a factor of 10^8–10^{10}. The acceleration may even exceed a value of 10^{12}, which is far above the values that chemical catalysts are capable of achieving; (2) enzymes are environmentally acceptable. Unlike heavy metals, for instance, biocatalysts are completely degraded in the environment; (3) enzymes act under mild conditions. Enzymes act in a temperature range of 20–40°C, under neutral aqueous, and in the absence of substrate functional group protection. This minimizes problems of undesired side reactions such as decomposition, isomerization, racemization, and rearrangement, which often plague traditional methodology; (4) enzymes display high chemoselectivity, regioselectivity, and enantioselectivity.

As a result, reactions that generally tend to be "cleaner" and laborious can largely be omitted; and (5) enzymes can catalyze a broad spectrum of reactions. There is an enzyme-catalyzed process equivalent to almost every type of organic reaction, such as oxidation, hydrolysis, addition, halogenation, alkylation, and isomerization. In addition, many enzymes accept unnatural substrates, and genetic engineering can further alter their stability, broaden their substrate specificity, and increase their specific activity. Thus, the application of enzymes in synthesis thus represents a remarkable opportunity for the development of industrial chemical and pharmaceutical processes [4–7].

Although it is well known that a given enzyme is able to catalyze a specific reaction efficiently, some unexpected experimental results have indicated that many enzymes have catalytic promiscuity [8–12]. Enzyme promiscuity is classified into three categories: (a) condition promiscuity, which is an enzyme's ability to work under unexpected condition; (b) substrate promiscuity, which is an enzyme's ability to work with unexpected substrates; and (c) catalytic promiscuity, which is an enzyme's ability to catalyze unexpected reactions. Among them, catalytic promiscuity has gained much attention as it opens a wide scope for the industrial application of enzymes.

During the past decade, biocatalytic promiscuity, as a new frontier extending the use of enzymes in organic synthesis, has received considerable attention and expanded rapidly. A classic example of promiscuous enzymatic behavior is pyru-vate decarboxylase, which not only decarboxylates pyruvate but also links acetal-dehyde and benzaldehyde to form R-phenylacetylcarbinol. The use of pyruvate decarboxylase to form carbon–carbon bonds, which does not occur in the natural reaction, was first studied in 1921 and was applied in industry today [13]. As one of the most rapidly growing areas in enzymology, multifunctional biocatalytic reac-tions not only highlights the existing catalysts but may provide novel and practical synthetic pathways which are not currently available. Miao et al. reviewed enzyme promiscuity for carbon-carbon bond-forming reactions like aldol couplings, Michael(-type) additions, Mannich reactions, Henry reactions, and Knoevenagel condensations [14]. Gotor-Fernández et al. also highlighted the hydrolase-catalyzed reactions for nonconventional transformations in the same year [15].

Hydrolases (such as lipase, protease, acylase) have received extensive attention as biocatalysts for a long time due to their many attractive properties like stability in

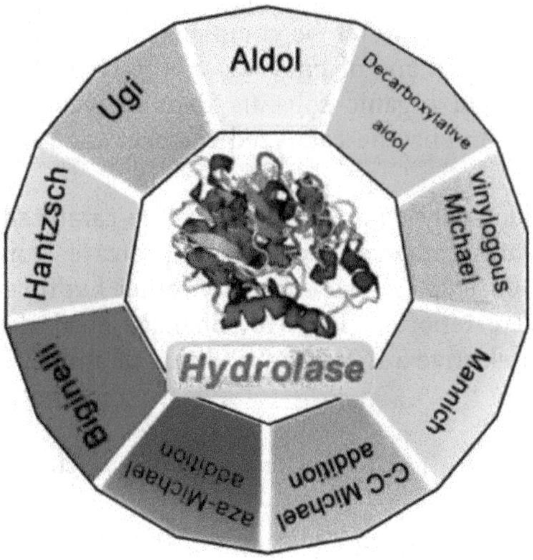

Figure 1.
Hydrolase-catalyzed promiscuous reactions.

organic solvents, not requiring cofactors, broad substrate tolerance, commercial availability, and high chemo-, regio-, and stereoselectivity. Hydrolases have demonstrated a great versatility in hydrolysis, transesterification, aminolysis reactions, etc. Some hydrolase-catalyzed promiscuous reactions have been done in the last decades (**Figure 1**). These research and other relevant reports encouraged us to believe that the catalytic activities for unconventional reactions rather than the well-known hydrolytic function may also have a natural role in hydrolase evolution.

The aim of the present chapter is to give a brief overview of the hydrolase-catalyzed C—C and C—N reactions and present some of the most recent applications in different fields for recent decade. The main work in our group will be disclosed, highlighting the catalytic properties of hydrolases to catalyze not only single processes but also multicomponent and tandem reactions. Consequently, the promiscuous hydrolase-catalyzed reactions are outlined with focus on Michael addition, aldol reaction, Mannich reaction, Biginelli reaction, etc.

2. Michael addition

Michael addition is a 1,4-addition of a nucleophile to α ,β-unsaturated compounds, and it is one of the most fundamental and important reactions for the formation of carbon-carbon bonds and carbon-heteroatom bonds in organic synthesis. Michael addition reactions are traditionally catalyzed under strong basic or acidic conditions, which can cause unwanted side reactions such as further condensation or polymerization of α ,β-unsaturated compounds. Thus, biocatalysis can afford a green and facile method for organic synthesis. Among different biocatalysts, hydrolases such as protease and lipase have been widely used as a green and efficient catalyst for Michael addition.

2.1 Carbon-heteroatom bond formation Michael addition

Michael addition is the early promiscuous reaction catalyzed by hydrolase. In 1986, Kitazume et al. reported the hydrolytic enzyme-catalyzed stereospecific Michael addition reactions in buffer solution (pH = 8.0) at 40–41°C (**Figure 2**) [16]. This discovery overthrows the long erroneous concept of enzymology that "biocatalysis must be carried out in aqueous solution," making many organic reactions that cannot be carried out in water be completed in organic solvents and greatly expanding the application scope of enzymes as catalysts. Moreover, enzymes are frequently more stable in organic solvents than in water. Thus, some research groups began to focus on enzyme-catalyzed Michael addition reactions in organic solvents.

Lin and Gotor et al. firstly reported the hydrolase-catalyzed Michael addition of imidazole with acrylates catalyzed by alkaline protease from *Bacillus subtilis* in organic solvent in 2004 [17, 18]. Subsequently, other hydrolase-catalyzed Michael addition reactions were reported. In 2010, Bhanage et al. developed an efficient protocol for the regioselective aza-Michael addition of amines with acrylates using

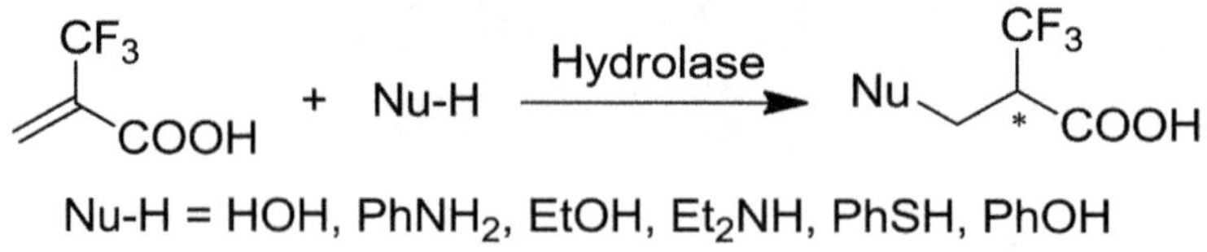

Figure 2.
The first hydrolase-catalyzed Michael addition in buffer.

Candida antarctica lipase B (CALB) as a biocatalyst at 60°C (**Figure 3**) [19]. The universality of the reaction, including the reactions of various primary and secondary amines with different acrylates, was also studied. Higher yields were obtained.

Gotor et al. have explored new synthetic possibilities of commercially available protease from *Bacillus licheniformis* immobilized as cross-linked enzyme aggregates (Alcalase-CLEA) in 2011, since the CLEA immobilization improves the stability of the protein toward denaturalization by heating, organic solvents, and autoproteolysis [20]. Alcalase-CLEA has achieved the best results in the aza-Michael addition of secondary amines to acrylonitrile (**Figure 4**). In all cases the formations of the corresponding Michael adduct were faster than in the absence of biocatalyst, but also in comparison with the inhibited enzyme, the reaction rates being highly dependent of the amine structure.

In 2012, Baldessari et al. firstly reported the synthesis of N-substituted β-amino esters by application of *Rhizomucor miehei* lipase in aza-Michael addition of mono- and bifunctional amines to α,β-unsaturated esters [21]. The authors selected ethyl acrylate and propyl acrylate as the Michael acceptors and different alkylamines, alkanolamines, and diamines as the Michael donors (**Figure 5**). The reactions were carried out in low-polarity solvents (hexane, toluene, and diisopropyl ether (DIPE)) at 30°C for 16 h with yields from 12 to 100%. Subsequently, the authors investigated the effect of the reaction conditions on the Michael addition systematically, such as commercially available enzyme sources, organic solvents, and the structure of the Michael acceptor and donor. The results showed that the alkanolamines in n-hexane were not selective and double Michael adducts could be obtained. Substrate concentration also plays an important role in enhancing the catalytic effect of enzymes on spontaneous reactions. High substrate concentration limits the efficiency of biocatalysts.

In 2013, Demeunynck et al. have optimized the lipase-biocatalyzed addition of benzylamine to ethyl propiolate. Immobilized *Candida antarctica* lipase B was

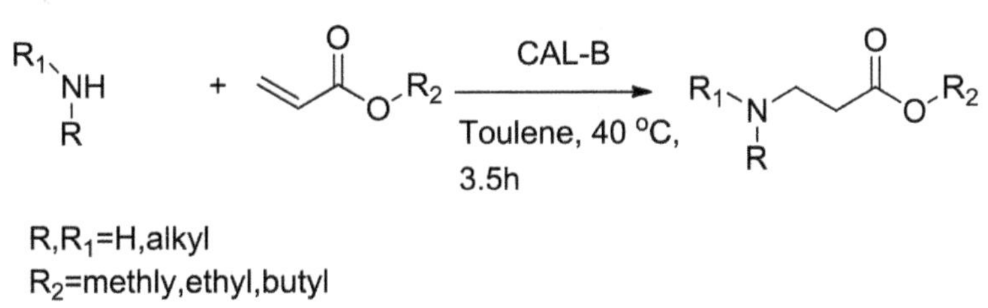

Figure 3.
CALB-catalyzed aza-Michael addition of amine to acrylate.

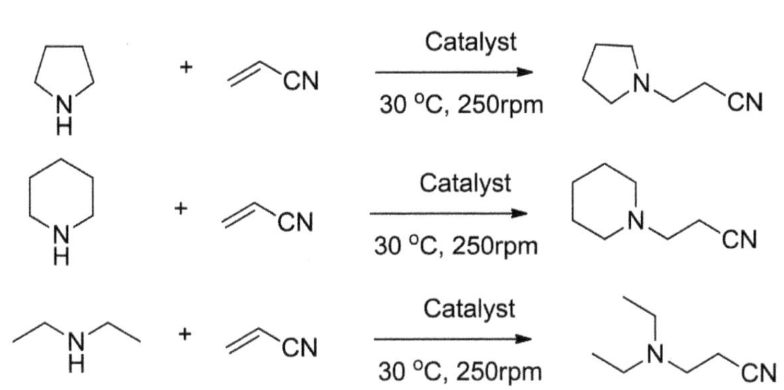

Figure 4.
Hydrolase-catalyzed Michael-type additions between secondary amines and acrylonitrile.

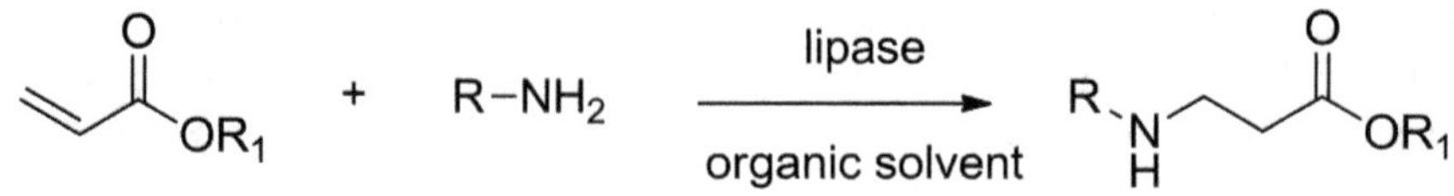

$R_1=CH_2CH_3, (CH2)_3CH_3$
$R=CH_3(CH2)_2$- , $(CH_3)_2CH$- ,$Ph(CH_2)_2$- ,$HO(CH_2)_2$-, $HO(CH_2)_3$-, $HO(CH_2)_4$-, $H_2N(CH_2)_3$, $H_2N(CH_2)_6$, $H_2N(CH_2)_8$,$H_2N(CH_2)_{12}$

Figure 5.
Lipase-catalyzed aza-Michael addition of amines to α,β-unsaturated esters.

beneficial to the chemoselective 1,2-addition, using TBME, dioxane, or toluene under overnight (15 h) gentle magnetic stirring at 50°C (**Figure 6**) [22]. Under these conditions, the yield of acrylamide was good at the Gram scale. S-trans-z and e-diphenylamine were formed as by-products. The reactions worked well with other primary amines but not with secondary amines that only gave the corresponding aminoacrylates. The chemoselectivity of CALB with N- and S-nucleophiles was also checked. The transesterification also worked in good yields in TBME, toluene, or dioxane. The best yields (near quantitative) were observed when the reactions were carried out in open vessels under gentle magnetic stirring at 50°C for 6 h.

In the same year, Franssen and co-workers demonstrated lipases from *Pseudomonas stutzeri* (PSL) and *Chromobacterium viscosum* (CVL) are excellent catalysts for the aza-Michael addition of amines to substituted or unsubstituted acrylates with high product selectivity and good yields (**Figure 7**) [23]. Comparative studies of other lipases, including Novoxin 435, have proven ineffective. The selective Michael addition of diamines to these substituted acrylates was also realized. In this paper, the catalytic effects of various lipases on aza-Michael addition reaction, especially on the lipase catalysis of *Pseudomonas aeruginosa* OM2 and PSL, are introduced. The 1,4-adducts of acrylate and benzylamine have high yield and selectivity.

Chemoselective synthesis of N-protected β -amino esters involving lipase-catalyzed aza-Michael additions is mainly hampered by the two electrophilic sites present on these compounds. In order to control the chemoselectivity, a solvent engineering strategy based on the thermodynamic behavior of products in media of different polarity was designed by Castillo et al. (**Figure 8**) [24]. This strategy can

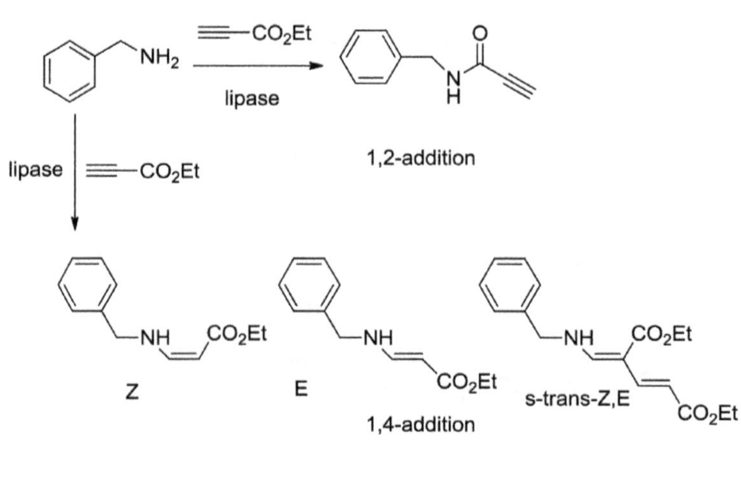

Figure 6.
Reactivity of benzylamine with ethyl propiolate.

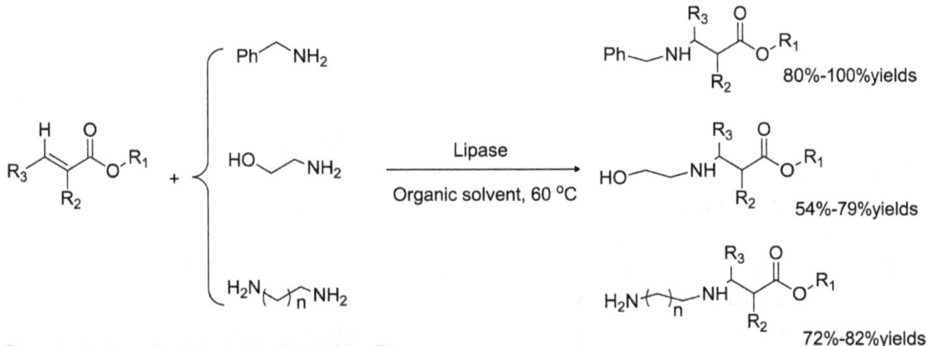

^{a}R1: Me, Et, Bu, t-Bu; R2: H, Me; R3: H, Me , Ph;

Figure 7.
Lipase-catalyzed aza-Michael addition of amines with different substituted acrylates.

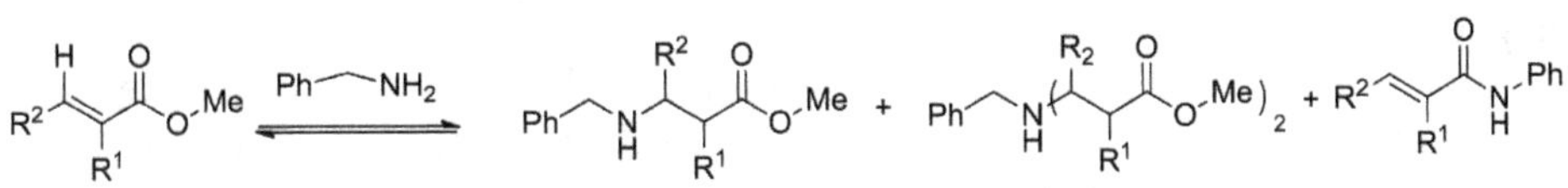

1.R^1=H, R^2=H; 2.R^1=H, R^2=Me; 3.R^1=Me, R^2=H; 4.R^1=Ph, R^2=H; 5.R^1=H, R^2=Ph;

Figure 8.
Chemoselectivity of the addition of benzylamine to α,β-unsaturated esters.

obtain highly selective aza-Michael adducts from benzylamine and different acrylates. Ammonia hydrolysis is avoided in almost all reactions with n-hexane (a nonpolar solvent) as solvent, while the corresponding Michael adduct is synthesized in 53–78% yield. On the contrary, if the reaction is carried out in polar solvents (e.g., 2-methyl-2-butanol (2M2B)), the product of ammonia hydrolysis will be advantageous. Thermodynamic analysis of these processes using the actual solvation conductor-like screening model (COSMO-RS) helps to understand some key factors affecting chemical selectivity and confirms that reliable estimates of the thermodynamic interactions between solutes and solvents allow adequate selection of reaction media that may lead to chemical selectivity.

Reaction media has an important effect on the yield and chemo- and enantioselectivity of biocatalytic reaction. Solvent engineering is an effective tool to direct chemo- and enantioselectivity of the aza-Michael addition and the subsequent kinetic resolution of the Michael adduct [25]. In the reaction of benzylamine and methyl crotonate catalyzed by CALB, three possible adducts can be isolated: aminolysis product, aza-Michael addition product, and double addition product (**Figure 9**). The authors selected n-hexane and 2-methyl-2-butanol (two solvents of opposite polarity) as solvents in the experiments. The Michael adduct is favored in more hydrophobic media, while the amide product in more polar solvents, and the best values of ee for aza-Michael adduct were obtained in almost 100% 2M2B. The experiment results clarify the origin of the enantiomeric excess of the aza-Michael

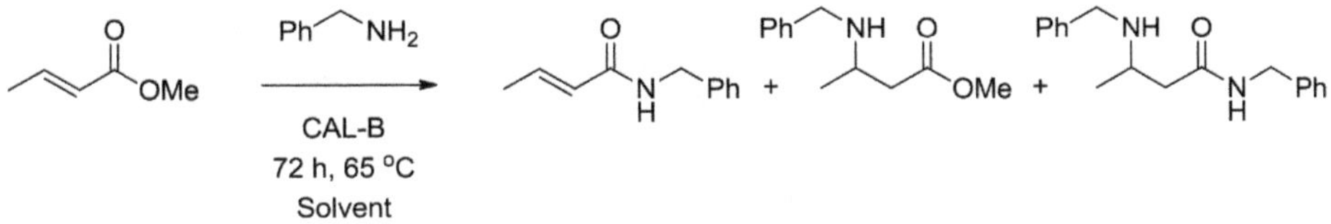

Figure 9.
Lipase-catalyzed addition and aminolysis reaction of benzylamine to methyl crotonate.

addition product, obtained in 2M2B, by a resolution process with CALB on the enantiomers of aza-Michael addition product.

Our group demonstrated that 3-substituted 2H-chromene derivatives were synthesized via biocatalytic domino oxa-Michael/aldol condensations (**Figure 10**) [26]. α-Amylase from *Bacillus subtilis* shows excellent catalytic activity and exerts good adaptability to different substrates in the reaction. The reaction conditions including organic solvents, water content, temperature, molar ratio of substrates, and enzyme loading were optimized.

It is generally believed that the hydrolysis site of hydrolase is also the active site of its miscible catalysis. On this basis, the possible mechanism of domino reaction catalyzed by hydrolase was proposed. First, salicylaldehyde was activated by amino residues and oxygen anions of amylase. Then methyl vinyl ketone is attacked by activated salicylaldehyde, and a new C—O bond is formed by oxa-Michael addition reaction. Next, an intramolecular aldol reaction begins to form carbon–carbon bonds. Finally, the adduct was dehydrated, and the required product was released (**Figure 11**).

Very recently, our group conducted an aza-Michael addition of aniline compounds and acrylate derivatives catalyzed by CALB and several mutants in order to investigate reaction mechanistic (**Figure 12**) [27]. The influence factors of the reaction were discussed systematically, including solvent, enzyme loading, temperature, and time of reaction. On this basis, dozens of substrates with different structures were conducted to occur aza-Michael addition on the optimized conditions. The results demonstrated that the structures of substrates had a great influence on the activity.

Four different reaction intermediates (Intermediate 1, 2, 3, and 4) were matched with the catalytic activity site of CALB to perform molecular docking simulation (**Figure 13A**). We can see that the binding mode of all the four intermediates with the active site is basically the same. The binding modes of four intermediates with CALB catalytic active sites were analyzed, in order to further study the blinding modes of aza-Michael addition intermediates and CALB and the driving forces of their mutual recognition. As shown in **Figure 13b**, it can be seen from the figure

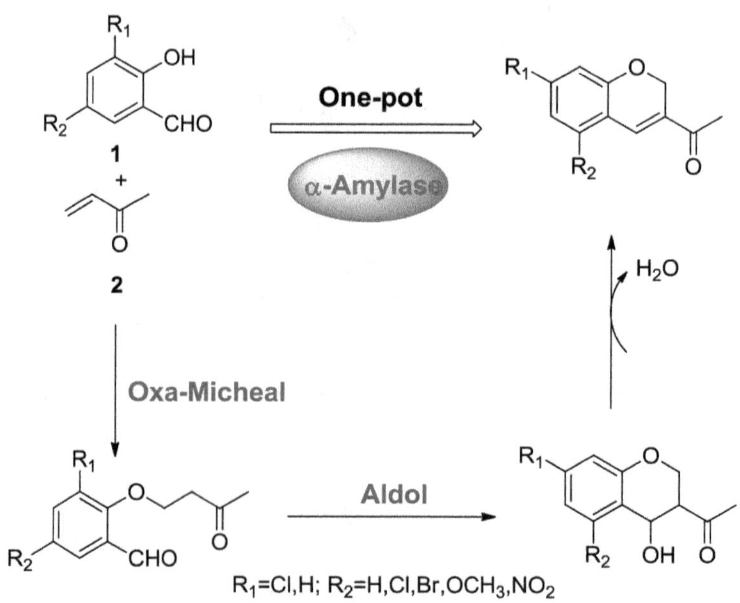

Figure 10.
α-Amylase-catalyzed synthesis of 3-substituted 2H-chromene derivatives.

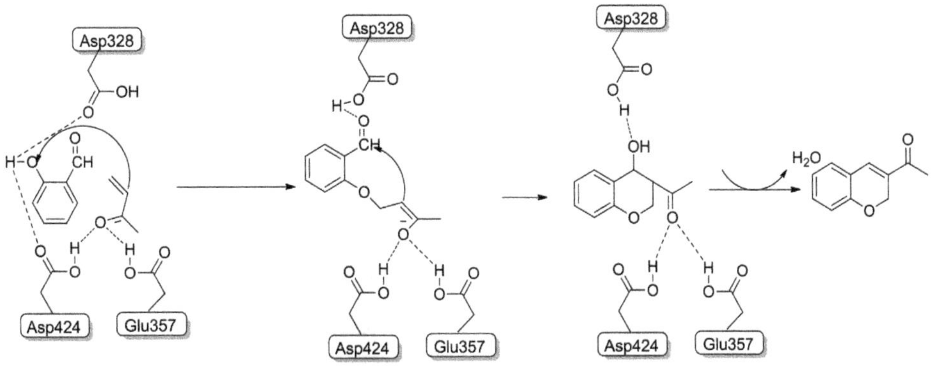

Figure 11.
The proposed mechanism for the α-amylase-catalyzed synthesis of 3-substituted 2H-chromene derivative.

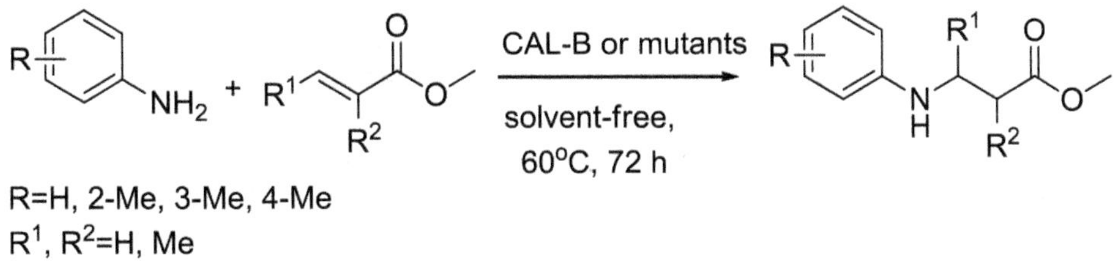

Figure 12.
CALB and mutants catalyzed aza-Michael addition.

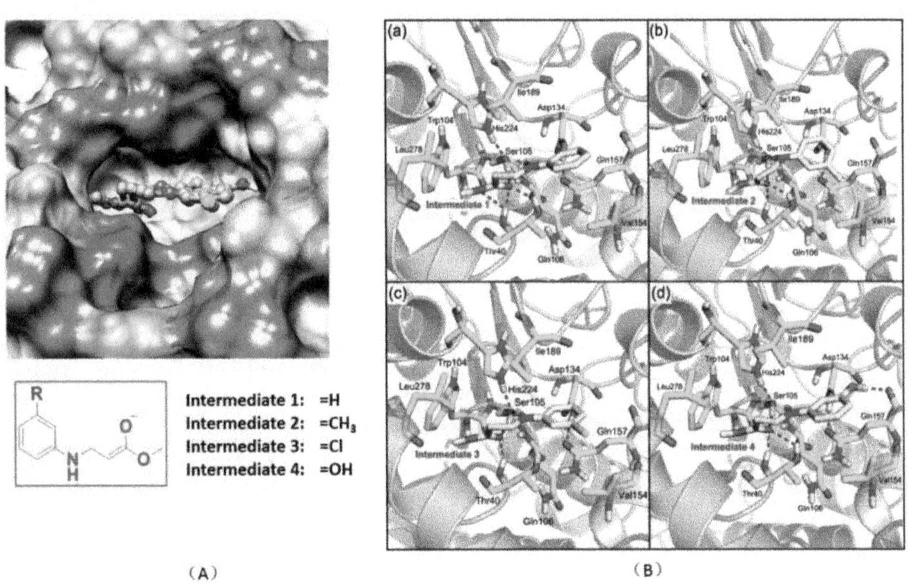

Figure 13.
Molecular docking simulation of CALB with four different reaction intermediates. (A)Hydrophobic matching of the four reaction intermediates with CAL B. The cavity represents the CAL B catalytic pocket which is able to bind and orient the substrates. The blue surface represents hydrophilic while the orange surface represents hydrophobic. The substrates are shown in the pocket in ball-and-stick representation with the atom of substrate coloured according to their atom types (carbon, grey; nitrogen, blue; oxygen, red; chlorine, green). (B)Three-dimensional model of the binding between four aza-Michael addition intermediates (1, 2, 3 and 4) and the CAL B active site. The protein is shown in grey with interacting residues shown as a sky blue stick model. The intermediate is shown as a yellow stick model, and the blue dotted lines indicate the hydrogen bonds between the intermediate and the active site of CAL B.

that the binding modes of the four intermediates with the CALB catalytic chamber are basically the same.

In order to determine the catalytic activity of CALB, three mutants and wild-type CALB were expressed in *E. coli* and were purified to catalyze the aza-Michael addition reaction. The results showed that aza-Michael activity could be dramatically decreased by the mutation of active sites: neither mutant S105 A nor mutant H224 could catalyze the reaction.

However, the mutant I189 A still had a weak catalytic effect on this reaction. Based on these experimental results, the molecular docking was carried out, and the mechanism of aza-Michael addition catalyzed by CALB was studied, and a reasonable reaction mechanism was proposed (**Figure 14**). This helped to explain the effect of substrate structure on the reaction. The substituents of substrates affect the interaction with CALB active sites. Some substituents enhance the binding of substrates and facilitate the reaction. In the whole process, the Ser105 and His224 residues played an important role in proton transfer. Without these two residues, the proton transfer would be blocked, and the aza-Michael addition could not be possible. Besides, the Ile189 residue forms hydrophobic interaction with the benzene ring of the substrate, which makes the substrate more stable in the active cavity.

The biocatalytic thia-Michael reaction is an attractive strategy to develop C—S bond-forming reactions. In 2012, Kiełbasinski and co-workers have reported the use of a number of lipases including PPL, MJL, CALB, and PSL in the addition of benzenethiol to racemic phenyl vinyl sulfoxide or 2-phosphono-2,3-didehydrothiolane S-oxide in organic solvents at room temperature (**Figure 15**) [28]. The addition of piperidine to phenyl vinyl sulfoxide in chloroform is carried out in both enzymatic and non-catalytic processes, while in the former, the reaction rate is 2.5 times faster. Conversely, the conjugate addition of phenylmercaptan with phenyl vinyl sulfoxide is only carried out in the presence of enzymes and ethanol as solvent. In any case, the product is not enantiomerically enriched. However, in the

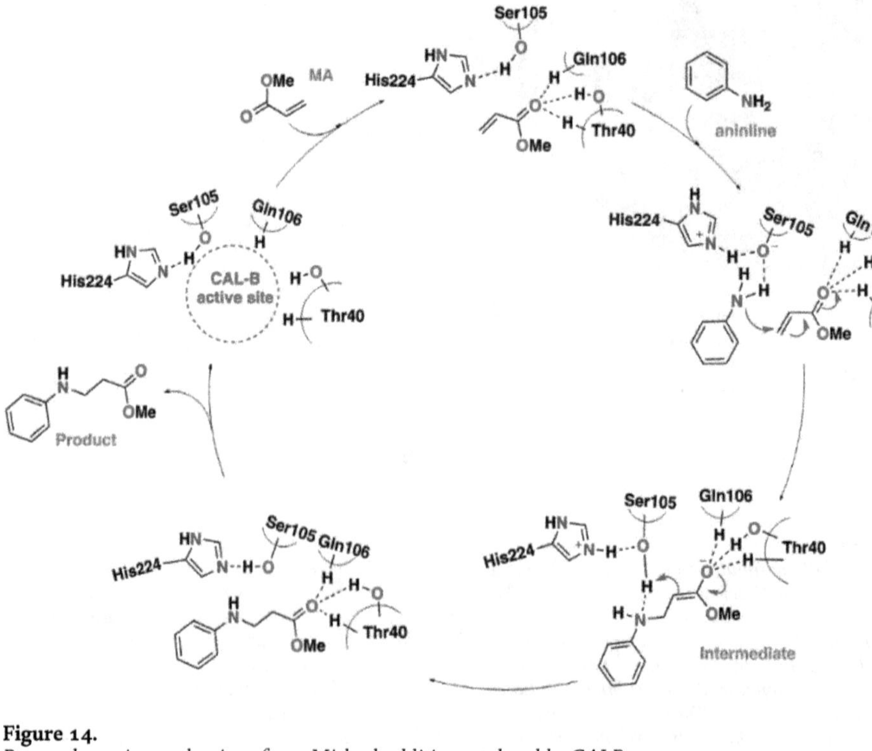

Figure 14.
Proposed reaction mechanism of aza-Michael addition catalyzed by CALB.

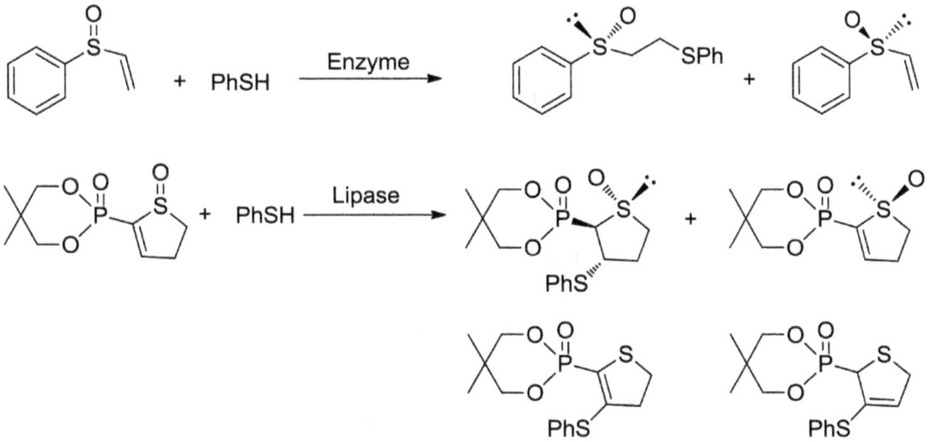

Figure 15.
The Michael addition of benzenethiol to racemic phenyl vinyl sulfoxide or 2-phosphono-2,3-didehydrothiolane S-oxide.

presence of various lipases, the addition of phenylmercaptan to a better Michael receptor, cyclic sulfonyl alkylphosphonate, in some cases resulted in up to 25% optical purity of the product and the recovered substrate.

Then, some mechanistic considerations are presented in the studies. The authors proposed sulfoxide oxygen atoms are bound to the "oxygen anion pore" of the enzyme activity site by hydrogen bond. Conversely, histidine catalyzed by binary enhances the nucleophilicity of sulfur centers in phenylmercaptan molecules. Although the interaction of the latter is the same as Michael's addition of mercaptan to enols, the H-binding of sulfoxide oxygen atom must be different from that of carbonyl oxygen atom, which results in the lower catalytic efficiency of the enzyme for the reaction. It is well known that oxygen anion holes bind to the transition state better than the ground state. When lipase catalyzes ester hydrolysis, the intermediate oxygen anion is tetrahedral. Although the sulfoxide group is tetrahedral, which indicates that the bonding of sulfoxide group should be uniform, compared with the oxygen anion, the sulfoxide group has no negative charge on the intermediate oxygen atom, which significantly reduces the strength of hydrogen bond. In addition, for the Michael addition of nucleophilic reagents, the intermediate oxygen anion is planar, which reduces the space requirement and makes it more suitable for the oxygen anion pore than the tetrahedral sulfoxide intermediate (**Figure 16**).

In 2014, Domingues and co-workers firstly reported the reaction between cinnamaldehyde and thiophenol. Several hydrolases such as PPL, lipozyme, chymosin, and papain have demonstrated different levels of activities, and PPL has found application on the multigram scale (**Figure 17**) [29]. These reactions were carried out at room temperature, and good or excellent sulfur Michael adducts were obtained. The scheme describes the use of EtOH as a solvent and fewer enzymes. The chymosin and papain were used as biocatalysts for organic reactions for the first time.

2.2 Carbon-carbon bond formation Michael addition

C—C bond-forming reactions are one of the mainstays of organic chemistry. In this field the hydrolase-catalyzed Michael reaction also has numerous applications in synthetic chemistry.

In 2011, the asymmetric C—C Michael addition catalyzed by lipozyme TLIM (immobilized lipase from *Thermomyces lanuginosus*) in organic medium in the presence of water was reported for the first time by Guan et al. The biocatalytic reaction

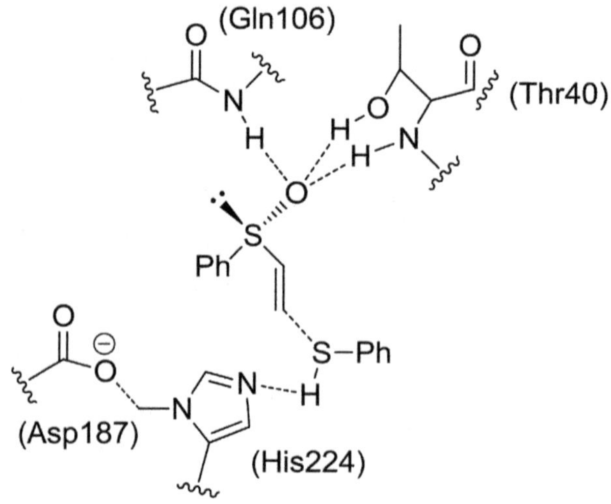

Figure 16.
Assumed mechanism of the conjugate addition of benzenethiol to racemic phenyl vinyl sulfoxide.

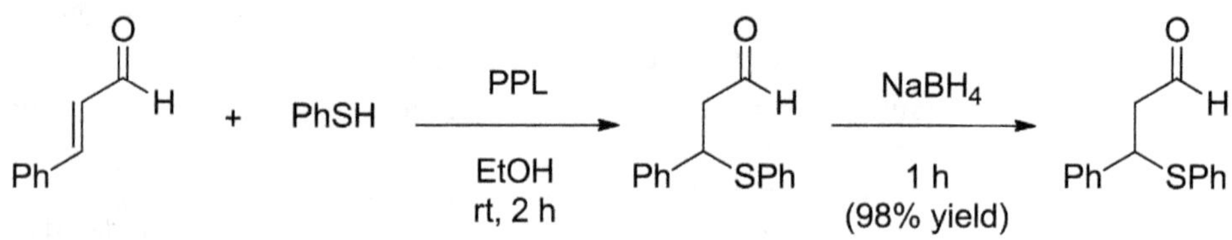

Figure 17.
PPL-catalyzed C—S bond-forming reaction between cinnamaldehyde and thiophenol.

is suitable for adding large amounts of 1,3-dicarbonyl compounds and cyclohexanone to aromatic and heteroaromatic nitroolefins and cyclohexenone in DMSO in the presence of water under mild reaction conditions (**Figure 18**) [30]. The enantioselectivities up to 83% ee and yields up to 90% were achieved.

Then, the same research group explored porcine pancreatic lipase (PPL) which was used as a biocatalyst to catalyze the Michael addition of 4-hydroxycoumarin to α,β-unsaturated enones in organic medium in the presence of water to synthesize warfarin and derivatives in 2012 (**Figure 19**) [31]. The products were obtained in moderate to high yields (up to 95%) with none or low enantioselectivities (up to 28% ee). The influence of reaction conditions including solvents, temperature, and molar ratio of substrates was systematically investigated. It was the first time warfarin and derivatives were prepared using a biocatalyst.

Sometimes Michael adducts are not the final targeted compounds. We studied lipase-catalyzed Michael addition between nitrostyrene and acetylacetone in DMSO in the presence of water under mild reaction conditions in 2011 [32]. Two possible adducts can be isolated: the routine Michael addition product and cyclic product

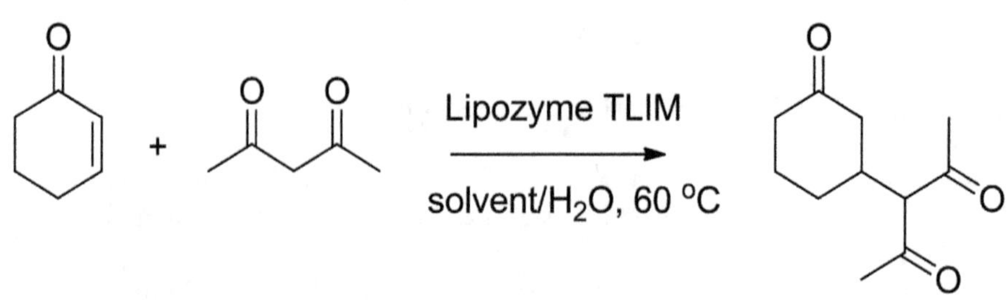

Figure 18.
Michael reaction of cyclohexenone and acetylacetone catalyzed by lipozyme TLIM in different solvents.

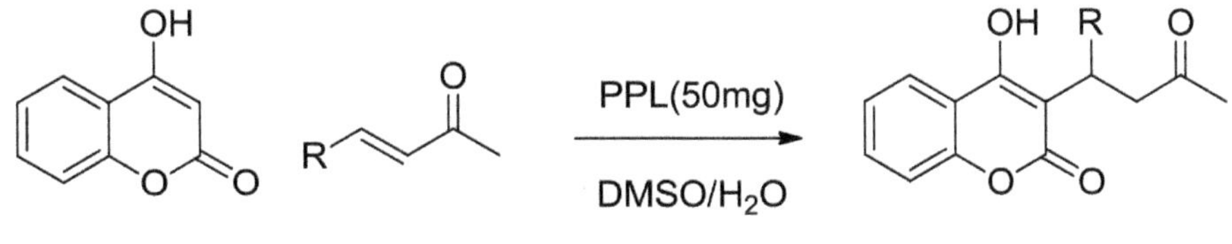

Figure 19.
The Michael addition of 4-hydroxycoumarin 1 to α,β-unsaturated enones for the formation of warfarin and derivatives.

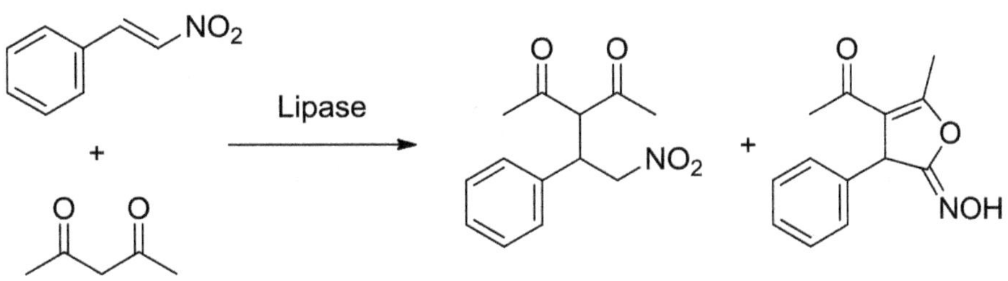

Figure 20.
Lipase-catalyzed reaction of nitrostyrene and acetylacetone.

(**Figure 20**). With the aim to get cyclic product in more efficient manner, the catalytic activities of several lipases were firstly tested in mixed ethanol/water solvents. Among the tested lipases, PPL showed the best activity. And according to the single-crystal X-ray diffraction analysis of cyclic products, the reaction was confirmed to give the product oximes with Z-stereoselectivity.

Then, our group reported for the first time lipase-catalyzed direct vinylogous Michael addition reactions of vinyl malononitriles to nitroalkenes (**Figure 21**) [33]. A series of nitroalkenes reacted with vinyl malononitriles to produce the corresponding products with moderate to high yields in the presence of Lipozyme® (immobilized lipase from *Mucor miehei*). The excellent diastereoselective products were produced in all reactions in acetonitrile at 30°C for 48 h. The enzyme has only a very slight loss of catalyst efficiency after being reused for seven consecutive cycles of the reaction in the previously determined optimized conditions.

The reaction mechanism was studied by computational simulation approach using dock. Based on the proposed catalytic mechanism of Michael reaction, two docking process of the substrates with the amino acids of the active site were performed. The calculation results explained the experimental results that the lipase possessed specific substrate selectivity. To further elucidate the different catalytic effects of RML, CALB, and CRL, structural characteristics of their active site were analyzed, respectively (**Figure 22**). The docking results showed that once vinyl malononitrile was occupied by nitrostyrene, it could not be docked with the active site of CRL. It indicates that the active site is too narrow to bind both two substrates at the same time, so it could not catalyze the direct Michael addition reaction. As for CALB, the active site seems big enough for both nitrostyrene and vinyl malononitrile, but the docking results showed that nitrostyrene blocked proton

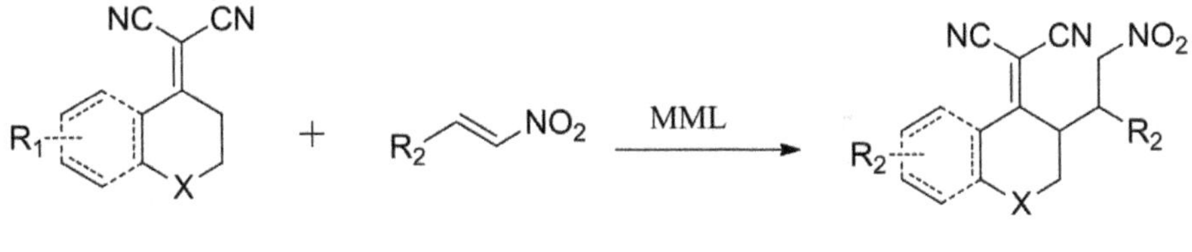

Figure 21.
Lipase-catalyzed direct vinylogous Michael addition reaction.

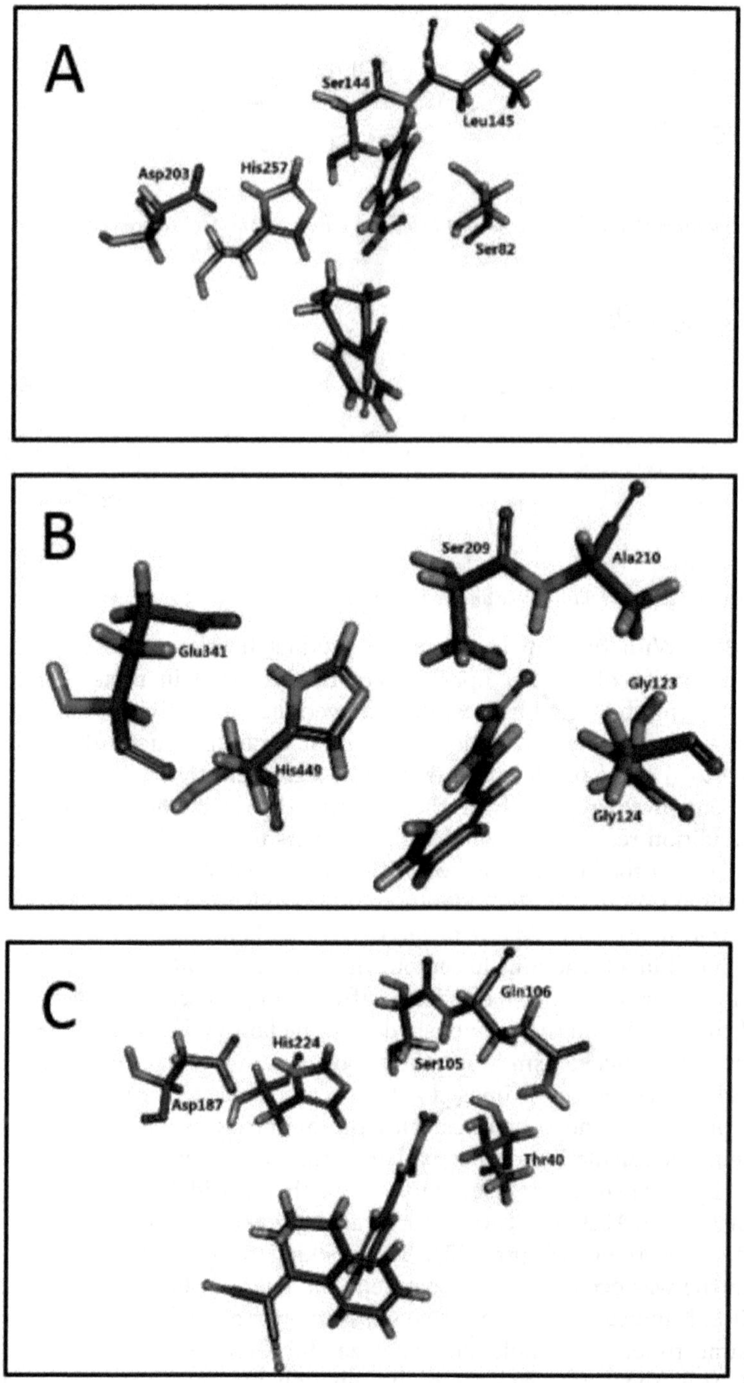

Figure 22.
The comparison of RML, CRL, and CALB: (A) the docking result of RML and substrates, (B) the docking result of CRL and substrates and (C) the docking result of CALB and substrates.

transferring from vinyl malononitrile to histidine, which may make it unable to catalyze the direct vinylogous Michael addition.

Reaction mechanism of the lipase-catalyzed direct vinylogous Michael addition reaction has been proposed (**Figure 23**). First, nitroalkenes bind to oxygen anion pores and were stabilized by three hydrogen bonds with Leu145 and Ser82.

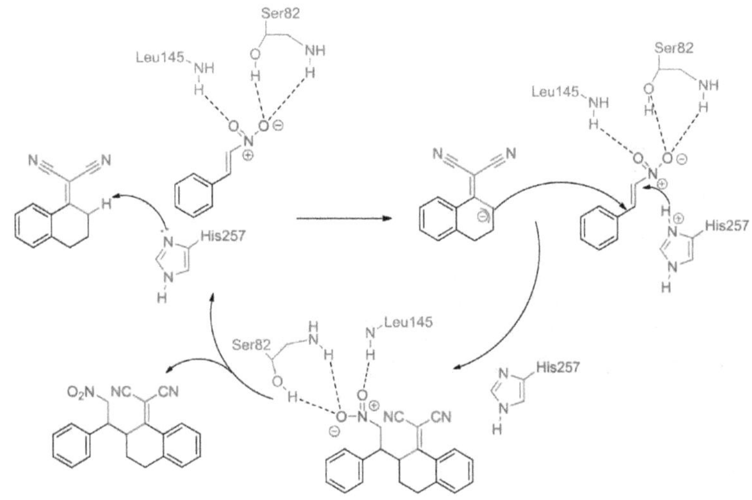

Figure 23.
The proposed mechanism of lipase-catalyzed vinylogous Michael addition.

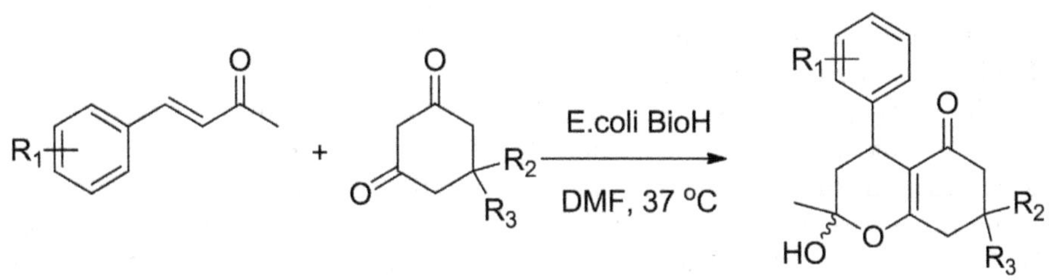

Figure 24.
BioH esterase-catalyzed Michael addition-cyclization cascade reaction.

The protons were then transferred from vinyl malononitrile to His 257 to form a transition state. Subsequently, the protons were transferred from the imidazole group of His 257 to nitroolefins, and the carbon of nitroolefins were attacked by nucleophilic carbon molecules to form products.

In 2014, Ye et al. reported the preparation of 2-hydroxy-2-methyl-4-(4-nitrophenyl)-3,4,7,8-tetrahydro-2H-chromen-5(6H)-one by Michael addition-cyclization cascade reaction of p-nitrobenzalacetone with 1,3-cyclohexanedione in anhydrous media, and control experiments were conducted (**Figure 24**) [34]. The high yield was observed with *Escherichia coli* BioH esterase in DMF at 37°C. In order to preliminarily explore the mechanism of the reaction, site-directed mutagenesis was performed on the hydrolysis catalytic triad of BioH, and the results indicated "alternate-site enzyme promiscuity." Using a series of substituted phenylacetone and 1,3-cyclodiketone as reactants, the yield could reach 76.3%.

3. Aldol reaction

The aldol reaction has long been recognized as one of the most useful tools for organic chemists. The ability to form carbon-carbon bonds can generate a broad range of both natural and novel poly-hydroxylated compounds. Thus, it is the most important and valuable reaction for the preparation of pharmaceuticals, fine chemicals, and natural products. Aldolases have evolved to catalyze the metabolism and catabolism of highly oxygenated metabolites and are found in many biosynthetic pathways of carbohydrates, keto acids, and some amino acids [35]. Aldolases bind their respective donor substrates with high specificity and generally will not accept any other donors, even if their structures are similar to the natural donor. The advantages of using aldolases are very high stereospecificity and environmentally benign reaction conditions [36]. However, the limited number of substrates as well as the high cost of these biocatalysts has led researchers to consider other more stable enzymes [37].

In 2003, Berglund and co-workers firstly reported the serine hydrolase *Candida antarctica* lipase B to have catalytic activity for aldol reactions [38]. Our group reported the first lipase-catalyzed asymmetric aldol reaction in 2008 [39]. However, these aldol reactions in earlier studies involving hydrolases just showed moderate activities and selectivities; some more efficient promiscuous aldol reaction have been researched and presented in the last decade.

3.1 Aldol reaction

In 2012, Guan et al. firstly demonstrated that lipase from porcine pancreas, type II (PPL II), has been observed to catalyze the direct asymmetric aldol reaction of heterocyclic ketones with aromatic aldehydes at 30°C in CH_3CN/H_2O (**Figure 25**) [40]. PPL II has good catalytic activity and good adaptability to different substrates. Its enantioselectivity can reach 87% ee and enantioselectivity 83:17 (anti/syn). Then PPL II has aldolase function in organic solvents.

In the same year, the same group also reported the similar asymmetric aldol reaction of aromatic and heteroaromatic aldehydes with cyclic and acyclic ketones in acetonitrile in the presence of a phosphate buffer by chymopapain, which is a cysteine proteinase isolated from the latex of the unripe fruits of *Carica papaya* [41]. Chymopapain exhibited the best catalytic activity and moderate stereoselectivity in DMSO, and the enzyme showed the best enantioselectivity of 79% ee in CH_2Cl_2 with low diastereoselectivity (**Figure 26**). In consideration of both diastereo- and enantioselectivities, the group chose MeCN as a suitable solvent for the asymmetric direct aldol reaction, which gave the best dr of 77:23 and a moderate ee of 76% among the tested solvents. Then, in order to further optimize the direct asymmetric aldol reaction catalyzed by papain, the effects of water content,

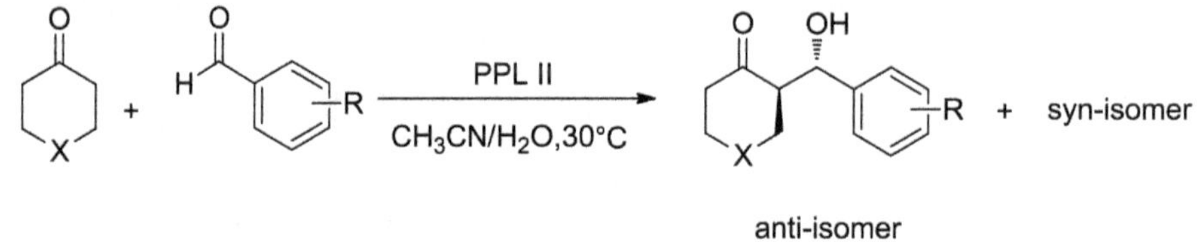

Figure 25.
Lipase-catalyzed direct asymmetric aldol reaction of heterocyclic ketones with aromatic aldehydes.

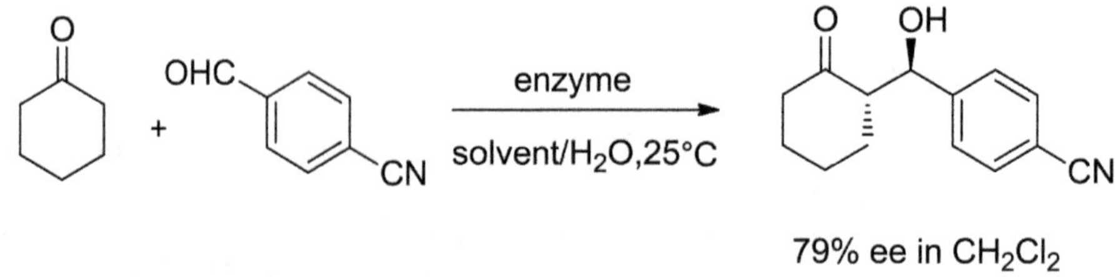

Figure 26.
The asymmetric aldol reaction of 4-cyanobenzaldehyde and cyclohexanone.

reaction temperature, and the amount of buffer on the enzymatic reaction were investigated. The reaction of 4-cyanobenzaldehyde with cyclohexanone was used as a model reaction.

The authors proposed a mechanism for the chymopapain-catalyzed aldol reaction (**Figure 27**). The catalytic triad of Cys, His, and Asn formed the active site of chymopapain. Firstly, the carbonyl of the substrate ketone is coordinated in the oxygen anion pore of Asn-His binary and active center. Secondly, a proton is transferred from the ketone to the His residue to form enolate ion. Thirdly, another substrate aldehyde accepts the proton from imidazolium cation and forms a new carbon-carbon bond with ketones. Finally, the product is released from the oxyanion hole and separates from the active site.

In 2013, our group firstly reported the asymmetric aldol reaction between aromatic aldehydes and cyclic ketones by PPL (**Figure 28**) [42]. The results showed that a small amount of water could promote the catalytic activity of PPL at 37°C. A wide range of aromatic aldehydes reacted with cyclic ketones to provide the corresponding aldol products with high yields (up to 99%) and moderate to good stereoselectivity (up to 90% ee and 99:1 dr).

In the same year, a simple and convenient synthesis route of series α,β-unsaturated aldehydes was formed by combining the two catalytic activities of the same enzyme with the one-pot method of aldehyde-alcohol reaction and in situ acetaldehyde formation (**Figure 29**) [43]. Lipase from *Mucor miehei* has conventional and promiscuous catalytic activities for the hydrolysis of vinyl acetate and aldol condensation with in situ-formed acetaldehyde.

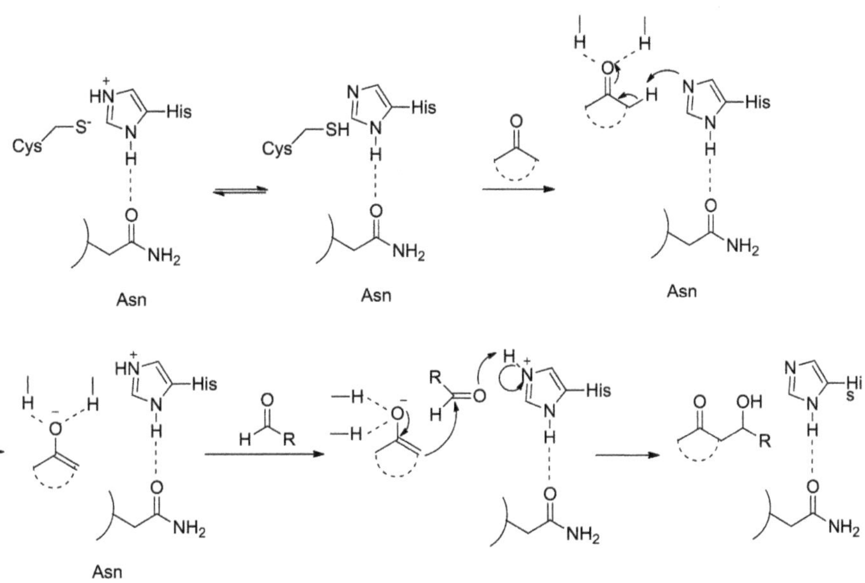

Figure 27.
Proposed mechanism for the chymopapain-catalyzed aldol reaction.

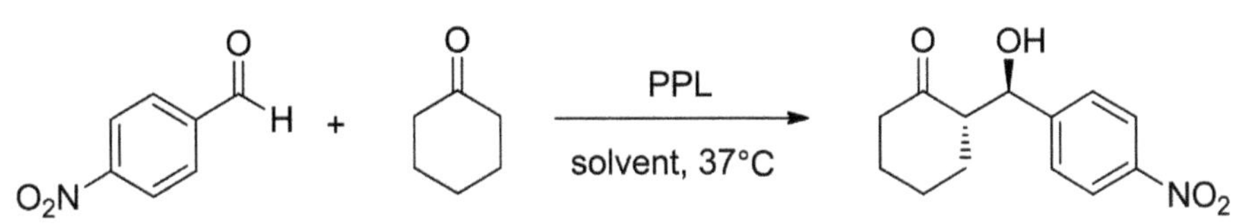

Figure 28.
The asymmetric aldol reaction between aromatic aldehydes and cyclic ketones by PPL.

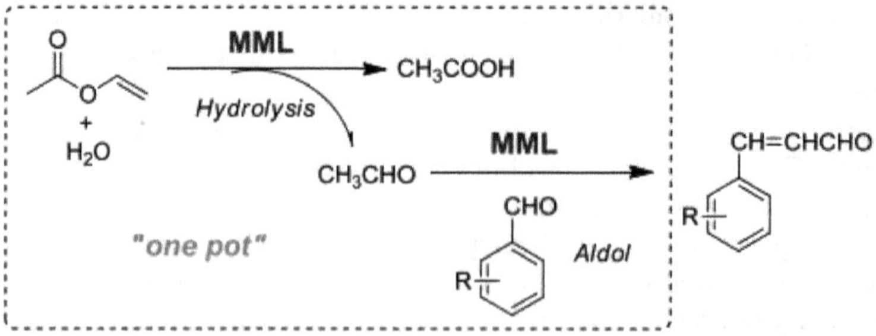

Figure 29.
MML-catalyzed aldol condensation using an in situ-generated acetaldehyde.

In 2014, Majumder and Gupta found that the properties of lipase-catalyzed reaction products of acetylacetone with 4-nitrobenzaldehyde depend on the source of lipase and reaction medium (**Figure 30**) [44]. *Mucor javanicus* lipase was found to give 70% aldol and 80% enantiomeric excess in anhydrous t-amyl alcohol.

Gao and Guo et al. demonstrated the catalytic promiscuity of an acyl-peptide releasing enzyme from *Sulfolobus tokodaii* (ST0779) for aldol addition reaction for the first time, and accelerated activity was observed at elevated temperature (**Figure 31**) [45]. The turnover number kcat (s^{-1}) of this thermostable enzyme at 55°C is 7.78-fold higher than that of PPL at its optimum temperature of 37°C, and the molecular catalytic efficiency kcat/Km ($M^{-1} s^{-1}$) adds up to 140 times higher than PPL.

The authors proposed a mechanism for the ST0779-catalyzed aldol reaction between acetone and 4-nitrobenzaldehyde (**Figure 32**). Based on the structure simulation of ST0779, the aldol reaction catalyzed by ST0779 with acetone and p-nitrobenzaldehyde as model reaction was proposed. Because of its thermodynamic superiority and high affinity, acetone first enters the active site and then is accommodated by the active site residues Ser439 and His 555. Proton transfer forms a transition state of enol salts, which is stable by Ser439. Asp523 is involved in stabilizing the positive charge of His 555-protonated imidazole ring. In the next

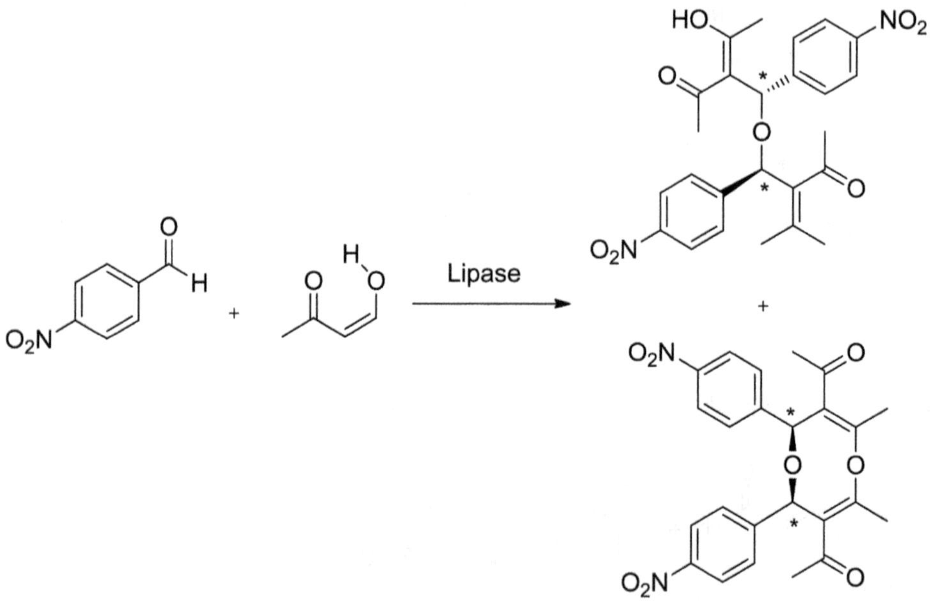

Figure 30.
The lipase-catalyzed promiscuous reaction: formation of acyclic and cyclic 2:2 adducts.

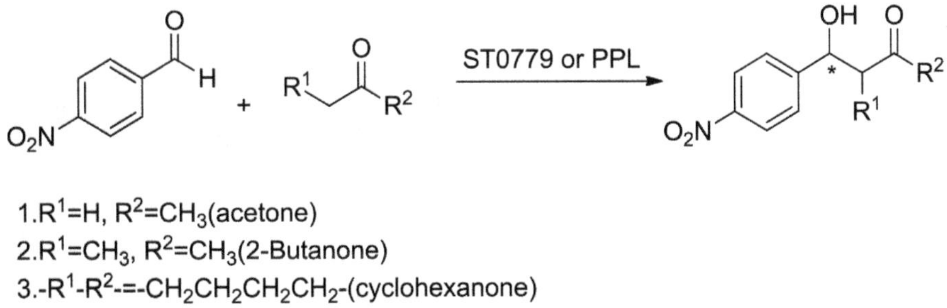

Figure 31.
Aldol reactions catalyzed by ST0779 or PPL.

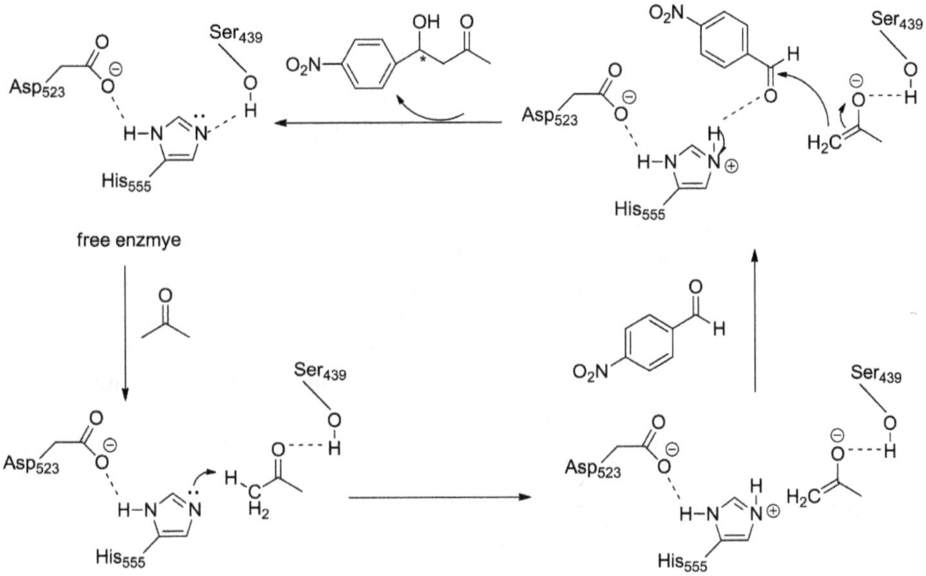

Figure 32.
Proposed mechanism for ST0779-catalyzed aldol reaction between acetone and 4-nitrobenzaldehyde.

step, the oxygen of carbonyl group in 4-nitrobenzaldehyde is protonated by protons from His 555 imidazole ring, and the carbon atoms in the same carbonyl group are neutrally attacked by oleic acid carbon to form a new C—C bond. Finally the aldol product is released from the enzyme, and the enzyme is freed for a new reaction.

In 2016, Wu et al. demonstrated a one-pot bienzymatic cascade in organic media to synthesize chiral β-hydroxy ketones for the first time (**Figure 33**) [46]. The decarboxylative aldol reaction catalyzed by an immobilized lipase from *Mucor miehei* (MML) and the synthesis of β-hydroxy ketone catalyzed by a lipase A or B from *Candida antarctica* (CALA or CALB) were combined in this one-pot protocol, reducing the purification step between the two reactions. (S)-β-hydroxy ketones and acylated (R)-β-hydroxy ketones could be obtained under mild reaction condition. The ee values of most chiral compounds were in a range of 94–99%, while the total yields of both chiral products were all above 85%. This enzymatic one-pot chain method is still very effective, not only can it be amplified to the level of grams but also the catalyst was recovered three times.

In 2019, Gao et al. demonstrated the construction of an unencapsulated remote-controlled nanobiocatalytic system. The system used three enzyme-conjugated gold nanorod composites (EGCs) to control reaction rates in real time by self-assembling

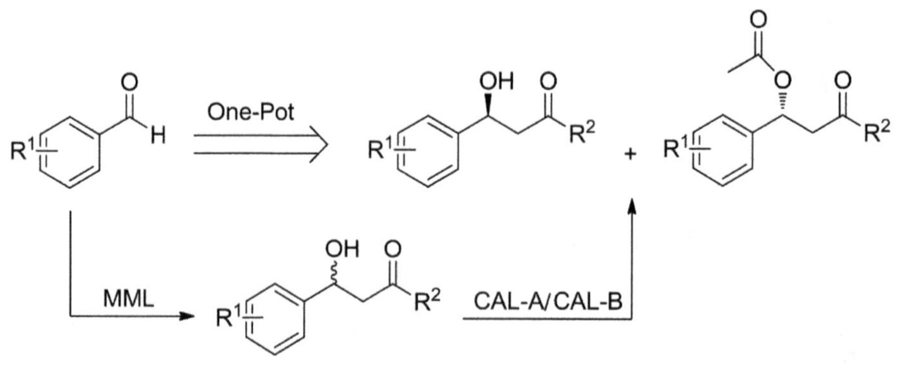

Figure 33.
One-pot cascade for synthesis of chiral β-hydroxy ketone derivatives.

enzymes formed by the combination of enzymes at different optimal temperatures and gold nanorods (GNRs) [47]. By using the photothermal effect of GRS to transfer energy quickly, coupled with the real-time and long-range regulation of enzyme activity, improving the thermal stability of the enzyme and effective catalysis of the aldol reaction can be achieved. The increase in energy inside GRS, stimulated by distant near-infrared (NIR) stimuli, leads to increased enzyme activity. The results show that the method of internal heating that transfers energy more directly to the enzyme-catalyzed site is a faster and more effective energy transfer method. The results also show that the catalytic effect of the remote-controlled nanocatalytic system at lower temperature is the same as that of the free enzyme at higher temperature, but it has the advantages of improving the stability of the enzyme and extending its service life. Specifically, PPL EGCs at room temperature exhibit the same catalytic effect as achieved by free PPL at 40°C, while ST0779EGCs and APE1547 EGCs at 33°C exhibit a higher catalytic effect than their corresponding free enzymes at 63°C. In addition, EGCs have superior catalytic efficiency and product yield compared with aldol addition in free enzyme systems.

3.2 Henry (nitroaldol) reaction

The nitroaldol or Henry reaction is one of the most useful carbon-carbon bond-forming reactions and has wide synthetic applications in organic chemistry. This reaction provides access to valuable racemic and optically active β-nitro alcohols, which are very useful in organic synthesis as precursors for pharmaceutical and biological purposes. In recent years, efficient nonconventional biocatalytic approaches have been reported [48].

Guan et al. firstly reported transglutaminase was used to catalyze Henry reactions of aliphatic, aromatic, and heteroaromatic aldehydes with nitroalkanes (**Figure 34**) [49]. The reactions were carried out at room temperature, and the corresponding nitroalcohols were obtained in yields up to 96%.

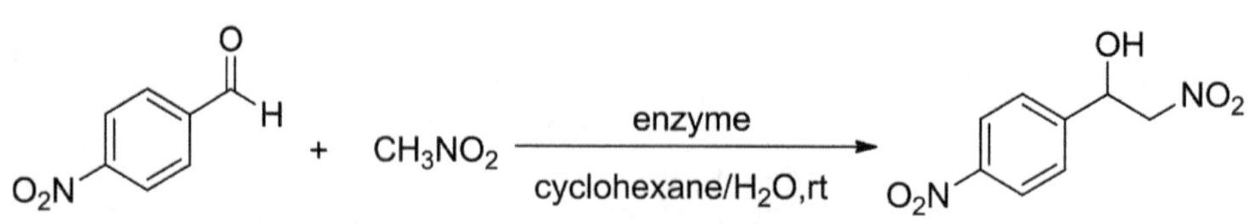

Figure 34.
Enzyme-catalyzed Henry reaction of 4-nitrobenzaldehyde and nitromethane.

Then, the same group reported glucoamylase from *Aspergillus niger* (AnGA) catalyze Henry reactions of aromatic aldehydes and nitroalkanes in 2013 [50]. The reactions were carried out at 30°C in the mixed solvents of ethanol and water, and the corresponding β-nitro alcohols were obtained in yields of up to 99%. Experiments demonstrated that AnGA could be inhibited by the product of the Henry reaction at 80°C. This enzymatic Henry reaction has a broad substrate scope and could be facilely enlarged to gram scale. Based on the experiments with denatured and inhibited AnGA, and the comparison of natural activity and promiscuous activity, the possible mechanism was also discussed (**Figure 35**). Glu400, as a base, deprotonates the α-carbon of the nitroalkane providing intermediate I. At the same time, Glu179, as an acid, donates a proton to the carbonyl oxygen of the aldehyde generating intermediate II. Then, the a-carbon of I, as a nucleophile, attacks the carbonyl of II forming a new carbon-carbon bond. Finally, the product (β-nitro alcohol) is released from the active site.

On the other hand, Lin and co-workers demonstrated the Henry reaction can also be catalyzed in a neat organic solvent. When using the D-aminoacylase from *Escherichia coli* as the promiscuous biocatalyst, DMSO was found to be the best solvent at 50°C [51]. Interestingly, the synthesis of optically active β-nitro alcohols was achieved by a two-step strategy combining the D-aminoacylase-catalyzed nitroaldol reaction with the PSL-catalyzed resolution of the so obtained racemic β-nitro alcohols (**Figure 36**) [52]. Both alcohols and acetates were isolated in good yields and high enantiomeric excess (>84% ee_s; >96% ee_p; E > 150).

In 2013, lipase A from *Aspergillus niger* was used in the Henry reaction between aromatic aldehydes and a large excess of nitroalkanes in an organic/water medium (**Figure 37**) [53]. The yield of corresponding β-nitro alcohols at 30°C reached 94%.

Gotor and co-workers reported the inexpensive carrier protein bovine serum albumin (BSA) as catalyst was firstly used in the condensation of an appropriate aldehyde with 1-nitroalkanes in aqueous media (**Figure 38**) [54]. By optimizing the reaction conditions, the yield of corresponding nitroalcohols at 30°C reached 91%.

Similarly, two other well-known lipases, *Pseudomonas cepacia* lipase and CALB, were found to catalyze the Henry reaction [55]. Nevertheless, spectroscopic experiments showed that the immobilization protocols contribute to the change in the secondary structure of the enzyme, which leads to improved conversion rates.

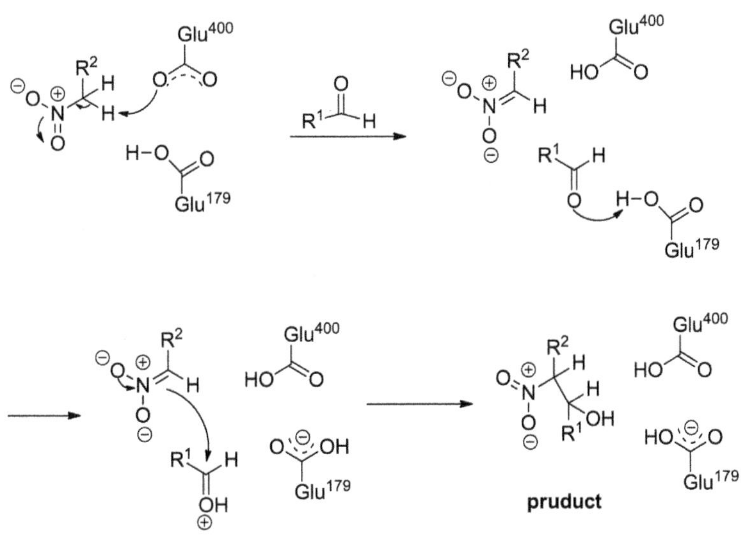

Figure 35.
Possible mechanism of the AnGA-catalyzed Henry reaction.

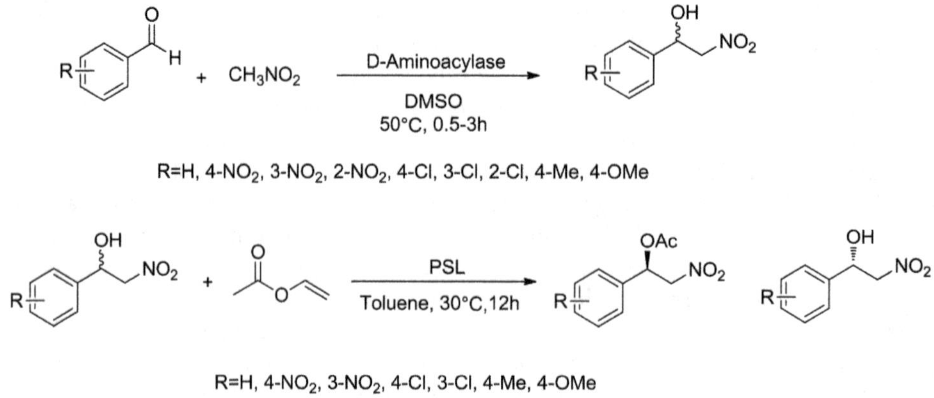

Figure 36.
Two-step method to obtain β-nitro alcohols and the corresponding acetates of both configurations based on a D-aminoacylase-catalyzed reaction and PSL-mediated kinetic resolution using vinyl acetate as acyl donor.

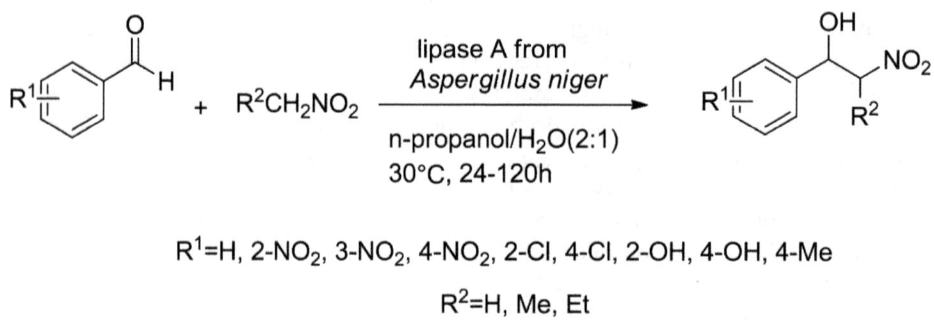

Figure 37.
Nitroaldol reaction between aromatic aldehydes and nitroalkanes catalyzed by lipase A from Aspergillus niger.

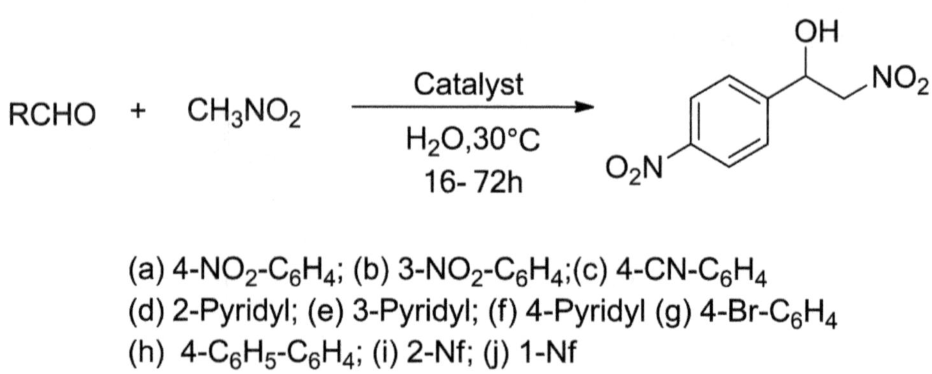

Figure 38.
Catalytic nitroaldol addition between different aromatic aldehydes and nitromethane.

3.3 Aldol (nitroaldol) reaction in untraditional solvent

Solvents for a biocatalysis reaction have experienced several generations of development. Traditional organic solvents (water miscible or water immiscible), in the form of cosolvents or second phase, can provide solutions for the above-described challenges. However, organic solvents inevitably face their own disadvantages, such as high volatility, difficulty in preparation, and inhibition of the activity of biocatalysts. So the untraditional solvents such as buffer solvent, ionic liquids (ILs), and deep eutectic solvents (DESs) have attracted the interest of many groups.

In 2013, our group demonstrated bovine pancreatic lipase (BPL) was first used to catalyze the aldol reaction and acidic buffer was first used for promiscuous enzymatic aldol reaction (**Figure 39**) [56]. The highest yield (99.0%), the best dr of 96:4, and a moderate ee of 66% were observed with aromatic aldehyde and ketone by BPL in phosphate-citrate buffer (pH 5.6, 5.0 mL) at 30°C.

Porto et al. demonstrated the lipase from *Rhizopus niveus* (RNL) catalyzed by unspecific protein catalysis the aldol reactions between cyclohexanone and aromatic aldehydes in organic solvents with water or aqueous buffer solution (**Figure 40**) [57]. The reactional conditions strongly influenced the yield (0–99%) and enantioselectivities in the anti-products (6–55% ee). The aldol products with enantioselectivities in the anti-product were observed for inactive enzyme and in denaturing conditions. Therefore, the reactions in the evaluated conditions were proceeded by unspecific protein catalysis with moderate enantioselectivities and not by promiscuous activity.

Ionic liquids are the first enzyme-compatible untraditional media developed by the green and sustainable concept (given their low vapor pressure). Numerous reactions, e.g., hydrolytic and redox reactions as well as formation of C—C bond, have been successfully performed in such ILs-containing media. We demonstrated PPL was used to catalyze asymmetric cross aldol reactions of aromatic and heteroaromatic aldehydes with various ketones in ionic liquid ([BMIM][PF_6]) for the first time in 2014 (**Figure 41**) [58]. PPL exhibited high catalytic activity and

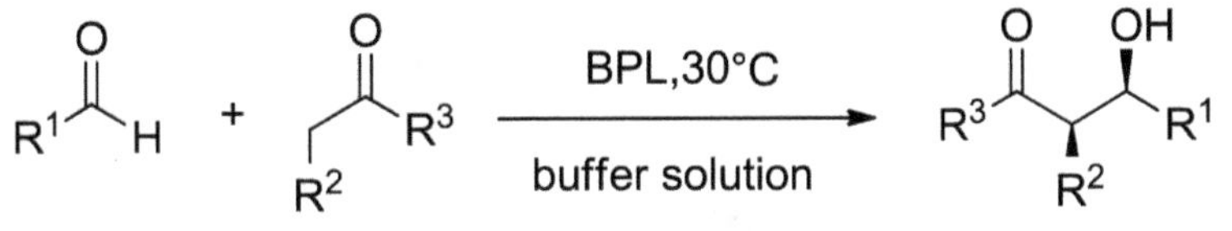

Figure 39.
The BPL-catalyzed asymmetric aldol reaction in buffer solution.

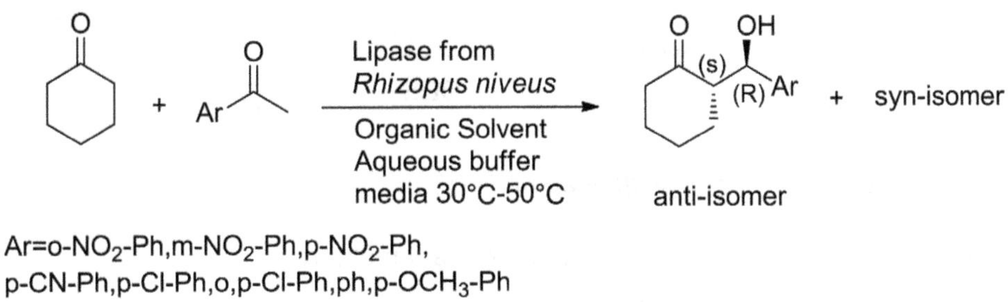

Figure 40.
Aldol reactions by lipase from Rhizopus niveus.

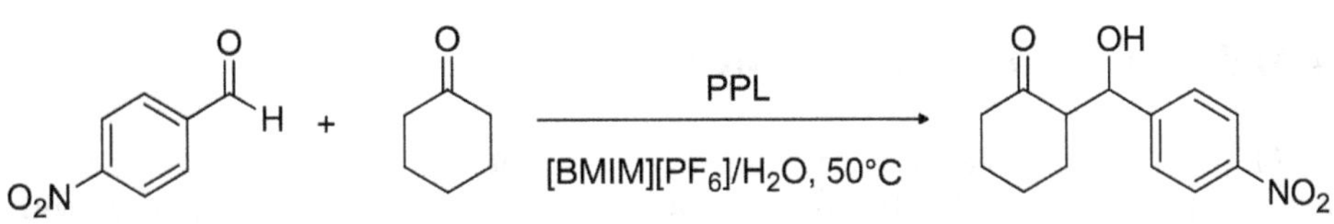

Figure 41.
The PPL-catalyzed asymmetric cross aldol reaction in ionic liquid.

excellent stereoselectivity in this efficient and recyclable room-temperature ionic liquid in the presence of moderate water. High yields of up to 99%, excellent enantioselectivities of up to 90% ee, and good diastereoselectivities of up to >99:1 dr were achieved.

Despite the excellent performance of ILs in biocatalysis, more doubts about their ungreenness and environmental influence have been gradually presented. Deep eutectic solvents, the recognized alternative of ILs, first came to the public vision in 2001. Since then, research on DESs faced a prosperous increase in many fields, such as extraction, materials synthesis and biotransformation, and biocatalysis.

Gotor-Fernández and co-workers reported a promiscuous lipase-catalyzed aldol reaction has been performed for the first time in DESs in 2016. The aldol reaction between 4-nitrobenzaldehyde and acetone was examined in-depth, with excellent compatibility being found between PPL and DESs (choline chloride/glycerol mixtures) for the formation of the aldol product in high yields (**Figure 42**) [59]. The system was compatible with a series of aromatic aldehydes and ketones including acetone, cyclopentanone, and cyclohexanone.

At the same year, Tian et al. explored the Henry reaction catalyzed by lipase AS using deep eutectic solvents as a reaction medium (**Figure 43**) [60]. The studies had shown that adding 30 vol% water to DES could increase the catalytic activity of enzymes. The final yield of the lipase AS-catalyzed Henry reaction was 92.2% in a DES-water mixture within only 4 h. In addition, the lipase AS activity was improved by approximately threefold in a DES-water mixture compared with that in pure water. The methodology was also extended to the aza-Henry reaction. The enantioselectivity of both Henry and aza-Henry reactions was not found.

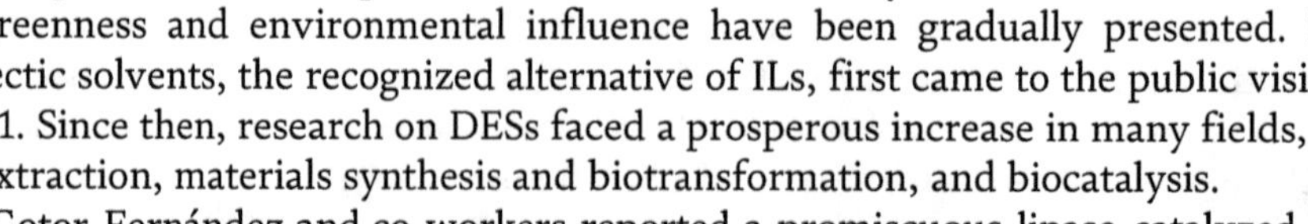

Figure 42.
The PPL-catalyzed asymmetric cross aldol reaction in DESs.

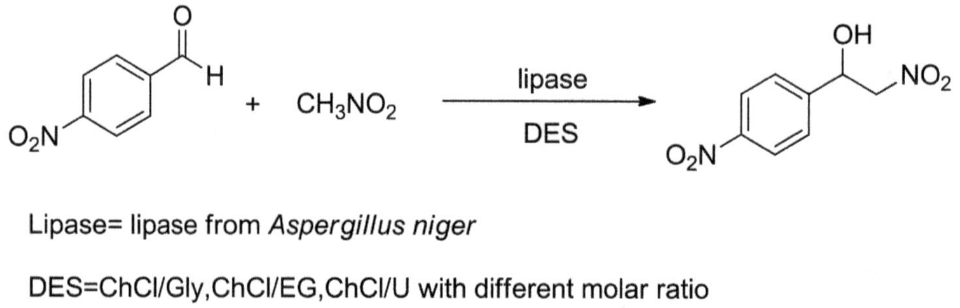

Figure 43.
Lipase AS-catalyzed Henry reaction in DES.

4. Multicomponent reactions (MCRs)

Multicomponent reactions have attracted sustained attention because they represent a powerful tool for the construction of complex molecular structures with evident advantages, such as simplified workup procedures, high overall yields, and versatile product libraries. Recently, hydrolases have allowed the development of multiple transformations and mainly served for the synthesis of heterocyclic compounds with high complexity in high yields. In this section, we will focus on the hydrolase-catalyzed multicomponent reaction in a one-pot transformation.

4.1 Mannich reaction

The Mannich reaction is a typical and the first example for the hydrolase-catalyzed multicomponent reaction, which is atom-economic and a powerful synthetic method for generating carbon-carbon bonds and nitrogenous compounds. An unprecedented "one-pot," direct Mannich reaction of ketone, aldehyde, and amine catalyzed by lipase was described first in 2009 [61]. Lipase from *Mucor miehei* (MML) efficiently catalyzed the Mannich reaction (**Figure 44**).

To assess the generality of the lipase-catalyzed Mannich reaction, we extended other substrates such as cyclohexanone, butanone, and 1-hydroxy-2-propanone in more benign reaction system (ethanol/water) catalyzed by the lipase from *Candida rugosa* (CRL) [62]. It was found that a wide range of aromatic aldehydes could effectively participate in the CRL-catalyzed Mannich reaction to give the corresponding β-amino carbonyl compounds (**Figure 45**). The reaction was favored by the electron-withdrawing substituents of the aldehydes.

In 2012, Guan et al. reported the enzyme-catalyzed, direct, three-component asymmetric Mannich reaction using protease type XIV from *Streptomyces griseus* (SGP) in acetonitrile (**Figure 46**) [63]. This characteristic makes it important to develop an enzyme-catalyzed asymmetric Mannich reaction as a more sustainable

Figure 44.
The first lipase-catalyzed direct Mannich reaction.

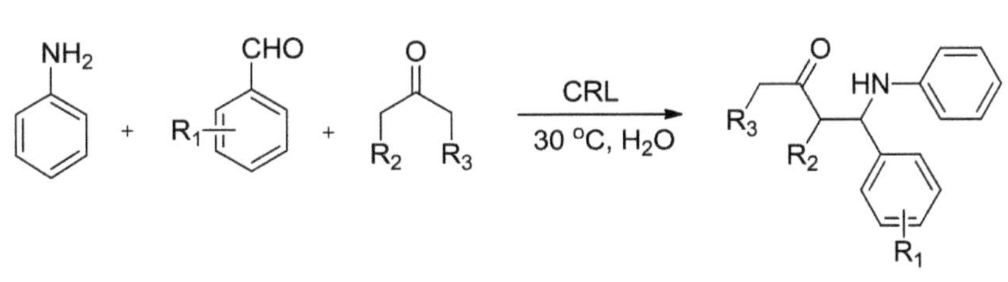

Figure 45.
Lipase-catalyzed direct Mannich reaction of various aryl aldehydes and ketones with aniline.

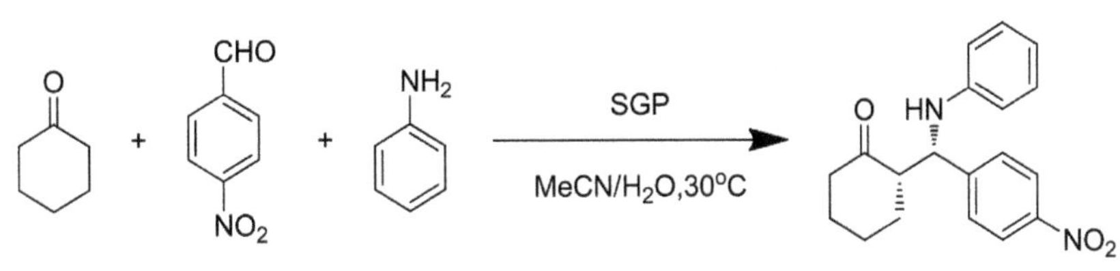

Figure 46.
SPG-catalyzed direct Mannich reaction.

complement to chemical catalysis. The control experiments with the denatured enzyme and non-enzyme proteins indicated that the specific natural fold of SGP was responsible for its stereoselectivity in the Mannich reaction. A wide range of substrates were accepted by the enzyme, and yields of up to 92%, enantioselectivities of up to 88% ee, and diastereoselectivities of up to 92:8 dr were achieved. As an example of enzyme catalytic promiscuity, this work broadens the scope of SGP-catalyzed transformations.

In the same year, Lin et al. inspired by chemical cofactors or mediators expect some small molecules to similarly improve the enzymatic Michael addition of unactivated carbon nucleophiles [64]. They found that the CALB/acetamide co-catalyst system can effectively catalyze the Michael addition between less-activated ketones and aromatic nitroolefins. This is of particular interest because neither CALB nor acetamide can independently catalyze the reaction to any significant extent. The CALB/acetamide catalyst system is also effective for other C—C bond-forming reactions with varying degrees of success, for example, CALB-catalyzed Mannich reaction (**Figure 47**). After adding acetamide as a co-catalyst, the yield increased by 50% (from 25 to 38%). The synergistic catalytic system of the lipase and the small molecule organic catalyst will greatly expand the application prospect of the enzyme in organic synthesis.

After 2 years, Guan et al. reported the use of acylase I from *Aspergillus melleus* in the asymmetric Mannich reaction (**Figure 48**) [65]. Compared to the current chemical technologies, this enzymatic reaction is more environmentally friendly and sustainable by using biocatalysts from inexpensive renewable resources. The activity and stereoselectivity of AMA can be improved by adjusting the solvent, pH, water content, temperature, substrate molar ratio, and enzyme loading. A wide range of substrates can be accepted by AMA, achieving enantioselectivities up to 89% ee, diastereoselectivities up to 90:10 dr, and yields up to 82% in the mixture of MeCN and phosphate buffer pH 8.1 (85:15 v/v 1 mL) at 30°C. This work not only provides new examples of enzyme-catalyzed reliability and potential synthetic methods of organic chemistry but may also help to better understand the metabolic pathways of nitrogen-containing compound biosynthesis.

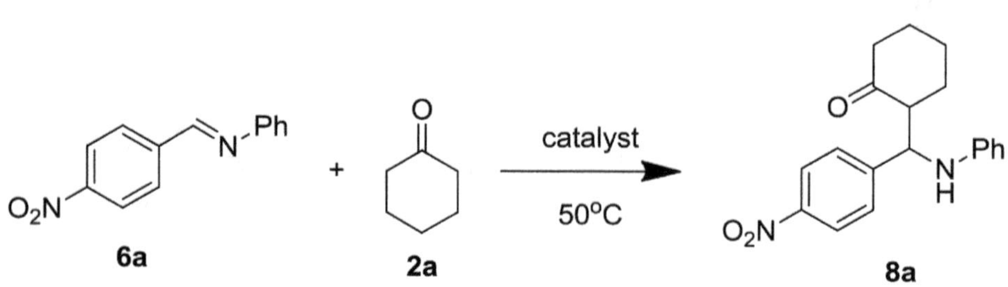

Figure 47.
Mannich reaction of (E)-N-(4-nitro-benzylidene)aniline with cyclohexanone.

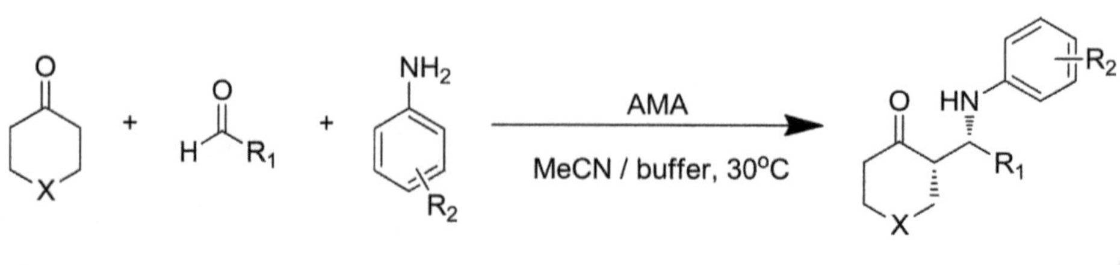

Figure 48.
The AMA-catalyzed Mannich reactions.

4.2 Biginelli reactions

In 2013, Sinha et al. reported bovine serum albumin promoted simple and efficient one-pot procedure for synthesis of 3,4-dihydropyrimidin-2(1H)-ones including potent mitotic kinesin Eg5 inhibitor monastrol under mild reaction conditions (**Figure 49**) [66]. After the reaction conditions are optimized, the yields reached up to 82% in EtOH at 60°C. The catalyst recyclability and gram-scale synthesis have also been demonstrated to enhance the practical utility of process.

Followed by our group, we reported trypsin as a multifunctional catalyst for synthesis of 3,4-dihydropyrimidin-2(1H)-ones by the Biginelli reaction of urea, β-dicarbonyl compounds, and in situ-formed acetaldehyde (**Figure 50**) [67]. This one-pot multistep reaction consists of two relatively independent reactions, both of which are catalyzed by trypsin. First is the transesterification of ethyl acetate and isobutanol at 60°C to produce in situ acetaldehyde, followed by in situ-generated acetaldehyde, urea, and β-dicarbonyl compounds for Biginelli reaction. The first reaction continuously provides a substance for the second reaction, effectively reducing the volatilization loss, oxidation, and polymerization of acetaldehyde and avoiding the negative influence of excess acetaldehyde on the enzyme. Under optimal conditions, a wide range of substrates participate in the reaction and provide the target product in high yield.

In 2017, CALB-catalyzed for synthesis of 3,4-dihydropyrimidin-2(1H)-ones by a tandem multicomponent reaction in one pot (**Figure 51**) has been reported [68]. Several control experiments were performed using acetaldehydes directly to explore the possible mechanism of this procedure. Moreover, owing to the distinct modularity and highly efficient features of the MCR, it assembles libraries of structurally diverse products and provides an exceptional synthesis tool for the discovery of the minimal deep-blue luminogen in the solid state, namely, a single ring. A few of the compounds show deep-blue emissions which only contain a single ring. This is an important application of green biocatalytic promiscuity for constructing a wide variety of new materials.

4.3 Hantzsch reaction

Lin et al. reported a three-component (aldehyde, 1,3-dicarbonyl compound, and acetamide) Hantzsch-type reaction in anhydrous solvent, which gave 1,4-dihydropyridines in moderate to good yields (**Figure 52**) [69]. The group used acetamide as a new source of ammonia. Initially the yield of the reaction with CALB

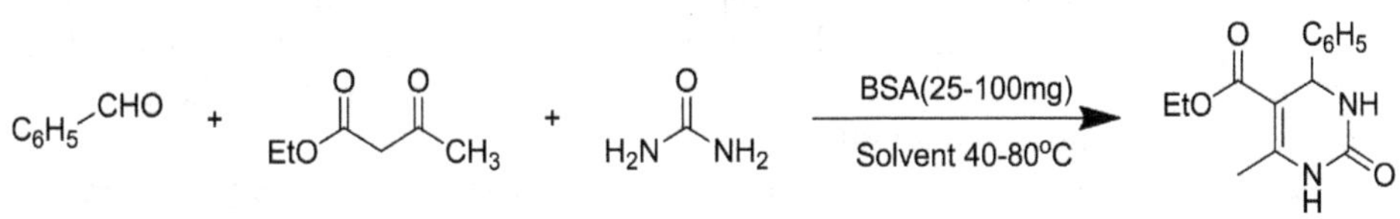

Figure 49.
BSA-catalyzed Biginelli reaction.

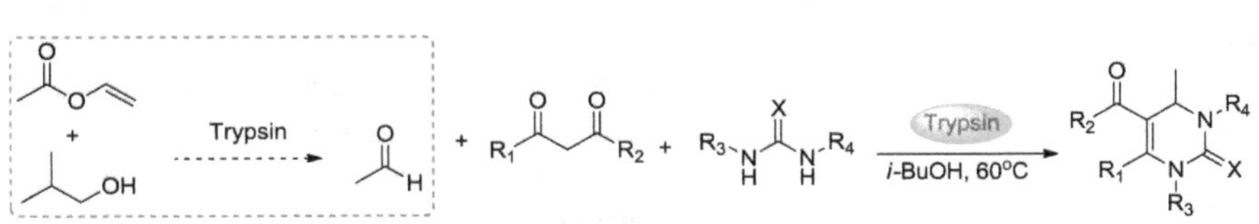

Figure 50.
Trypsin-catalyzed Biginelli reaction using an in situ-generated acetaldehyde.

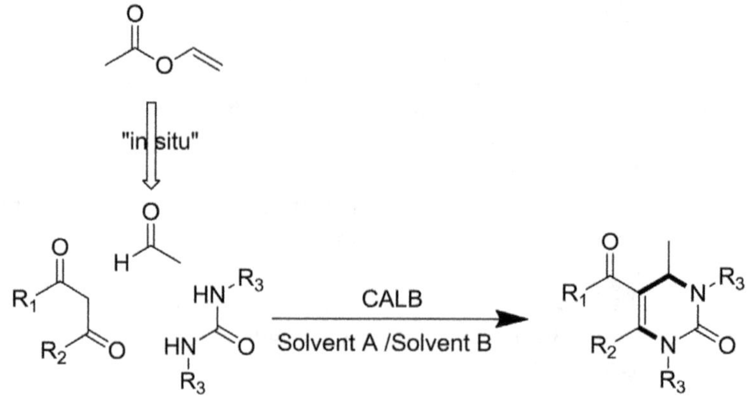

Figure 51.
CALB-initiated tandem Biginelli reaction.

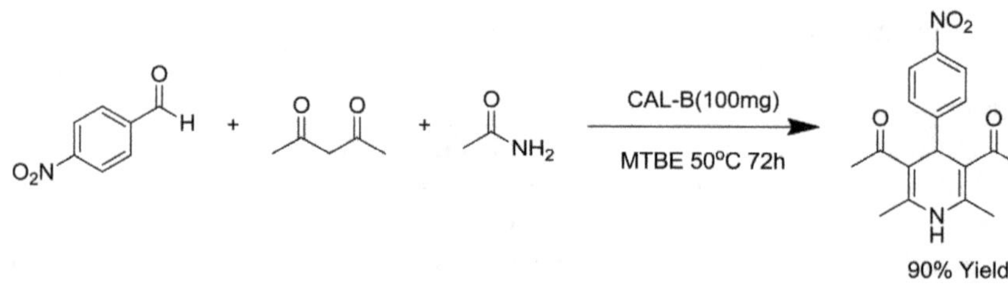

Scheme

Figure 52.
Lipase-catalyzed Hantzch-type reaction.

was very low (only 25%). The yield of the product was slightly improved using a mixed solvent (the ratio of MTBE to acetylacetone was 6:4), and the molar ratio of 4-nitrobenzaldehyde acetamide was 1:4 at 50 mg/ml lipase. When the lipase concentration (100 mg/mL) was increased, the yield increased sharply.

They proposed a reasonable mechanism of the reaction, wherein Asp-His dyad and oxyanion hole in the active site stabilized acetamide (**Figure 53**). This activated acetamide reacted with 1,3-dicarbonyl compounds to form an intermediate, which upon subsequent hydrolysis by CALB formed an enamine intermediate. During this period, a CALB-catalyzed Knoevenagel condensation reaction of the 1,3-dicarbonyl compound with aldehyde formed a separate intermediate (α,β-unsaturated carbonyl compound). Subsequently, the intermediate that is stabilized by the catalytic center of the lipase forms the final product (1,4-dihydropyridine) by Michael addition and intramolecular condensation.

In 2017, our group reported a series of 1,4-dihydropyridines was produced via facile enzymatic Hantzsch reactions in one pot, using acetaldehydes/aromatic aldehydes prepared in situ (**Figure 54**) [70]. After screening several parameters on a model reaction, the tandem process afforded 1a in 80% yield. This approach provided an opportunity to discover novel libraries of AIEEgens that contain the minimum requirement necessary for AIEE behavior, namely, a single ring.

Meanwhile, we found that certain 1,4-DHPs could stain the mitochondria in live cells with high selectivity but without obvious guiding units (such as cationic groups). Taking one of the 1,4-DHPs as an example, we found that it exhibited excellent photostability and storage stability and that it could be utilized in applications such as real-time imaging, long-term tracking of mitochondrial morphological changes, and viscosity mapping (**Figure 55**). We believe that the use of biocatalysis

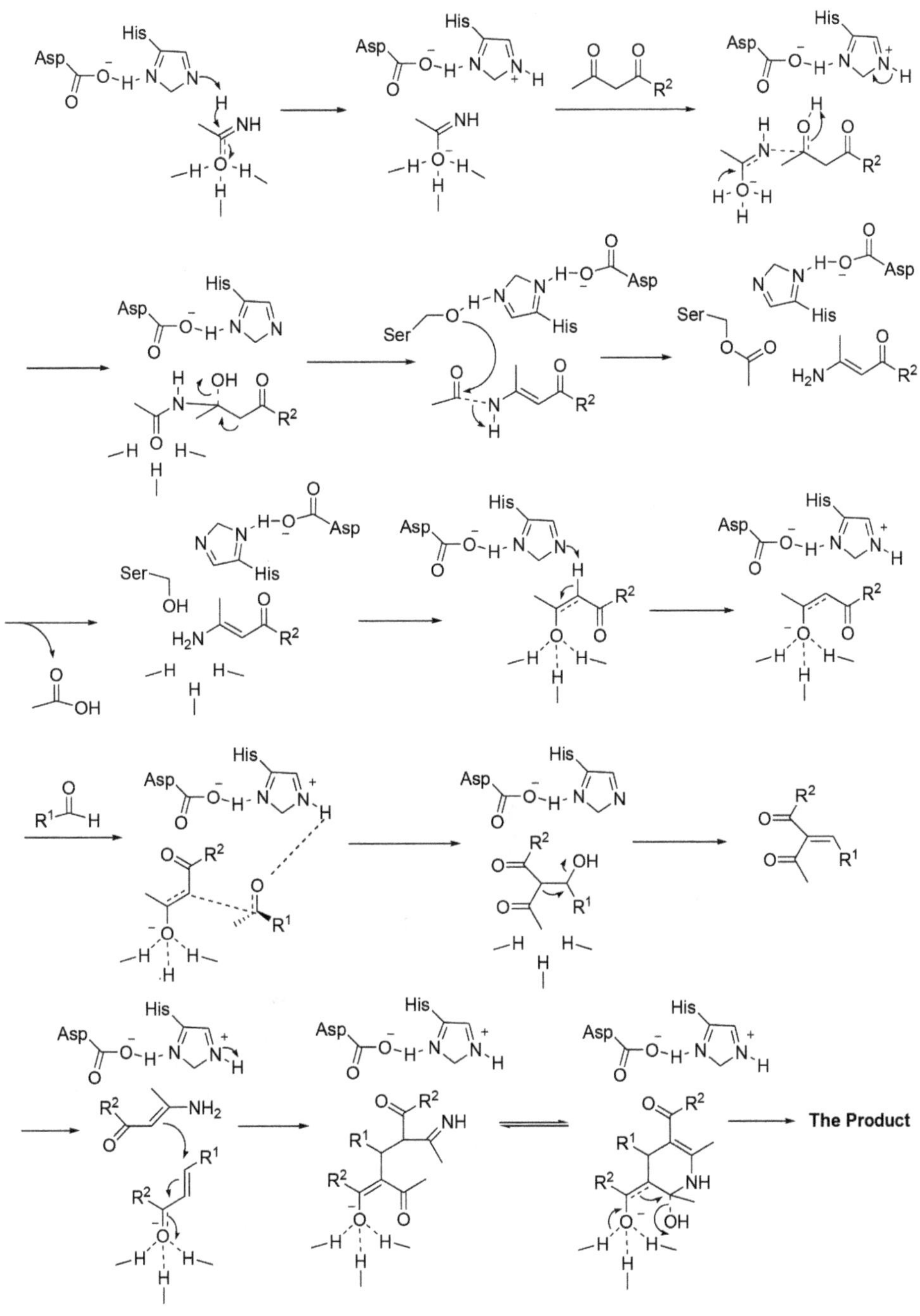

Figure 53.
Proposed mechanism of lipase-catalyzed Hantzsch-type reaction of an aldehyde with acetamide and 1,3-dicarbonyl compounds.

could be simplified with workup procedures and could provide avenues for discovering a wide variety of new materials.

In recently, Ye et al. reported solvent-free quick synthesis of 1,4-DHP calcium antagonists felodipine, nitrendipine, nifedipine, and nemadipine B and their derivatives by Lipozyme® RM IM-catalyzed multicomponent reactions of aromatic aldehyde, alkyl acetoacetate, and alkyl 3-aminocrotonate under ball milling conditions (**Figure 56**) [71]. The product was obtained in moderate yield (up to 86.8%), and the effects of the reaction conditions were investigated, including catalyst loading,

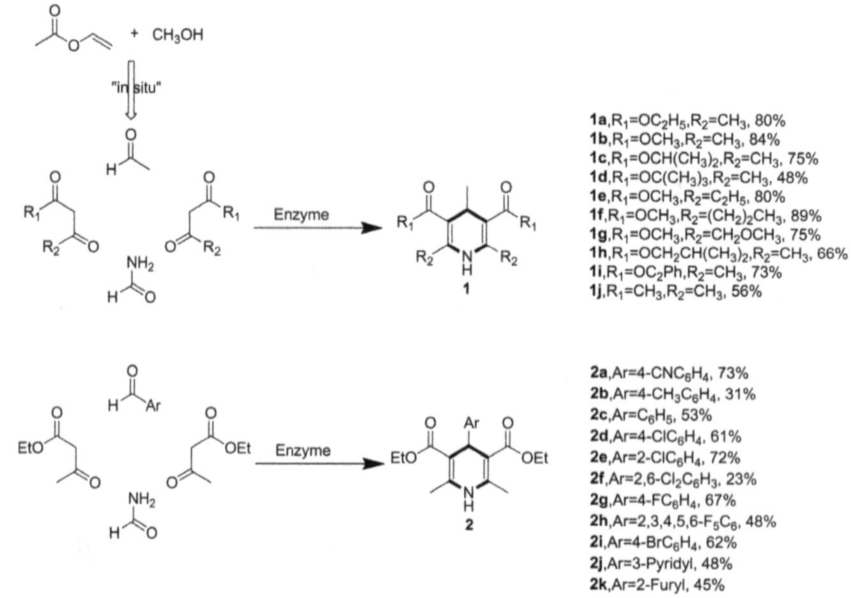

Figure 54.
Synthesis of 1 and 2 through one-pot multicomponent reactions.

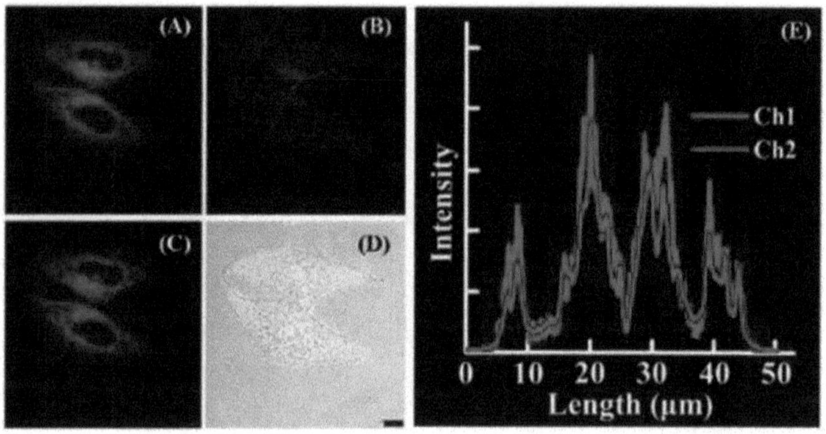

Figure 55.
Confocal fluorescence images of HeLa cells stained with 2 h and MitoTracker® Deep Red FM (MTDR). (A) Fluorescent image of 2h (10.0 μM) in HeLa cells, collected at 410-460 nm and λex = 405 nm. (B) Fluorescent image of MTDR (1.0 μM), collected at 660-740 nm and λex = 633 nm. (C) Merged image of A and B. (D) Overlay of fluorescence and bright-field image. (E) Intensity profiles of linear regions of interest (ROI) across the HeLa cells. Scale bar: 7.5 μm.

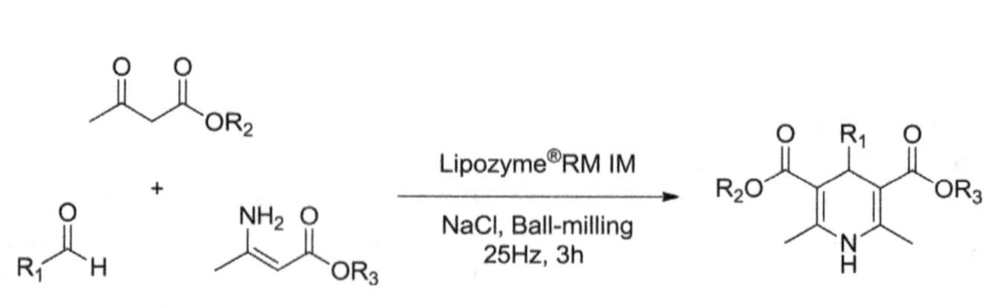

Figure 56.
Lipozyme® RM IM-catalyzed rapid synthesis of 1,4-DHP calcium antagonists and derivatives under ball milling conditions.

grinding aid, and milling frequency. The reaction features environmentally friendly, simple, and efficient operation. A major feature distinguishing this enzyme promiscuity from previously reported work is the use of mechanochemical ball milling techniques that overcome disadvantages such as long reaction times and the use of hazardous organic solvents. This work demonstrates the potential application of mixed enzyme-catalyzed reactions for drug synthesis under ball milling conditions.

4.4 Ugi reaction

In 2013, Berlozecki et al. reported for the first time an enzyme-catalyzed Ugi reaction that has many advantages over previous reactions, such as good reaction at room temperature and extensive solvent selection (**Figure 57**) [72]. In this three-component reaction, the aldehyde, amine, and isocyanide are condensed to form a dipeptide. Of all the selected lipases, Novozym 435 had the highest yield of 75%.

Recently, Thomas et al. reported the concatenation of the Ugi four-component synthesis, and the CALB-catalyzed aminolysis of the intermediary formed Ugi methyl ester products furnishes a novel consecutive five-component reaction for the formation of triamides (**Figure 58**) [73]. This one-pot method is compatible with metal catalysis methods such as copper-catalyzed alkyl azide ring addition and Suzuki cross-coupling or both in a one-pot process.

The mild reaction conditions make this sequence superior to the stepwise process with isolation of the Ugi product and even more favorable than other base-catalyzed or microwave-assisted aminolyses. This efficient scaffold forming

Figure 57.
CALB-catalyzed Ugi reaction.

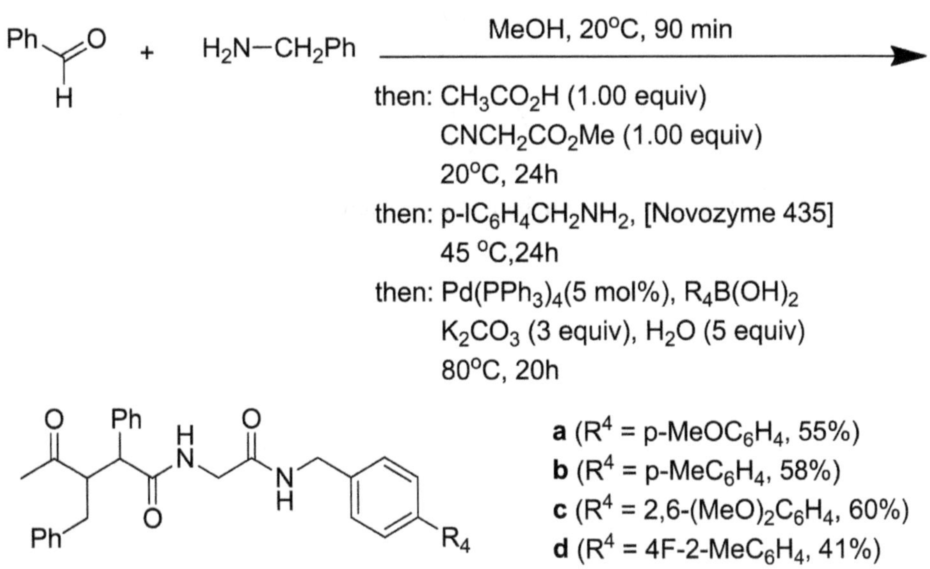

Figure 58.
Consecutive six-component U-4CR-CALB-catalyzed aminolysis-Suzuki cross-coupling sequence of biaryls.

processes are particularly favorable for creating compound libraries for medicinal chemistry lead finding and for functional chromophores in materials sciences.

5. Summary

This chapter has reviewed some examples of various types of hydrolase catalytic promiscuous reactions and their applications in the past decade. Several different types of hydrolases catalyzed carbon-carbon or carbon-heteroatom formation reactions have been discussed: aldol reactions, Michael reactions, and multicomponent reactions.

From these examples, it is clear that enzymes that display catalytic promiscuity can provide new opportunities for organic synthesis. Exploiting enzyme catalytic promiscuous reactions might lead to new, efficient, and stable catalysts with alternative activity and could provide more promising and green synthetic methods for organic chemistry. The development of protein engineering and enzyme engineering can extend the application of metagenome libraries and find enzymes with specific promiscuous behavior. We believe the progress in the area of biocatalytic promiscuity will greatly extend the useful applications of enzymes.

Acknowledgements

The authors gratefully acknowledge funding of our research in this area by grants from the Natural Science Foundation of Guangdong Province (Grant No. 2018A030307022) and the Special Innovation Projects of Common Universities in Guangdong Province (Grant No.2018KTSCX126).

Author details

Yun Wang[1] and Na Wang[2*]

1 School of Chemistry and Chemical Engineering, Lingnan Normal University, Zhanjiang, China

2 Key Laboratory of Green Chemistry and Technology, Ministry of Education, College of Chemistry, Sichuan University, Chengdu, P.R. China

*Address all correspondence to: wnchem@scu.edu.cn

References

[1] Faber K. Biotransformations in Organic Chemistry. 7th ed. Berlin Heidelberg: Springer-Verlag; 2018. 434 p. DOI: 10.1007/978-3-319-61590-5

[2] van Rantwijk F, Sheldon RA. Biocatalysis in ionic liquids. Chemical Reviews. 2007;**107**(6):2757-2785. DOI: 10.1021/cr050946x

[3] Yadav JS, Reddy GSKK, Sabitha G, Krishna AD, Prasad AR, Hafeez URR, et al. Daucus carota and baker's yeast mediated bio-reduction of prochiral ketones. Tetrahedron: Asymmetry. 2007;**18**(6):717-723. DOI: 10.1016/j.tetasy.2007.03.009

[4] Schmid A, Dordick JS, Hauer B, Kiener A, Wubbolts M, Witholt B. Industrial biocatalysis today and tomorrow. Nature. 2001;**409**:258-268. DOI: 10.1038/35051736

[5] Ran N, Zhao L, Chen Z, Tao J. Recent applications of biocatalysis in developing green chemistry for chemical synthesis at the industrial scale. Green Chemistry. 2008;**10**(4): 361-372. DOI: 10.1039/b716045c

[6] Koeller KM, Wong CH. Enzymes for chemical synthesis. Nature. 2001;**409**: 232-240. DOI: 10.1038/35051706

[7] Liese A, Seelbach K, Wandrey C, editors. Industrial Biotransformations, 2nd, Completely Revised and Extended Edition. Weinheim: Wiley-VCH; 2006. 536 p. DOI: 10.1002/3527608184

[8] O'Brien PJ, Herschlag D. Catalytic promiscuity and the evolution of new enzymatic activities. Chemistry & Biology. 1999;**6**:R91-R105. DOI: 10.1016/S1074-5521(99)80033-7

[9] Copley S. Enzymes with extra talents: Moonlighting functions and catalytic promiscuity. Current Opinion in Chemical Biology. 2003;**7**(2):265-272. DOI: 10.1016/S1367-5931(03)00032-2

[10] Bornscheuer UT, Kazlauskas RJ. Catalytic promiscuity in biocatalysis: Using old enzymes to form new bonds and follow new pathways. Angewandte Chemie International Edition. 2004; **43**(45):6032-6040. DOI: 10.1002/anie.200460416

[11] Kazlauskas RJ. Enhancing catalytic promiscuity for biocatalysis. Current Opinion in Chemical Biology. 2005; **9**(2):195-201. DOI: 10.1016/j.cbpa.2005.02.008

[12] Hult K, Berglund P. Enzyme promiscuity: Mechanism and applications. Trends in Biotechnology. 2007;**25**(5):231-238. DOI: 10.1016/j.tibtech.2007.03.002

[13] Neuberg C, Hirsch J. An enzyme which brings about union into carbon chains. Biochemische Zeitschrift. 1921; **115**:282-310

[14] Miao YF, Rahimi M, Geertsema EM, Poelarends GJ. Recent developments in enzyme promiscuity for carbon–carbon bond-forming reactions. Current Opinion in Chemical. 2015;**25**:115-123. DOI: 10.1016/j.cbpa.2014.12.020

[15] López-Iglesias M, Gotor-Fernández V. Recent advances in biocatalytic promiscuity: Hydrolase-catalyzed reactions for nonconventional transformations. The Chemical Record. 2015;**15**:743-759. DOI: 10.1002/tcr.201500008

[16] Kitazume T, Ikeya T, Murata K. Synthesis of optically active trifluorinated compounds: Asymmetric Michael addition with hydrolytic enzymes. Chemical Communications. 1986;**17**:1331-1333. DOI: 10.1039/c39860001331

[17] Cai Y, Yao SP, Wu Q, Lin XF. Michael addition of imidazole with acrylates catalyzed by alkaline protease from Bacillus subtilis in organic media. Biotechnology Letters. 2004;**26**:525-528. DOI: 10.1023/b:bile.0000019562.21256.39

[18] Torre O, Alfonso I, Gotor V. Lipase catalysed Michael addition of secondary amines to acrylonitrile. Chemical Communications. 2004;**15**:1724-1725. DOI: 10.1039/b402244k

[19] Dhake KP, Tambade PJ, Singhal RS, Bhanage BM. Promiscuous candida antarctica lipase B-catalyzed synthesis of b-amino esters via aza-Michael addition of amines to acrylates. Tetrahedron Letter. 2010;**51**:4455-4458. DOI: 10.1016/j.tetlet.2010.06.089

[20] López-Iglesias M, Busto E, Gotor-Fernández V, Gotor V. Use of protease from bacillus licheniformis as promiscuous catalyst for organic synthesis: Applications in C-C and C-N bond formation reactions. Advanced Synthesis & Catalysis. 2011;**353**: 2345-2353. DOI: 10.1002/adsc.201100347

[21] Monsalve LN, Gillanders F, Baldessari A. Promiscuous behavior of Rhizomucor miehei lipase in the synthesis of N-substituted β -amino esters. European Journal of Organic Chemistry. 2012;**6**:1164-1170. DOI: 10.1002/ejoc.201101624

[22] Bonte S, Ghinea IO, Baussanne I, Xuereb JP, Dinica R, Demeunynck M. Investigation of the lipase-catalysed reaction of aliphatic amines with ethyl propiolate as a route to N-substituted propiolamides. Tetrahedron. 2013;**69**: 5495-5500. DOI: 10.1016/j.tet.2013.04.093

[23] Steunenberg P, Sijm M, Zuilhof H, Sanders JPM, Scott EL, Franssen MCR. Lipase-catalyzed aza-Michael reaction on acrylate derivatives. The Journal of Organic Chemistry. 2013;**78**:3802-3813. DOI: 10.1021/jo400268u

[24] Rivera-Martınez JD, Escalante J, Lopez-Munguıa A, Marty A, Castillo E. Thermodynamically controlled chemoselectivity in lipase-catalyzed aza-Michael additions. Journal of Molecular Catalysis B: Enzymatic. 2015; **112**:76-82. DOI: 10.1016/j.molcatb.2014.12.009

[25] Ortega-Rojas MA, Rivera-Ramírez JD, Ávila-Ortiz CG, Juaristi E, González-Muñoz F, Castillo E, et al. One-pot lipase-catalyzed enantioselective synthesis of (R)-(−)-N-benzyl-3-(benzylamino)butanamide: The effect of solvent polarity on enantioselectivity. Molecules. 2017;**22**:2189-2196. DOI: 10.3390/molecules22122189

[26] Zhou LH, Wang N, Zhang W, Xie ZB, Yu XQ. Catalytical promiscuity of a-amylase: Synthesis of 3-substituted 2H-chromene derivatives via biocatalytic domino oxa-Michael/aldol condensations. Journal of Molecular Catalysis B: Enzymatic. 2013;**91**:37-43. DOI: 10.1016/ j.molcatb.2013.02.001

[27] Gu B, Hu ZE, Yang ZJ, Li J, Zhou ZW, Wang N, et al. Probing the mechanism of CAL-B-catalyzed aza-Michael addition of aniline compounds with acrylates using mutation and molecular docking simulations. ChemistrySelect. 2019;**4**:3848-3854. DOI: 10.1002/slct.201900112

[28] Madalinska L, Kwiatkowska M, Cierpał T, Kiełbasinski P. Investigations on enzyme catalytic promiscuity: The first attempts at a hydrolytic enzyme-promoted conjugate addition of nucleophiles to a,b-unsaturated sulfinyl acceptors. Journal of Molecular Catalysis B: Enzymatic. 2012;**81**:25-30. DOI: 10.1016/j.molcatb.2012.05.002

[29] Rizzo PVS, Boarin LA, Freitas IOM, Gomes RS, Beatriz A, Rinaldi AW, et al. The study of biocatalyzed thio-Michael

reaction: A greener and multi-gram protocol. Tetrahedron Letter. 2014;**55**: 430-434. DOI: 10.1016/j.tetlet.2013. 11.047

[30] Cai JF, Guan Z, He YH. The lipase-catalyzed asymmetric C–C Michael addition. Journal of Molecular Catalysis B: Enzymatic. 2011;**68**:240-244. DOI: 10.1016/j.molcatb.2010.11.011

[31] Xie BH, Guan Z, He YH. Promiscuous enzyme-catalyzed Michael addition: Synthesis of warfarin and derivatives. Journal of Chemical Technology and Biotechnology. 2012;**87**: 1709-1714. DOI: 10.1002/jctb.3830

[32] Wu MY, Li K, He T, Feng XW, Wang N, Wang XY, et al. A novel enzymatic tandem process: Utilization of biocatalytic promiscuity for high stereoselective synthesis of 5-hydroxyimino-4,5-dihydrofurans. Tetrahedron. 2011;**67**:2681-2688. DOI: 10.1016/j.tet.2011.01.060

[33] Zhou LH, Wang N, Chen GN, Yang Q , Yang SY, Zhang W, et al. Lipase-catalyzed highly diastereoselective direct vinylogous Michael addition reaction of a, a-dicyanoolefifins to nitroalkenes. Journal of Molecular Catalysis B: Enzymatic. 2014;**109**:170-177. DOI: 10.1016/j. molcatb.2014.09.001

[34] Jiang L, Wang B, Li RR, Shen S, Yu HW, Ye LD. Catalytic promiscuity of Escherichia coli BioH esterase: Application in the synthesis of 3,4-dihydropyran derivatives. Process Biochemistry. 2014;**49**:1135-1138. DOI: 10.1016/j.procbio.2014.03.020

[35] Dean SM, Greenberg WA, Wong CH. Recent advances in aldolase-catalyzed asymmetric synthesis. Advanced Synthesis and Catalysis. 2007;**349**(8–9):1308-1320. DOI: 10.1002/adsc.200700115

[36] Gijsen HJM, Qiao L, Fitz W, Wong CH. Recent advances in the chemoenzymatic synthesis of carbohydrates and carbohydrate mimetics. Chemical Reviews. 1996;**96**: 443-474. DOI: 10.1021/cr950031q

[37] Palomo C, Oiarbide MJ, Garcia M. Current progress in the asymmetric aldol addition reaction. Chemical Society Reviews. 2004;**33**:65-75. DOI: 10.1002/chin.200421285

[38] Branneby C, Carlqvist P, Magnusson A, Hult K, Brinck T, Berglund P. Carbon–carbon bonds by hydrolytic enzymes. Journal of the American Chemical Society. 2003;**125**: 874-875. DOI: 10.1021/ja028056b

[39] Li C, Feng XW, Wang N, Zhou YJ, Yu XQ. Biocatalytic promiscuity: The first lipase-catalysed asymmetric aldol reaction. Green Chemistry. 2008;**10**(6): 616-618. DOI: 10.1039/b803406k

[40] Guan Z, Fu JP, He YH. Biocatalytic promiscuity: Lipase-catalyzed asymmetric aldol reaction of heterocyclic ketones with aldehydes. Tetrahedron Letters. 2012;**53**:4959-4961. DOI: 10.1016/j.tetlet.2012.07.007

[41] He YH, Li HH, Chen YL, Xue Y, Yuan Y, Guan Z. Chymopapain-catalyzed direct asymmetric aldol reaction. Advanced Synthesis & Catalysis. 2012;**354**:712-719. DOI: 10.1002/adsc.201100555

[42] Xie ZB, Wang N, Zhou LH, Wan F, He T, Le ZG, et al. Lipase-catalyzed stereoselective cross-aldol reaction promoted by water. ChemCatChem. 2013;**5**:1935-1940. DOI: 10.1002/ cctc.201200890

[43] Wang N, Zhang W, Zhou LH, Deng QF, Xie ZB, Yu XQ. One-pot lipase-catalyzed aldol reaction combination. Applied Biochemistry and Biotechnology. 2013;**171**:1559-1567. DOI: 10.1007/s12010-013-0435-4

[44] Majumder AB, Gupta MN. Lipase-catalyzed condensation reaction of

4-nitrobenzaldehyde with acetyl acetone in aqueous-organic cosolvent mixtures and in nearly anhydrous media. Synthetic Communications. 2014;**44**:818-826. DOI: 10.1080/00397911.2013.834059

[45] Li R, Perez B, Jian H, Gao R, Dong MD, Guo Z. Acyl-peptide releasing enzyme from Sulfolobus tokodaii (ST0779) as a novel promiscuous biocatalyst for aldol addition. Catalysis Communications. 2015;**66**:111-115. DOI: 10.1016/j. catcom.2015.03.030

[46] Xu F, Xu J, Hu YJ, Lin XF, Wu Q. One-pot bienzymatic cascade combining decarboxylative aldol reaction and kinetic resolution to synthesize chiral b-hydroxy ketone derivatives. RSC Advances. 2016;**6**: 76829-76837. DOI: 10.1039/c6ra12729k

[47] Li W, Liu DN, Geng X, Li ZQ , Gao RJ. Real-time regulation of catalysis by remote controlled enzyme-conjugated gold nanorod composites for aldol reaction-based applications. Catalysis Science & Technology. 2019;**9**: 2221-2230. DOI: 10.1039/c9cy00167k

[48] Milner SE, Moody TS, Maguire AR. Biocatalytic approaches to the Henry (nitroaldol) reaction. European Journal of Organic Chemistry. 2012;**16**:3059-3067. DOI: 10.1002/ejoc.201101840

[49] Tang RC, Guan Z, He YH, Zhu W. Enzyme-catalyzed Henry (nitroaldol) reaction. Journal of Molecular Catalysis B: Enzymatic. 2010;**63**:62-67. DOI: 10.1016/j.molcatb.2009.12.005

[50] Gao N, Chen YL, He YH, Guan Z. Highly efficient and large-scalable glucoamylase catalyzed Henry reactions. RSC Advances. 2013;**3**: 16850-16856. DOI: 10.1039/c3ra41287c

[51] Wang JL, Li XH, Xie Y, Liu BK, Lin XF. Hydrolase-catalyzed fast Henry reaction of nitroalkanes and aldehydes in organic media. Journal of Biotechnology. 2010;**145**:240-243. DOI: 10.1016/j.jbiotec.2009.11.022

[52] Xu F, Wang J, Liu B, Wu Q , Lin XF. Enzymatic synthesis of optical pure b-nitroalcohols by combining D-aminoacylase-catalyzed nitroaldol reaction and immobilized lipase PS-catalyzed kinetic resolution. Green Chemistry. 2011;**13**:2359-2361. DOI: 10.1039/c1gc15417f

[53] Le ZG, Guo LT, Jiang GF, Yiang XB, Liu HQ. Henry reaction catalyzed by lipase a from *Aspergillus niger*. Green Chemistry Letters and Reviews. 2013;**6**: 277-281. DOI: 10.1080/17518253.2013.818721

[54] Busto E, Gotor-Fernández V, Gotor V. Protein-mediated nitroaldol addition in aqueous media. Catalytic promiscuity or unspecific catalysis?Organic Process Research & Development. 2011;**15**:236-240. DOI: 10.1021/op100130c

[55] Izquierdo DF, Barbosa O, Burguete MI, Lozano P, Luis SV, Fernández-Lafuente R, et al. Tuning lipase B from Candida antarctica C–C bond promiscuous activity by immobilization on polystyrene-divinylbenzene beads. RSC Advances. 2014;**4**:6219-6225. DOI: 10.1039/c3ra47069e

[56] Xie ZB, Wang N, Jiang GF, Yu XQ. Biocatalytic asymmetric aldol reaction in buffer solution. Tetrahedron Letters. 2013;**54**:945-948. DOI: 10.1016/j.tetlet.2012.12.022

[57] Birolli WG, Fonseca LP, Porto ALM. Aldol reactions by lipase from Rhizopus niveus, an example of unspecific protein catalysis. Catalysis Letters. 2017;**147**: 1977-1987. DOI: 10.1007/s10562-017- 2121-6

[58] Zhang Y, Wang N, Xie ZB, Zhou LH, Yu XQ. Ionic liquid as a

recyclable and efficient medium for lipase-catalyzed asymmetric cross aldol reaction. Journal of Molecular Catalysis B: Enzymatic. 2014;**110**:100-110. DOI: 10.1016/j.molcatb.2014.10.008

[59] González-Martínez D, Gotor V, Gotor-Fernández V. Application of deep eutectic solvents in promiscuous lipase-catalysed aldol reactions. European Journal of Organic Chemistry. 2016;**8**: 1513-1519. DOI: 10.1002/ejoc.201501553

[60] Tian XM, Zhang SQ, Zheng LY. Enzyme-catalyzed Henry reaction in choline chloride-based deep eutectic solvents. Journal of Microbiology and Biotechnology. 2016;**26**(1):80-88. DOI: 10.4014/jmb.1506.06075

[61] Li K, He T, Li C, Feng XW, Wang N, Yu XQ. Lipase-catalysed direct Mannich reaction in water: Utilization of biocatalytic promiscuity for C–C bond formation in a "one-pot" synthesis. Green Chemistry. 2009;**11**(6):777-779

[62] He T, Li K, Wu MY, Feng XW, Wang N, Wang HY, et al. Utilization of biocatalytic promiscuity for direct Mannich reaction. Journal of Molecular Catalysis B: Enzymatic. 2010;**67**(3–4): 189-194. DOI: 10.1016/j.molcatb.2010.08.004

[63] Xue Y, Li LP, He YH, Guan Z. Protease-catalysed direct asymmetric Mannich reaction in organic solvent. Scientific Reports. 2012;**2**:761-764. DOI: 10.1038/srep00761

[64] Chen XY, Chen GJ, Wang JL, Wu Q, Lin XF. Lipase/acetamide-catalyzed carbon-carbon bond formations: A mechanistic view. Advanced Synthesis & Catalysis. 2013; **355**:864-868. DOI: 10.1002/adsc.201201080

[65] Guan Z, Song J, Xue Y, Yang DC, He YH. Enzyme-catalyzed asymmetric Mannich reaction using acylase from Aspergillus melleus. Journal of Molecular Catalysis B: Enzymatic. 2015; **111**:16-20. DOI: 10.1016/j.molcatb.2014.11.007

[66] Sharma UK, Sharma N, Kumar R, Sinha AK. Biocatalysts for multicomponent Biginelli reaction: Bovine serum albumin triggered waste-free synthesis of 3,4-dihydropyrimidin-2-(1H)-ones. Amino Acids. 2013;**44**: 1031-1037. DOI: 10.1007/s00726-012-1437-1

[67] Xie ZB, Wang N, Wu WX, Le ZG, Yu XQ. Trypsin-catalyzed tandem reaction: One-pot synthesis of 3,4-dihydropyrimidin-2(1H)-ones by in situ formed acetaldehyde. Journal of Biotechnology. 2014;**170**:1-5. DOI: 10.1016/j.jbiotec.2013.10.031.

[68] Zhang W, Wang N, Yang ZJ, Li YR, Yu Y, Pu XM, et al. Lipase-initiated tandem Biginelli reactions via in situ-formed acetaldehydes in one pot: Discovery of single-ring deep blue luminogens. Advanced. Synthesis & Catalysis. 2017;**359**:3397-3406. DOI: 10.1002/adsc.201700599

[69] Wang JL, Liu BK, Yin C, Wu Q, Lin XF. *Candida antarctica* lipase B-catalyzed the unprecedented three-component Hantzsch-type reaction of aldehyde with acetamide and 1,3-dicarbonyl compounds in non-aqueous solvent. Tetrahedron. 2011;**67**:2689-2692

[70] Zhang W, Wang N, Liu YH, Jiao SY, Zhang WW, Pu XM, et al. 1,4-Dihydropyridines: Discovery of minimal AIEEgens and their mitochondrial imaging applications. Journal of Materials Chemistry B. 2017;**5**:464-469. DOI: 10.1039/c6tb02135b

[71] Jiang L, Ye LD, Gu JL, Su WK, Ye WT. Mechanochemical enzymatic synthesis of 1,4-dihydropyridine calcium antagonists and derivatives. Journal of Chemical Technology Biotechnology. 2019;**94**:2555-2560. DOI: 10.1002/jctb.6051

[72] Klossowski S, Wiraszka B, Berlozecki S, Ostaszewski R. Model studies on the first enzyme-catalyzed Ugi reaction. Organic Letters. 2013;**15**: 566-569. DOI: 10.1021/ol3033829

[73] Gesse P, Müller TJJ. Consecutive five-component Ugi-4CR-CAL B-catalyzed aminolysis sequence and concatenation with transition metal catalysis in a one-pot fashion to substituted triamides. European Journal of Organic Chemistry. 2019;**11**: 2150-2157. DOI: 10.1002/ejoc.2019 00198

Permissions

All chapters in this book were first published by InTech Open; hereby published with permission under the Creative Commons Attribution License or equivalent. Every chapter published in this book has been scrutinized by our experts. Their significance has been extensively debated. The topics covered herein carry significant findings which will fuel the growth of the discipline. They may even be implemented as practical applications or may be referred to as a beginning point for another development.

The contributors of this book come from diverse backgrounds, making this book a truly international effort. This book will bring forth new frontiers with its revolutionizing research information and detailed analysis of the nascent developments around the world.

We would like to thank all the contributing authors for lending their expertise to make the book truly unique. They have played a crucial role in the development of this book. Without their invaluable contributions this book wouldn't have been possible. They have made vital efforts to compile up to date information on the varied aspects of this subject to make this book a valuable addition to the collection of many professionals and students.

This book was conceptualized with the vision of imparting up-to-date information and advanced data in this field. To ensure the same, a matchless editorial board was set up. Every individual on the board went through rigorous rounds of assessment to prove their worth. After which they invested a large part of their time researching and compiling the most relevant data for our readers.

The editorial board has been involved in producing this book since its inception. They have spent rigorous hours researching and exploring the diverse topics which have resulted in the successful publishing of this book. They have passed on their knowledge of decades through this book. To expedite this challenging task, the publisher supported the team at every step. A small team of assistant editors was also appointed to further simplify the editing procedure and attain best results for the readers.

Apart from the editorial board, the designing team has also invested a significant amount of their time in understanding the subject and creating the most relevant covers. They scrutinized every image to scout for the most suitable representation of the subject and create an appropriate cover for the book.

The publishing team has been an ardent support to the editorial, designing and production team. Their endless efforts to recruit the best for this project, has resulted in the accomplishment of this book. They are a veteran in the field of academics and their pool of knowledge is as vast as their experience in printing. Their expertise and guidance has proved useful at every step. Their uncompromising quality standards have made this book an exceptional effort. Their encouragement from time to time has been an inspiration for everyone.

The publisher and the editorial board hope that this book will prove to be a valuable piece of knowledge for researchers, students, practitioners and scholars across the globe.

List of Contributors

Andivelu Ilangovan and Thumadath Palayullaparambil Adarsh Krishna
School of Chemistry, Bharathidasan University, Tiruchirappalli, Tamilnadu, India

Badri Nath Jha and Nishant Singh
University Department of Chemistry, T. M. Bhagalpur University, Bhagalpur, India

Abhinav Raghuvanshi
Department of Chemistry, Indian Institute of Technology Indore, Indore, India

David Joshua Ferguson
Taylor University, Upland, Indiana, United States of America

Tangali Ramanaik Ravikumar Naik
Vijayanagara Srikrishna Devaraya University (VSKU), Ballari, Karnataka

Chebolu Naga Sesha Sai Pavan Kumar
Division of Chemistry, Department of Sciences and Humanities, Vignan's Foundation for Science, Technology and Research, Guntur, Andhra Pradesh, India

Jamel Mejri
Higher School of Engineers of Medjez el Bab (ESIM), University of Jendouba, Tunisia

Youkabed Zarrouk
Field Crops Laboratory, National Agronomic Research Institute of Tunisia (INRAT), Tunisia

Majdi Hammami
Aromatic and Medicinal Plants Laboratory, Biotechnology Center of Borj-Cedria Hammam-Lif, Tunisia

Vitaliy Viktorovich Libanov, Alevtina Anatolévna Kapustina and Nikolay Pavlovich Shapkin
Department of General, Inorganic and Elementorganic Chemistry, School of Natural Sciences, Far Eastern Federal University, Vladivostok, Russia

Sakthivel Pandaram and Tamilselvan Duraisamy
School of Chemistry, Bharathidasan University, Tiruchirappalli, Tamilnadu, India

Yan-Chao Wu, Yun-Fei Cheng and Hui-Jing Li
School of Marine Science and Technology, Harbin Institute of Technology, Weihai, P. R. China

Tejpal Singh Chundawat
Department of Applied Sciences, The NorthCap University, Gurugram, Haryana, India

Yun Wang
School of Chemistry and Chemical Engineering, Lingnan Normal University, Zhanjiang, China

Na Wang
Key Laboratory of Green Chemistry and Technology, Ministry of Education, College of Chemistry, Sichuan University, Chengdu, P.R. China

Index

Printed in the USA
CPSIA information can be obtained
at www.ICGtesting.com
JSHW051354091023
49903JS00006B/145

9 781647 265991